D0899175

WITHDRAWN

Key Topics in Conservation Biology

Dedication

This book celebrates the move of Oxford's WildCRU to its own centre, Tubney House, and is thus dedicated to the memory of Miles and Briony Blackwell and to the Trustees of the Tubney Trust who made this possible.

The Student Panel
(whose members commented on drafts of every chapter)

Ewan Macdonald, Edinburgh University, UK
Michael Mills, Cape Town University, RSA
Stephanie Pimm, Carleton College, USA
Ryan Waples, Wesleyan University, USA
Ross Wrangham, Colorado University, USA

When I was a boy of fourteen, my father was so ignorant I could hardly stand to have the old man around. But when I got to be twenty-one, I was astonished at how much he had learned in seven years.
(Mark Twain, 'Old Times on the Mississippi', *Atlantic Monthly*, 1874)

Key Topics in Conservation Biology

Edited by

David W. Macdonald

of the Wildlife Conservation Research Unit,
University of Oxford

Katrina Service

of the University of East London

Blackwell
Publishing

© 2007 by Blackwell Publishing Ltd

BLACKWELL PUBLISHING
350 Main Street, Malden, MA 02148-5020, USA
9600 Garsington Road, Oxford OX4 2DQ, UK
550 Swanston Street, Carlton, Victoria 3053, Australia

The right of David Macdonald and Katrina Service to be identified as the Authors of the Editorial Material in this work has been asserted in accordance with the UK Copyright, Designs, and Patents Act 1988.

All rights reserved. No part of this publication may be reproduced, stored in a retrieval system, or transmitted, in any form or by any means, electronic, mechanical, photocopying, recording or otherwise, except as permitted by the UK Copyright, Designs, and Patents Act 1988, without the prior permission of the publisher.

First published 2007 by Blackwell Publishing Ltd

1 2007

Library of Congress Cataloging-in-Publication Data

Key topics in conservation biology / edited by David W. Macdonald & Katrina Service.
    p. cm.
  Includes index.
  ISBN-13: 978-1-4051-2249-8 (pbk. : alk. paper)
  ISBN-10: 1-4051-2249-8 (pbk. : alk. paper) 1. Conservation biology. I. Macdonald, David W.
  (David Whyte) II. Service, Katrina.

QH75.K47 2006
333.95'16—dc22

                                        2006001711

A catalogue record for this title is available from the British Library.

Set in 10/12.5pt Meridian
by SPI Publisher Services, Pondicherry, India
Printed and bound in the UK
by TJ International, Padstow, UK

The publisher's policy is to use permanent paper from mills that operate a sustainable forestry policy, and which has been manufactured from pulp processed using acid-free and elementary chlorine-free practices. Furthermore, the publisher ensures that the text paper and cover board used have met acceptable environmental accreditation standards.

For further information on
Blackwell Publishing, visit our website:
www.blackwellpublishing.com

Contents

Contributors

Alonzo C. Addison is interested in the application of sensor technology to everything from natural and cultural heritage to architecture and engineering. He currently serves as Special Advisor to the Director of the United Nations Educational Scientific and Cultural Organization World Heritage Centre. President of the Virtual Heritage Network, he founded the University of California at Berkeley's Center for Design Visualization and in the early 1990s helped create the first high-accuracy long-range laser scanner as Vice President of Cyra Technologies (now Leica Geosystems). He holds degrees in engineering, architecture and computing from Princeton and Berkeley and resides between Paris and San Francisco with his wife and daughter.

H. Resit Akçakaya is Senior Scientist at the Applied Biomathematics Institution long Island, New York. where he works on models for ecological risk analysis, population viability analysis, integrating metapopulation and landscape dynamics, and incorporating uncertainty into criteria for threatened species.

Steve Albon is Head of Science at the Macaulay Institute and is best known for his research on the population ecology of ungulates, in particular, red deer on the Isle of Rum, and Soay sheep on St Kilda. He is currently Chair of the Scottish Biodiversity Forum's, Action Plan and Science Group (APSG), which advises the Scottish Executive on the science underpinning Scotland's Biodiversity Strategy with the overall aim to conserve biodiversity for health, enjoyment and well being of the people of Scotland.

Lise Albrechtsen studied for her DPhil with Oxford University's Wildlife Conservation Research Unit focusing on the economics of bushmeat within central Africa, with particular focus on Equatorial Guinea. She received a BA in Political Science from Texas Lutheran University (Seguin, USA), a MSc in Environmental Change and Management from the Environmental Change Institute of Oxford University (UK), a MA in Environmental Economics from Scuola Mattei (Milano, Italy), and has attended a Master's course in System Dynamics Modelling at the University of Bergen (Norway). Currently she is a junior economist with the UN Food and Agriculture Organization (FAO) working on food security analyses.

Elizabeth J. Asteraki is a senior research scientist with CAB International, Malaysia. With a background in entomology and community ecology, her research has focused on biodiversity conservation in agro-ecosystems. More recently she has been active in developing ustainable production systems and integrated pest management programmes for developing countries in South East Asia.

Sandra Baker is a behavioural ecologist with the at Oxford University Wildlife Conservation Research Unit at Oxford University. Her main research interest is in exploiting (and manipulating) natural animal behaviour to resolve wildlife management and conflict situations – in particular, those involving carnivores. This fascination evolved during her doctoral thesis in which she developed learned food aversions for the non-lethal control of European badgers and red foxes.

Cristian Bonacic, DVM, MSc, DPhil is a wildlife conservation scientist. He is Director of Fauna Australis (a research group in Chile which tackles practical conservation problems), a member of the Wildlife Trust Alliance and associate researcher of the Wildlife Conservation Research Unit, University of Oxford. He is an Associate Professor in the School of Agriculture and Forestry at Pontificia Universidad Católica de Chile.

Mark S. Boyce is Professor of Biological Sciences and the Alberta Conservation Association Chair in Fisheries and Wildlife at the University of Alberta. His research interests include population ecology of vertebrates, especially relative to conservation and wildlife management. He has served as Editor in Chief of the Journal of Wildlife Management and is currently an editor for Ecology and Ecological Monographs.

David Brown is a Research Fellow of the Overseas Development Institute inLondon. His research focuses on the livelihoods dimensions of forest and conservation policy. He works throughout the tropics, but particularly in the bushmeat heartlands of west and central Africa. He co-authored the UK Submission to CITES 'Bushmeat as a Trade and Wildlife Management Issue' (CITES Decision 11/166), which led to the setting up of the CITES Central African Bushmeat Working Group.

Marcel Cardillo is a NERC postdoctoral fellow at Imperial College London. His primary research interests are in the application of phylogenetic comparative methods in conservation biology, macroecology and community ecology.

Stephen Cobb leads The Environment and Development Group, an international consulting firm based in Oxford. Specializing in tropical conservation, he spent the first half of his career living in remote parts of Africa, reconciling the needs of wildlife conservation and rural development. Latterly he has drafted laws and policies, reformed ailing institutions and created new financial mechanisms for keeping conservation wheels turning in perpetuity.

N. Mark Collins is Director of the Commonwealth Foundation in London. Until March 2005 he was with the United Nations Environment Programme, most recently in the Division of Environmental Conventions in Nairobi, Kenya, and previously as Director of the UNEP World Conservation Monitoring Centre in Cambridge, UK. Mark's earlier career included ecological research in Kenya, Nigeria, Malaysia and elsewhere in the tropics. He serves on many international committees, advisory groups and voluntary bodies, and in 2000 received the Royal Geographical Society Busk Medal in recognition of his work.

Chris R. Dickman is a Professor in Ecology at the University of Sydney, and Director of the University's Institute of Wildlife Research. He has particular interests in the ecology of arid environments and in vertebrate conservation, and is a Past President of the Australian Mammal Society and Chair of the New South Wales Government's Scientific Committee on threatened biota.

C. Patrick Doncaster is in the School of Biological Sciences at the University of Southampton. His research covers population biology and evolutionary ecology, mainly developing and testing conceptual models. His conservation work focuses particularly on the responses of animal populations to habitat degredation at local and regional scales.

Stephen A. Ellwood studied for his doctorate at University of Oxford's Wildlife Conservation Research Unit. Although his thesis focused on aspects of density dependence and independence in wild deer, and validating some methods commonly used to count them, he has a parallel interest in the development and application of technologies used for behavioural observation. He is also a founder member of, and Scientific Adviser to, OxLoc Ltd, a company specialising in the miniaturization of GPS–GSM technology for use in animal and asset tracking.

John E. Fa is Director of Conservation Science at Durrell Wildlife Conservation Trust in Jersey, UK, responsible for research and conservation projects on a variety of species and habitats worldwide. He has undertaken research in a number of conservation biology topics in Europe, Africa and South America, with special emphasis in biodiversity assessment and endangered species management.

Ruth E. Feber is at the Wildlife Conservation Research Unit, University of Oxford. Her research interests include the ecology and conservation of species inhabiting agricultural habitats and the wider impacts of farming practice on biodiversity.

Les Firbank is head of Land Use Systems at the Centre for Ecology and Hydrology. His research addresses interactions between agriculture and biodiversity, and he has looked at biodiversity impacts of organic farming, set-aside, agri-environment schemes, and led the Farm Scale Evaluations of GM crops. His current research focuses on assessing trends in biodiversity resulting from land-use change at national and European scales.

Eli Geffen is a behavioural and molecular ecologist in the Department of Zoology at Tel Aviv University. He uses molecular markers to address evolutionary biology questions at levels varying from the individual to the species, in a number of mammalian model systems.

Joshua Ginsberg is the Vice-President for Conservation Operations at the Wildlife Conservation Society where he also served as Director of the Asia and Africa Programmes. He serves as Chairman of the NOAA/NMFS Hawaiian Monk Seal Recovery Team. A Professor (adjunct) at Columbia University, he supervizes a research group that focuses their efforts on questions of population persistence and effectiveness of protected areas.

Susanna Hecht is Professor in the School of Public Affairs, University of California at Los Angeles. She is an analyst of the dynamics of forest trends in the New World Tropics, alternatives to deforestation and tropical environmental history

Peter Hudson is The Willaman Chair in Biology and the Director of The Center for Infectious Disease Dynamics at Penn State University, Pennsylvania. He has worked extensively on the dynamics of infectious diseases in wild animal populations, particularly a long-term study of the effects of parasites on red grouse population dynamics and more recently parasite community dynamics of mice and rabbits.

Carolyn King comes from Kent, took her DPhil (on weasels) from Oxford in 1971, and has spent her professional life in New Zealand, specializing on the ecology of introduced mustelids. She is a senior lecturer in the Department of Biological Sciences at the University of Waikato, Hamilton, New Zealand.

Nigel Leader-Williams is Professor of Biodiversity Management at the Durrell Institute of Conservation and Ecology at the University of Kent, Canterbury, UK. He is interested in a wide range of conservation issues, including sustainable use of natural resources, law enforcement and illegal use in protected areas, conservation on private and communal land, and human–animal conflict.

Diana Liverman is the Director of the Environmental Change Institute at University of Oxford. Her research interests include the human dimensions of global environmental change and environmental governance in Latin America.

A.J. Loveridge completed a DPhil at Oxford in 1999 on the behavioural ecology of southern African jackals. He is currently a Junior Research Fellow at Lady Margaret Hall, Oxford and a post-doctoral researcher in the Wildlife Conservation Research Unit. He works on the impacts of sport hunting on a population of African Lions in western Zimbabwe and has an interest in carnivore behavioural ecology and the sustainable use of wildlife.

Gordon Luikart has general research interests in conservation biology, population genetics and evolutionary biology. His research focuses primarily on the development and application of molecular and statistical methods for use in conservation genetics and wildlife management, and for inferring the evolutionary history of wild and domestic mammal species (especially ungulates). He is currently a visiting professor at the Centre for Investigation of Biodiversity and Genetic Resources at the University of Porto (CIBIO-UP) in Portugal, and a research associate professor in the Division of Biological Sciences at the University of Montana, USA.

Tim Lynam is a systems ecologist who has spent much of his professional life working at the interface of human and ecological systems in southern Africa. He has built and worked with models designed to capture the details of complex social and ecological systems as well as those built with users to reflect their current understandings of the essential components of the systems they manage. He is particularly interested in developing and using the theory of resilience in linked social and ecological systems, and has recently moved to join CSIRO in Australia.

David W. Macdonald, D.Sc. is Professor of Wildlife Conservation at University of Oxford, and is founder and Director of Oxford's Wildlife Conservation Research Unit, and a Fellow of Lady Margaret Hall, Oxford. He was, until recently, A.D. White Professor at Cornell University. He specializes in mammals, especially carnivores and was for 25 years Chairman of the IUCN/SSC Canid Specialist Group, and winner of the 2004 Dawkins Prize for Conservation. He is involved in conservation policy nationally and internationally, and has produced prize-winning books and TV documentaries on wildlife.

Georgina M. Mace is Director of Science at the Zoological Society of London, and head of the Institute of Zoology, a research institute specializing in conservation biology. Her research focuses on the assessment and prediction of threatened status in wild species, and more generally on measures of and responses to biodiversity loss. She has served as a member of the IUCN/SSC Steering Committee for over 10 years, working on methods to improve Red List assessments and biodiversity indicators, and she is the next President of the Society for Conservation Biology.

Graeme McLaren is a conservation biologist with the UK Environment Agency. His research examines the animal welfare implications of conservation practices such as trapping and handling.

Gus Mills is Research Fellow for SANParks, based in the Kruger National Park, Head of the Carnivore Conservation Group of the Endangered Wildlife Trust and Extraordinary Professor at the Mammal Research Institute,

University of Pretoria. As Chairman of the Wild Dog Advisory Group, South Africa he has promoted and guided the South African Wild Dog Metapopulation Programme.

E.J. Milner-Gulland is Reader in Conservation Science in the Division of Biology at Imperial College, London. Her interests are in the interactions between people and wildlife, including population dynamics, sustainability of hunting and the incentives that people face to conserve and use natural resources.

Chris Newman is a senior researcher with the Wildlife Conservation Research Unit at Oxford University. Having completed his DPhil with the Wildlife Conservation Research Unit looking at the population demography of the Eurasian badger, he has gone on to co-ordinate the Unit's Mammal Monitoring and Volunteer projects. Chris serves on the committee of the UK's Tracking Mammal's Partnership and is the Wildlife Officer for the University's Ethical Review Committee. His current research is focused on climatic impacts on wildlife, and on engaging the public in conservation.

David Pearce was the founding Director of the Centre for Social and Economic Research on the Global Environment at the University of East Anglia. He was responsible for initiating and encouraging a huge number of successful careers in both academia and the public sector, and he was undoubtedly one of the modern founding fathers of environmental economics and its policy applications. His contribution was recognized by a lifetime achievement award from the European Association of Environmental and Resource Economists' Association. His OBE is also testament to the high regard in which he was held by the wider civil society in the UK. David died after a very short illness on 8 September 2005.

Stuart L. Pimm is the Doris Duke Professor of Conservation Ecology at the Nicholas School of the Environment and Earth Sciences at Duke University and also holds the title of Extraordinary Professor at the Conservation Ecology Research Unit at the University of Pretoria, South Africa. He is an expert on endangered species conservation, biodiversity, species extinction and habitat loss. Working in southern Africa, South America, Central America and the Everglades, his work has contributed to new practices and policies for species preservation and habitat restoration in many of the world's most threatened ecosystems. The Institute of Scientific Information recognized him in 2002 as being one of the world's most highly cited scientists. In 2004 he was elected a Fellow of the American Academy of Arts and Sciences.

Hugh Possingham born in 1962, completed his DPhil. in Biomathematics at University of Oxford in 1987 as an Australian Rhodes Scholar. In 1995 he was appointed Foundation Chair and Professor of the Department of Environmental Science at The University of Adelaide and in 2000 he became Professor of Ecology at The University of Queensland, Brisbane. His laboratory has a unifying interest in environmental applications of decision theory and Hugh has the privilege of chairing three Federal Government committees. He was recently elected to the Fellowship of The Australian Academy of Science in 1995 and he suffers from obsessive bird watching.

Jonathan Reynolds is a senior scientist at The Game Conservancy Trust, where he is responsible for research into predation control. This research has included studies on the impact of culling on predator populations at different geographical scales; the consequences of predator control for prey species; and the trade-offs between effectiveness, target accuracy and humaneness that are inherent among different control methods.

Philip Riordan is at the University of Oxford's Wildlife Conservation Research Unit, researching wildlife disease issues. In addition to examining the impacts of behavioural and ecological influences on disease transmission within and between wildlife species and assemblages, he is also interested in the wider human context of wildlife disease.

Terry L. Root a Senior Fellow at the Center for Environmental Science and Policy in Freeman Spogli Institute for International Studies at Stanford University, primarily works on large-scale ecological questions with a focus on impacts of global warming. She actively works at making scientific information accessible to decision makers and the public (e.g. being a Lead Author for IPCC Third and Fourth Assessment Reports). In 1999 she was chosen as an Aldo Leopold Leadership Fellow, in 1992 as a Pew Scholar in Conservation and the Environment, and in 1990 received a Presidential Young Investigator Award from the National Science Foundation.

Andrew Rowan was educated as a biochemist at Cape Town and Oxford Universities and subsequently also dabbled in philosophy, sociology, anthropology, toxicology and several other disciplines in a 30 year career in animal welfare science working for FRAME, The Humane Society of the USA and Tufts University School of Veterinary Medicine.

Steve P. Rushton is an ecological modeller based in the Institute for Environmant and Sustainability at the University of Newcastle. He specializes in modelling the distribution and dynamics of animal populations.

Katrina Service completed a BSc in Zoology at the University of Glasgow before joining the Mammal Research Unit at the University of Bristol, where she studied the scent marking behaviour of Eurasian badgers for her PhD. This was followed by 4 years of post-doctoral research on carnivore ecology at the University of Oxford's Wildlife Conservation Research Unit, during which time she was also involved in running an MSc course in conservation. She

is now a lecturer in conservation biology at the University of East London, and continues to research carnivore ecology and behaviour.

Claudio Sillero-Zubiri grew up on a farm in Argentina before studying Zoology in La Plata and taking a doctorate at Oxford University. He has researched widely on conservation biology in Africa and South America, and in 1998 was awarded the Whitley Award for Conservation for his work on Ethiopian wolves. He specializes in human–wildlife conflict, in 2005 he became Chairman of the IUCN/SSC Canid Specialist Group and is currently a senior researcher in Oxford's Wildlife Conservation Research Unit and Bill Travers Fellow for Wildlife Conservation at Lady Margaret Hall.

Grant Singleton's main interest is in wildlife management, particularly experimental field studies on small mammals. He is a leader in ecologically based rodent management and has vast experience in studying the biology and management of rodents in Australia, Indonesia, Vietnam, Lao PDR, Myanmar, Bangladesh and Philippines. He also has a strong interest in rodent diseases and led an 8 year project on immunocontraception of house mice.

Rob Smith has recently moved from Leicester University to become Dean of Applied Sciences at the University of Huddersfield and a member of the Ecology and Evolution Group at the University of Leeds. He has worked on the ecology and evolution of a variety of pest species, mostly insects but also various rodents. He has worked on rodents internationally as a consultant for a World Bank project on cacao in Equatorial Guinea, and more recently in Mexico. Rob was one of three editors of the *Journal of Animal Ecology* for 6 years.

Robert Strachan is a Biodiversity Technical Specialist with Environment Agency Wales. He advises on issues relating to riparian mammals and invasive non-native species. He heads the species recovery programme for the water vole in the UK in close collaboration with the Wildlife Conservation Research Unit.

Raman Sukumar is Professor of Ecology at the Indian Institute of Science, Bangalore, India. He is a leading expert on the ecology and conservation of Asian elephants, and was Chair of the IUCN/SSC Asian Elephant Specialist Group (1996–2004). A pioneer in the study of wildlife-human conflicts, he is the author of 3 books on elephants and over 70 scholarly papers on wildlife ecology, tropical forest ecology and climate change.

Jorgen Thomsen is an ornithologist by training and has held leadership positions in several international conservation organizations, including TRAFFIC International, World Wide Fund for Nature and IUCN – The World Conservation Union. A native of Denmark, he now resides in Washington, DC, where he is currently Senior Vice president, Conservation International, overseeing its Conservation Funding Division, and is the Executive Director of the Critical Ecosystem Partnership Fund.

Adrian Treves is a visiting professor of environmental planning, research and monitoring in the Conservation Biology Programme of Makerere University in Kampala, Uganda. He also directs 'COEX: Sharing the Land with Wildlife, Inc.' (www.coex-wildlife.org) a non-profit organization dedicated to promoting coexistence of wildlife and people with minimal conflict. His speciality is human–carnivore conflict (livestock loss and crop damage) and the human dimensions of wildlife conservation. He works in the field in East Africa and the USA (Wisconsin).

Frank Vorhies is a sustainability economist. He has worked for environmental NGOs including the African Wildlife Foundation, IUCN-The World Conservation Union and the Earthwatch Institute. Frank is particularly interested in making capitalism work for conservation. Currently he is a consultant to the UNCTAD BioTrade Initiative and the IUCN Global Marine Programme in Geneva.

Robin S. Waples is a Senior Research Scientist at the NOAA Northwest Fisheries Science Center in Seattle. For over a decade he headed a group charged with developing the scientific basis for listing determinations and recovery planning for Pacific salmon under the US Endangered Species Act. His research interests include: integrating genetic, ecological and life history information to identify conservation units; evaluating genetic interactions of hatchery and wild fish; theoretical and empirical evaluations of effective population size; genetic mixture analysis; and the analysis of gene flow in marine and anadromous species.

Rory Wilson currently holds the position of Chair in Aquatic Biology at the University of Wales Swansea, UK. He has been working with recording devices attached to animals for over 25 years and is particularly interested in foraging decisions in air-breathing marine animals.

Richard Wrangham is a Professor of Biological Anthropology at Harvard University, Chair of the Great Ape World Heritage Species Project and President of the International Primatological Society. He has been studying chimpanzee behavioural ecology in Kibale National Park, Western Uganda, since 1987.

Boxes

Preface

It's easy to think that as a result of the extinction of the dodo we are now sadder and wiser, but there's a lot of evidence to suggest that we are merely sadder and better informed.

(Douglas Adams and Mark Carwardine, *Last Chance To See*, 1990.)

Why bother?

It seems only fair to the reader, at the start of any book, to explain why the trouble was taken to write it. In the case of *Key Topics in Conservation Biology*, the question has answers at two different levels – the first explains why the topic itself is rivettingly relevant for everyone who gives even a jot, not just about Nature, but about the future of the human enterprise worldwide (and surely that makes it relevant to just about everybody), whereas the second explains why we tackled it in this particular way – an answer which reveals, unusually, that in this case the process is almost as interesting as the product.

At the first level, the reason why the key topics of wildlife conservation are relevant are not only because we are in the midst of an extinction crisis, but also because countless species not yet facing extinction, and their

habitats, are nonetheless facing grave change (almost always for the worse), invariably due ultimately to the hand of Man and often with consequences that also affect people. The extinction crisis itself is the topic of the first essay, by Pimm, Dickman and Cardillo, so there is no need to repeat the detail here. Similarly, issues such as bushmeat, hunting, pest control, agriculture and other forms of conflict are each the topic of other essays, as are such issues as infectious disease, invasive species and climate change. Again, other than drawing attention to the breadth of these topics, our purpose here is not to summarize these essays, but rather to direct the reader to them.

Like medicine, conservation biology is a mission-driven science. Physicians take it for granted that we all care about saving and extending human lives. Thus motivated, they study the pathology of ill health and practice methods to prevent or minimize it. Although death is an inevitable part of life, we deem a high number of premature deaths – from disease or accident – to be a particular concern. Likewise, conservation biology is about biodiversity loss and the methods to minimize it. Essays in this collection introduce some of the tools of this trade – spanning the ingenious gadgetry reviewed by Ellwood, Wilson & Addison, through the computer models explained by Boyce, Rushton & Lynam, to the institutional structures described by Cobb, Ginsberg & Thompsen. Others introduce the biological framework within which the natural environment can be understood, for example through the genetics (Geffen, Luikart & Waples) and the spatial organization (Akçakaya, Mills & Doncaster) of populations.

Why should we care about the loss of biodiversity? It is conventional to couch the answer in terms of economics, ethics and aesthetics (which, with the neologism of American spelling, can catchily be labelled the 'three e's'). These three resonate with the elements of triple bottom-line accounting (economics, environment and social responsibility) that has rightly become fashionable in reporting the impacts of corporations, and are also the basis of accounting in any conservation debate. Both trios emphasize that costs and benefits are measured in many different, and often awkwardly incommensurate currencies. You might value a species on the basis of its direct market worth, or its indirect value (e.g. in persuading people to go on holiday to watch it – calculated by so-called hedonic valuation), or in more abstract terms by the value you put on its existence (fuzzily quantified by so-called contingent valuation). The revelation of 50 years of conservation biology is that every issue is complicated, and every solution must be interdisciplinary – biology is a necessary component, but not a sufficient one, for understanding and thus solving conservation problems. This reality, which makes clear that there is a 'human dimension' to every conservation issue, and that this dimension is generally unavoidably central to the solution, reverberates through every essay in this book – it is the

entire topic of the essay on environmental economics by Pearce, Hecht & Vorheis, and a central message of the concluding overview in Macdonald, Collins & Wrangham's postscript. In short, whether or not an individual happens to realize it, or to be interested in biodiversity, everybody's life is affected by, and affects its conservation.

Turning to the more nuts-and-bolts question of how, and why, we produced this book the way we have, the answer lies in the invitation, in 2000, to create a module in Conservation Biology within the University of Oxford's Master of Science course entitled *Integrative Biology*, which is organized by the University's Department of Zoology. Believing that there was little merit in cajoling lecturers to prepare, and then compelling students to listen to, lectures that rehearsed conventional material that could more efficiently be gleaned from textbooks, we decided instead to organize the course as a series of workshops at which front-line specialists of international standing led discussions on their experiences at the cutting-edge of conservation. These sessions took the form of day-long Think Tanks, in which not only the Masters students and our invited guests, but also researchers from the Wildlife Conservation Research Unit pitched in together. Rather than wearisome essays, the course assignments involved snappy thought-pieces on emergent issues – the key topics in wildlife conservation. The formula was so energizing – flatteringly, the students repeatedly voted it their favorite module – that we thought to develop the approach as a book. Close to the front of our minds, and it was a thought that found favor with many of our visiting speakers, was a growing disquiet that the very welcome rise to prominence of conservation biology was tainted by an occasional and unwelcome tendency towards bluster! Specifically, our Think Tank sessions became vigilant to such refrains as 'it is really important that we study … such and such', to which the probing chorus of 'why?' sometimes revealed that although the topic might indeed be interesting, it was less obvious why it was operationally important. The notion grew, therefore, of assembling teams to write the essays that now comprise this book, and of selecting for each team a trio of renowned authors, each with a different perspective, and urging them to work together on the difficult task of stripping down to the essentials the issues that really are important in their topic. It is for that reason that the working title of this book has been 'Conservation Without Crap' – although the proposal that this should be the cover title was one from which the publisher politely demurred.

Because discussion had been such a prominent strength of the workshops that had catalysed this book, we sought to emulate this by subjecting each essay to the equivalent of a roomful of discussants. Not only did all the authors review each other's essays, but all members of the Wildlife Conservation Research Unit reviewed them too. The result was that most essays received over a dozen reviews, and went through many drafts in response. In conversation amongst the authors it emerged that a surprising number of our sons and daughters were biology students, and who more critical than an offspring to savage that which flows from the parental pen! Therefore, with something of a family feel, we assembled the Student Panel (listed on preliminary p.ii of this book) to cast a critical consumer's eye over each essay (although we hope these essays will fascinate, inform and entertain a wide readership from the loftiest authority to the aspirant Sixth Former, from interested layman to policy-maker to naturalist, our imagined modal reader might well be a Masters student). Each essay represents a hill – a vantage point from which a particular trio of specialists views the conservation landscape. Having been assaulted by the assembled army of over 70 reviewers, almost all the authors commented that the toughest comments to deal with came from the Student Panel – tellingly, perhaps their views of the hills were unencumbered by the baggage that older reviewers had accumulated while climbing them – who as yet know nothing of things

like the Research Assessment Exercise (RAE) and why it matters, or does not (for any readers also fortunate enough to be in this position, the RAE is a performance indicator that encourages scientists in the pursuit of a high score rather than of wisdom or usefulness). One of the most inescapable realizations drawn from the process of producing these essays, and something perhaps felt most keenly by the five of us reading the comments of our offspring, is just how radically the conservation landscape has changed in just one professional generation. As a practical aside, a lesson that might assist editors and authors as they recruit reviewers: as we five fathers watched our offspring toil over early drafts of these essays during the 2004 Christmas vacation, we also learnt how fiercely one cares about the quality of a script that presents to one's children the subject to which we have devoted our lives! Anyway, the quality of the final essays owes much to the diligent reviews not only of the authorial team and the Student Panel but equally to the following members of the Wildlife Conservation Research Unit: Christina Buesching, Ruthi Brandt, Zeke Davidson, Harriet Davies-Mostert, Carlos Driscoll, Hannah Dugdale, Adam Dutton, Paul Johnson, Jan Kamler, Kerry Kilshaw, Steven Gregory, Lauren Harrington, Donna Harris, Jorgelina Marino, Fiona Mathews, Tom Moorhouse, Inigo Montes, Jed Murdoch, Deborah Randall, Greg Rasmussen, Lucy Tallents, Hernan Vargas, Nobby Yamaguchi and Zinta Zommers.

Finally, each essay in this collection is intended to stand alone, but the collection as a whole is more than the sum of its parts, together introducing the nature of the problem, the framework in which it can be understood, some tools that can be used in the quest for solutions, and various of the issues that are topical. As such, it has no pretensions to compendiousness – there are many more than 18 key topics in wildlife conservation – nor even balance (although they drift variously over every type of organism, most of the authors have greater expertise in animals than in plants, and most specialize in vertebrates).

Nonetheless, this collection of essays does give a representative insight across the landscape of conservation. Just as it was spawned by days of debate between a diverse assemblage of people in our discussion groups, we hope that *Key* *Topics in Conservation Biology* will be the catalyst for countless fruitful discussions amongst those to whom it will fall to deliver the solutions that are required if Nature is to survive as more than a poor shadow of its former glory.

Truths would you teach, and save a sinking land?
All fear, none aid you, and few understand.

(Alexander Pope, in *Essay on Man*, 1994.)

David W. Macdonald
Wildlife Conservation Research Unit, Zoology Department, University of Oxford

Katrina Service
University of East London

The pathology of biodiversity loss: the practice of conservation

Chris R. Dickman, Stuart L. Pimm and Marcel Cardillo

Don't it always seem to go, that you don't know what you've got 'til it's gone... They paved paradise and put up a parking lot

(Joni Mitchell, *Big Yellow Taxi*, Siquomb Publishing Co. 1969.)

Introduction – what is biodiversity?

In this essay we start with the definitions of biodiversity and the problems of measuring it. These problems are significant, but not so insurmountable that we cannot quantify the timing and geographical distribution of biodiversity loss. We show that the loss of biodiversity is now hundreds to thousands of times faster than it should be because of human actions involving a variety of mechanisms. Some places, however, are very much more vulnerable to biodiversity loss than others; i.e. biodiversity loss is variable geographically.

Over the past 25 years the concept of biodiversity has been studied, reviewed and debated passionately by increasing numbers of scientists and resource managers, and has exploded into the public consciousness so pervasively that it underpins national agendas in many parts of the world. A search for the term on Internet websites yields far more hits than for many icons of popular culture (Norse

& Carlton 2003). So, what is biodiversity, why is it so important, and why has it become 'mainstream' only recently?

The term 'biodiversity' is commonly used to connote the 'variety of life', or 'God's Creation' to some, whereas others have proposed that it encompasses nothing less than the 'irreducible complexity of the totality of life' (Williams et al. 1994).

People have studied the variety of life for millennia, as hunter–gatherers harvesting food and other products of the natural world for their immediate survival, as settlers in agro-economies, as curiosity-driven natural historians, and as bioprospectors who seek new medicines and genetic improvements for agriculture. Studies of biodiversity are clearly not new. They have, however, become more urgent owing to concern that life's variety is being eroded by human activity. Warnings of impending 'extinction cascades' or 'biodiversity crises' are becoming increasingly common. In the current climate, 'biodiversity' appears to be moving beyond being a neutral term to one that additionally conveys emotion and value. Indeed, for some authors 'biodiversity' and 'nature conservation' are interchangeable (Bowman 1993).

Our definition of biodiversity is that provided by Elliott Norse for a report produced for the US Congress Office of Technology Assessment (OTA 1987): 'Biological diversity refers to the variety and variability among living organisms and the ecological complexes in which they occur.... (T)he term encompasses different ecosystems, species, genes...'

This three-part definition 'genes, species, ecosystems' – along with their evolutionary and ecological histories – produces a comprehensive value-free definition. It is also a **practical** one. We can measure the numbers of species and map their distributions. Maps of different ecosystems – forests or grasslands, for example, have been familiar for 100 years or more. Although more difficult, we can sometimes quantify the variety within a species. The diversity of genetic varieties of crop plants is one example.

This three-part definition forms the core of the ideas in UNEP's (United Nations Environment Programme) Convention on Biodiversity (signed by 150 government leaders at the 1992 Rio Earth Summit; SCBD 2005) and the Global Biodiversity Assessment (Heywood 1995). As one might expect, the easy-to-measure numbers of species provide these documents with most of their examples.

Equally, scientists wish to make the meaning of biodiversity more complex. The term sometimes means not just species and their genes, but the evolutionary history they represent and the ecological communities and processes that they create. Several authors have argued that 'biodiversity' should also include behavioural, ecological, physiological and life-form variation between individual organisms of the same species (e.g. Soulé 1991; Reich et al. 2003).

Biodiversity is thus a multifaceted concept that we can be measure in a variety of ways, though no single measure can capture all of its aspects (Purvis & Hector 2000). For practical purposes, we need a surrogate measure that allows biodiversity to be assessed effectively and that identifies major patterns and changes.

In practice, the measures most commonly used are simple counts of species (species richness) or counts that are weighted by the relative abundances and representation of species (species diversity) in samples. Species-based assessments have several advantages over possible alternatives. The primary one is that species are usually easier to count than genes, ecological interactions or other processes (Gaston 1996). Use of species measures can also be problematic.

First, species boundaries are sometimes difficult to define, especially in sibling taxa and in small, cryptic species that are morphologically conservative. Resolution of species is usually possible if small portions of the genome are characterized, but this adds cost and time to any assessment of biodiversity.

Second, even for conspicuous, well-differentiated species, taxonomists have described relatively few of the likely total. Taxonomists have named just over one and a half million species, but estimates of the total number of insects alone vary from 10 million to 100 million (Stork 1998; May 2000). Discoveries of 'extremophile' organisms deep in the soil profile, in underground lakes and around oceanic vents with no access to sunlight suggest further that much life remains to be inventoried. Despite such stocktaking problems, species remain the primary currency of biodiversity measurement, and lists of threatened species provide triggers for conservation action at local, national and international levels (Heywood, 1995; Burgess 2001).

An alternative approach, gaining in popularity, is to use measures of phylogenetic or phenotypic disparity among species. The philosophy underlying this approach is that it is preferable to conserve, for example, a member of a monotypic genus with no close living relatives than a species with numerous members of the same genus, because loss of the first species represents the loss of a far greater amount of unique evolutionary history. Phylogenetic diversity (PD) quantifies evolutionary history by measuring the summed lengths of the phylogenetic branches that separate species, either in terms of time since separation or the amount of evolutionary divergence (Faith 1992). This relies, of course, on availability of phylogenetic information, which is still non-existent for the great majority of species. The approach also represents an interesting value judgment. In diametric opposition one might argue that a large genus represents a lineage that is producing many new species and is thus one that merits priority for its evolutionary dynamism.

Measuring the loss of biodiversity

Most of this essay will be about species loss, for the practical reasons already noted. At least two major research efforts take exception to this emphasis. The first, mounted by scientists at the World Wildlife Fund (WWF), categorizes global ecosystems, then produces more finely divided continental **ecoregions** and assesses the threats to them. The second is from Paul Ehrlich's group at the Centre for Conservation Biology at Stanford University, California. If present trends continue, although many species may be saved in protected areas, these survivors will merely be remnants of their once geographically extensive and genetically diverse selves. The emphasis, they argue, should be on measuring the loss of local populations, for the 'services' biodiversity provide depend on what is present locally. (We shall return to this idea at the essay's end.)

The loss of ecosystems

The WWF has classified terrestrial ecosystems into 825 ecoregions, has another 500 for freshwater ecosystems and is working on classifying marine ecosystems (Ricketts et al. 1999; see http://www.worldwildlife.org/science/ecoregions.cfm). As an example, the first ecoregion listed for South America is the forest type dominated by *Araucaria* ('monkey puzzle') trees. They occur in the coastal mountains of Brazil, extending into northern Argentina. Of an original area of c.200,000 km^2, only c.13% remains.

As this example illustrates, one can immediately rank ecoregions by the fraction of their former extent that remains. Those with the least fraction remaining represent priorities for conservation action. How do such priorities match those based on species? Ecosystems such as tropical dry forests, deserts, tundra, temperate grasslands, lakes, polar seas and mangroves all contain characteristic species. Although conservation justifiably prioritizes tropical moist forests because they hold such a large fraction of the world's species (see below), a comprehensive strategy should also save distinctive ecosystems. Ecoregions also house distinctive ecological and evolutionary phenomena – they are, in part, defined by them. Given that we know such a small fraction of the world's species, it is at least possible that ecoregions provide a better clue to where distinctive species live than areas defined only on what we know about the few well-known taxa such as birds.

The loss of within-species variety

Other estimates of biodiversity loss focus on populations. Populations supply genetic diversity, because different populations across a species' range will differ to varying degrees in their genetic composition. Thus, as populations disappear locally, genes may become globally extinct. Hughes et al. (1997) defined population diversity as the number of populations on the planet. They estimated that an average species consists of 220 populations, suggesting that there may be more than 2 billion populations globally, of which 160 million populations (8%) are lost each decade. This is a much higher rate that the loss of species (below) because many populations are often lost before the species itself expires. Large areas of North America and Europe, for example, have lost almost all their large birds and mammals.

An obvious example is our own species. While our numbers are expanding rapidly, our cultural diversity – as measured by the number of languages that we speak – is shrinking rapidly. Few languages spoken by fewer than 500,000 people are taught to children on contact with western culture. Than means that about 90% of the world's > 6000 present languages will disappear in a generation or so (Pimm 2000).

At issue here is a matter of scale. Although much of the concern over the loss of biodiversity focuses on the global loss of species, most of the benefits conferred by biodiversity arise from large numbers of local populations of species (Hughes et al. 1997). An obvious example is the loss of forests that provide protection to a town's watershed. Although no species might become extinct globally, the forest trees provide a local service in preventing the soil erosion that would follow if the forests were cleared. Even if a species is not in danger of global extinction, it is 'ecologically extinct' if it has disappeared from most of its former distribution, and hence no longer performs any ecological role there.

The loss of species: what should we expect?

The arguments for measuring biodiversity as populations or ecosystems are compelling, but so too are measures of species numbers. That generations to come might not experience 'lions and tigers and bears, Oh my!' (Wizard of Oz) probably motivates public opinion in a direct way that the loss of (say) *Araucaria* forest does not. (Of course, species will continue to go extinct precisely because they lose their habitats.) Yet, how can we make sensible statements about species loss, if we do not know how many species there are?

We cannot estimate how many extinctions there are per year without making extravagant guesses from better known species groups. Absolute estimates of the numbers of extinctions must be extrapolated from the 100,000 well-known species to the one and a half million described species, to the likely grand total of a few to tens of millions of species (May 2000). Statements of how many species become extinct per year, or per day, can vary 100-fold because of uncertainties about total numbers of species (Pimm et al. 1995).

We can derive more confident **relative** estimates of extinction rates using the proportions of species that become extinct over time (Pimm et al. 1995). Such estimates beg the obvious question: are these proportions, which are based inevitably on well-known species, typical of the great majority of species groups that are not well-known? They are likely to be so if extinction rates in widely different groups and regions are broadly similar.

There is another way in which we must make estimates of extinctions relative. Extinctions have always been a part of Earth's history, so we scale any claims of massive extinctions now or in the future to past extinctions. The fossil data suggest that species last for one to a few million years except for the major upheavals, such as the one that eliminated dinosaurs (but not birds) at the end of the Cretaceous Period.

These background rates of extinction derive from the abundant and widespread species that dominate the fossil record (Pimm et al. 1995). Species most prone to current extinction are rare and local, so fossil data may still underestimate past extinction rates.

Recent work supplements these assessments of fossils by using the rapidly expanding knowledge of speciation rates based on molecular estimates of the evolutionary divergence of species. The argument has two parts. The first is that speciation and extinction rates cannot be very different. Were the latter higher than the former, the variety of life would have shrunk. If speciation rates were higher, we would often observe very 'bushy' evolutionary trees. There are some – that for human female mitochondrial DNA is an example – showing that all variation arose recently and in Africa. Such examples are rare, however.

The second part of the argument is that we can date some speciation events from well-timed geological events. They can be old, such as the division of the Caribbean from the Pacific, when the Panamanian land bridge was formed, or very recent, such as the isolation of populations on mountain tops as the last glaciation retreated. From these events, we can produce a time-calibrated scale of molecular divergence and so predict the time when other species diverged. It is this process that estimates the divergence of the human line from the chimpanzee line at a few million years ago.

These estimates now include a wide variety of species, including those that are rare and local (Pimm 2001). Molecular estimates are broadly compatible with the fossil data in suggesting a benchmark value for species longevity of a million years (perhaps more). It follows, that each year about one in a million species will expire from natural causes. Any more than that indicts human actions as responsible for their cause.

Box 1.1 presents several case studies demonstrating that current rates of extinction exceed the background rate by orders of magnitude. These examples demonstrate that extinctions can take place quickly, over large areas, in a wide variety of habitats, and involve very different kinds of species. In the next section we ask the obvious questions of what (if any) are the common patterns in what causes extinctions and which species and places are most vulnerable.

Causes of biodiversity loss

In his overview of recent extinctions, Diamond (1989) succinctly described four processes – the 'Evil Quartet' – that exterminate species. They are (i) habitat destruction, (ii) overexploitation, (iii) introduced species and (iv) secondary extinctions – the loss of a species that follows from the extermination of another species.

Habitat loss

Habitat loss through destruction and fragmentation is the predominant cause of extinction (WCMC 1992). On land, perhaps three-quarters of all well-known species live in tropical moist forests. Within the past 100 years (and often much less), human actions have shrunk these forests by half (Pimm 2001). The rates of deforestation are probably increasing. Other ecosystems are also shrinking, some, such as prairies and some tropical drywoodlands, at rates faster than tropical moist forests.

Habitat loss also has a significant impact on oceanic and freshwater habitats, with human activities such as damaging fishing techniques, exploitation, pollution and coastal development threatening 58% of the world's coral reefs (Bryant et al. 1998). Riverine habitats are similarly affected through extensive physical modifications such as damming and channelling. The seas cover more than two-thirds of the planet's surface yet only 250,000 to 300,000 marine species have been described, compared with more than one million on land. As on land, the peak of marine biodiversity lies in the tropics. Coral reefs account for almost 100,000 of these species, perhaps as much as 40% of the worlds' marine fishes, yet comprise just 0.2% of the ocean surfaces (Roberts et al. 2000). Although damage to coral reefs is important for the loss of species, by area, trawling does the greatest physical damage to ocean ecosystems. These effects occur across larger areas of the planet than tropical deforestation and involve even greater, more frequent disturbances. Watling & Norse (1998) estimated that $15 \times 10^6 \, km^2$ of the world's sea floor is ploughed each year by bottom trawling. Almost all the world's fisheries are concentrated in the $30 \times 10^6 \, km^2$ of nutrient-rich waters that are on the continental shelf, plus a few upwellings. On average, the ocean floor of these productive waters is trawled every 2 years. In reality, although a few areas escape trawling, others may be trawled five or even 50 times a year. Regrowth of animals is slow,

Box 1.1 Case histories

Freshwater mussels

In North American freshwater mussels, approximately 21 out of 297 species have become extinct since 1900 owing to habitat modification (Williams et al. 1992). Divide the number of extinctions (21) into the regional total (297) and multiplying it by the number of years over which the extinctions have occurred (c.100). That is, there have been 21 extinctions per 297 species in 100 years. Extrapolated to a million species, this regional rate is approximately 714 extinctions per million species per year – compared with the expectation of a single extinction per million species per year. Of course, we have selected these mussels as a special case of rapid extinction.

We can also generate a conservative estimate of **global** extinction rates, by supposing that these were the **only** freshwater mussel extinctions worldwide. By dividing the known extinctions per year by the worldwide total of species of freshwater mussels (c.1000), the global extinction rate of freshwater mussels over the past 100 years is approximately 200 extinctions per year per million species.

Freshwater fish

Of the approximately 950 species of freshwater fish in the USA, Canada and Mexico, 40 have become extinct in the past 100 years (Miller et al. 1989). The northern lakes, southern streams, wetlands and desert springs are very different habitats, but all have lost species. The arid region of south-western North America has lost most species, mainly from physical habitat changes and introduced species at springs, which are highly sensitive to disturbance. Some 50 species of Cyprinidae are threatened, including 14 species that inhabit spring systems in Nevada and 14 species in the Colorado River system. Impoundments, ground-water extraction, channelization and irrigation schemes appear to be contributory factors in 18 extinctions. Of 488 species of freshwater fish in south-eastern USA, four have become extinct and 80 more are threatened. Increasing development and chemical alteration of Appalachian and Cumberland mountain streams pose serious threats to many species (Miller et al. 1989).

Australian mammals

Australia and its surrounding islands are home to a unique mammal fauna: 85% of species are endemic, and it is the only region with all three major divisions of mammals (marsupials, monotremes and placentals) extant. Australian mammals have suffered two recent waves of extinction. The first was the 'megafauna' extinction event in the late Pleistocene between 50,000 and 10,000 years ago, when at least 20 genera of Australian mammals were lost. This wave of extinction has been attributed to ice-age climate change (Main 1978) and to human impact (Martin 1984; Flannery 1994). Some evidence suggests that the timing of megafaunal extinctions corresponds more closely with the arrival and spread of humans, around 50,000 years ago, than with the period of most extreme aridity around the Last Glacial Maximum 20,000 years ago (Miller et al. 1999; Roberts et al. 2001). The human overkill hypothesis is widely, but not universally, accepted (Wroe et al. 2004).

The second wave of Australian mammal extinctions began with European settlement in the late eighteenth century. Twenty-two mammal species have become extinct in that period, more than on any other continent (Australian Terrestrial Biodiversity Assessment 2002). A further eight species persist in tiny populations on offshore islands but are extinct in mainland Australia or Tasmania. If we include these eight species in the calculation, the recent rate of extinction of Australian terrestrial mammals is around 1400 species per million per year – twice that of the North American mussels. Of the Australian mammal species remaining, 59 are threatened with extinction (IUCN 2004). There is little doubt that the impacts associated with European settlement are to blame, although scientists debate the exact mechanism. The main causes seem to be habitat clearance for agriculture, habitat degradation by domestic stock and introduced rabbits, predation by introduced cats and foxes, and the breakdown of indigenous land-management regimes. Moreover, it is clear that this wave of extinctions is continuing. Within the past decade, substantial declines in abundance of several mammal species have been recorded in the relatively intact tropical savannas of northern Australia (Woinarski et al. 2001). Around 5 years ago, foxes were introduced into Tasmania and have rapidly established a breeding population, posing severe threats to the native mammal fauna as they have in mainland Australia.

particularly on the outer continental shelf and its slope, where natural storm damage is negligible, and many areas do not even start to recover before they are ploughed again

Sediment pollution from terrestrial run-off, another form of habitat destruction, causes severe reef degradation worldwide. The problem is concentrated in areas with rapid rates of land clearing and high rainfall, which causes swift erosion of exposed soils. Areas worst affected include Southeast Asia, East Africa, the Eastern Pacific and the Caribbean (Bryant et al. 1998), which are also richest in marine biodiversity. It is truly ironic that destruction of the most diverse terrestrial ecosystems – tropical rainforests – is causing the destruction of coral reefs, the most diverse marine system.

Overexploitation

Overkill results in the hunting of animals and the cutting of plants at rates faster than they can reproduce. Current rates of hunting for 'bush meat' are unsustainable in most areas where this activity occurs. (Bush meat is almost any vertebrate, often small ones, but obviously, the larger species are preferred.) Overexploitation occurs both in terrestrial ecosystems – deforestation often results in overharvesting of species such as mahogany, *Swientenia mahoganii* (Oldfield 1984) – and, perhaps more famously, in marine systems. Overfishing has resulted in valuable resources being driven to such low levels that exploitation is no longer sustainable and, in some cases, species have been driven to extinction. For example, since the 1990s, to supply a growing international market, many sharks are declining and are unable to recover due to their low reproduction rates (Manire & Gruber 1990; Waters 1992).

Introduced species

The translocation of alien species to new environments has caused mass extinctions of endemic faunas and floras, especially on islands where the biota were naïve to the effects of the invaders. Rats, rabbits, goats, pigs and predators such as cats have been among the most widely translocated and destructive of alien species (see Chapter 13).

Secondary extinctions

Finally, 'chains of extinction' (or 'extinction cascades') describe situations where the loss of one species causes the extinction of others that depend on it. For example, a specialized parasite would disappear if its specific host became extinct, as would plant species that lost their specific pollinators or seed dispersers. Other changes can be quite complicated. Once a species is lost, the species that fed upon, were fed upon, benefited or competed with that species will be affected. In turn, these species will affect yet other species. Food-web theory suggests that the pattern of secondary extinction may be quite complicated and thus difficult to predict (Pimm 1991). It also predicts that following the removal of particular species – often called 'keystones' – that the community of species that remain may change dramatically.

Which species are vulnerable?

Quantifying extinction risk

To understand which species are particularly vulnerable to these causes of extinction, we can obviously examine the characteristics of species that have already become extinct. It also makes sense to study those that we deem more or less close to extinction to glean insights from them.

Unfortunately, the information needed to do this is available for only a small fraction of the world's taxa, and estimates of extinction probability are usually little more than guesswork. The most comprehensive attempt to quantify extinction risk for large numbers of species is the Red List, compiled by the International Union for

Box 1.2 The IUCN Red List categories and criteria

Categories

The general aim of the Red List categories is to 'provide an explicit, objective framework for classifying the broadest range of species according to extinction risk' (IUCN 2004). Species categorized as Vulnerable, Endangered or Critically Endangered are grouped as 'Threatened'.

Extinct: there is no reasonable doubt that the last individual of this taxon has died.

Extinct in the Wild: the taxon is known to survive only in captivity or cultivation.

Critically Endangered: the taxon faces an extremely high risk of extinction in the wild. *Endangered*: the taxon faces a very high risk of extinction in the wild.

Vulnerable: the taxon faces a high risk of extinction in the wild.

Near Threatened: the taxon is close to being threatened in the near future.

Least Concern: the taxon is not at risk.

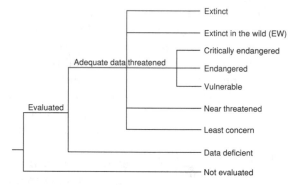

Criteria

The five Red List criteria (A–E), each with several subcriteria, are a set of objective guidelines for the classification of Threatened species (or subspecific taxa) as Vulnerable, Endangered or Critically Endangered. The criteria are based on: observed or estimated population or range reduction (A, B and C), extent of distribution (B and C), total population size (C and D), degree of population fluctuation (B and C) or fragmentation (B), geographical location (D) and quantitative modelling of extinction risk (E).

A rapid population or range decline
B small distribution and decline or fluctuation
C small population and decline
D very small or restricted population
E quantitative analysis

the Conservation of Nature and Natural Resources (IUCN 2004). Species are evaluated under a common set of quantitative criteria and assigned to one of seven ranked categories (see Box 1.2). Species can be categorized under several criteria, but the highest category specified by any criterion is taken as the species' extinction risk. Species for which too little is known to enable them to be placed in a category are labelled data deficient. To date, the only three major taxa for which all species have been assessed are birds, mammals and amphibians. Targets for the complete assessment of several other major taxa (reptiles, freshwater fish, sharks, rays and chimaeras, freshwater molluscs and plants) have been set (IUCN 2004). For other groups, however, the number of species currently assessed is a tiny fraction of the known species numbers; for example, fewer than 800 of the several million known insect species appear in the Red List.

The selectivity of extinction risk

Examining the lists of threatened and recently extinct species shows that by far the most common vulnerability is a small geographical range. Most recent extinctions have been on islands that, by definition, are small. One might think that island species might be unusually vulnerable because they are also ecologically naïve – they have not met the number of predators, for example, present on mainlands. Interestingly, this is not the case: for the same range size, island species are often less likely to be threatened than species on continents (Manne et al. 1999). The explanation is probably the second leading cause of threat – local scarcity. For a given range size, locally scarce species are much more likely to be threatened than species that are locally common. Island species, though geographically restricted, are often unusually common in their small ranges.

The explanation for these major vulnerabilities is obvious: other things being equal, the four major causes of extinction are likely to be greater threats to scarce, geographically restricted species than to common, widespread ones. Habitat destruction, for example, can more easily destroy a species if it has a small range encompassed by that destruction than if it has a larger one.

An unfortunate feature of global human impacts is that they disproportionately affect centres of endemism, where concentrations of geographically restricted species occur. Range-restricted species tend to have lower population densities and higher risks of extinction than widespread species. Myers (1988, 1990) defined these areas, centres of endemism combined with unusual levels of habitat destruction, as 'hotspots'. Species ranges are so concentrated that roughly half of all species on land are found in only 25 'hotspots', occupying only about 10% of the world's land surface. In 2000, approximately 12% of the original habitat of these 25 hotspots remained (Myers et al. 2000), a mere 37% of which is protected in any

way. Sixteen of these hotspots are forests and almost all are tropical forests. As a consequence of these high levels of habitat loss, these 25 hotspots are where the majority of threatened and recently extinct species are to be found.

Other factors are involved in extinction risk and the picture becomes more complex as one looks at smaller sets of species and particular regions. Species with small and declining populations, restricted geographical ranges and large area requirements are likely to be more at risk than common species, but traits such as body size, intrinsic rate of population increase and ecological specialization can all be important (McKinney 1997). Comparative studies of contemporary extinction risk (which typically use the Red List categories as a measure of risk) largely confirm this. In birds, for example, large body size, low fecundity and habitat specialization are associated with high extinction risk (Bennett & Owens 1997; Owens & Bennett 2000). In mammals, species at higher risk tend be at high trophic levels and have small geographical ranges, low population densities and slow life histories (Purvis et al. 2000b; Cardillo 2003; Cardillo et al. 2004).

It is clear, then, that extinction risk is determined not only by where a species lives and the external conditions it is exposed to, but also by its intrinsic, biological attributes. So far, we know little about whether external or intrinsic risk-promoting factors are more important (Fisher et al. 2003; Cardillo et al., 2004). There is evidence, however, that the two interact to determine extinction risk. In the mammal order Carnivora, there is an interaction between species' biological traits and degree of exposure to human populations. Slow life histories, low population densities and restricted distributions have a more acute influence on extinction risk among species that inhabit regions of high human population (Cardillo et al. 2004). Across mammals generally, body size has important interactive effects: many external and intrinsic factors that affect extinction risk do so more strongly for mammal species of larger size. Moreover, external factors seem to be more important in determining extinction risk among small species, whereas for large species, both external and intrinsic factors are important (Cardillo et al. 2005).

Intriguingly, there is also evidence that species in small, ancient or distinct lineages are more at risk of extinction than more recently evolved taxa (Johnson et al. 2002). As May (1990) has noted, some of the best-known threatened species, such as giant pandas (*Ailuropoda melanoleuca*) and tuataras (*Sphenodon* spp.), or recently extinct species such as the thylacine (*Thylacinus cynocephalus*), are phylogenetically old and distinct. The actual loss of bird and mammal species results in greater loss of genetic and evolutionary diversity than if extinct species were distributed randomly among higher taxonomic groups (Russell et al. 1988; Purvis et al. 2000a).

The ecological consequences of biodiversity loss

Do extinctions matter?

When species interfere with human endeavours we often suppress their numbers and – in some quarters – celebrate their demise. Such is the case with many carnivorous species such as the thylacine in Tasmania or wolves (*Canis lupus*) and brown bears (*Ursus arctos*) in the UK. Potential competitors or pests such as the passenger pigeon (*Ectopistes migratorius*), and disease organisms such as smallpox, have also been deliberately eliminated. Even when species do not compete with *Homo sapiens*, their passing often elicits little comment; good recent examples include many species of tropical rainforest frogs, small freshwater fish and the spectacular but ill-fated Miss Waldron's red colobus (*Procolobus badius waldroni*) from Ghana. So do extinctions matter?

The answer is in an emphatic 'yes' for a wealth of reasons. The effects of extinctions are obvious and matter profoundly. There are

practical ones about the species itself. Since the invention of bread, housewives may have wished the demise of the mould that spoiled baking. Only in the past 50 years or so have so many of us owed our very lives to *Penicillin* and other antibiotics. Not only are species useful, but so too are the ecosystems of which they are part. Species rarely go extinct as carefully excised members of some ecological community. Most commonly, a hillside is clear-cut, a reef dynamited or a wetland cleared, taking with it species and ecosystem benefits in one fell swoop.

Do extinctions matter ecologically?

Playing Devil's Advocate, we might argue that species are rare and often sparsely scattered before they finally expire, so their impact on energy and nutrient flows, use of resources and interactions with other species should be hardly noticeable after they have gone. We could also point to species that contribute little to community function and hence appear to be functionally redundant (Walker 1992). One problem with such arguments is that they often view species as static entities with fixed roles, and do not consider times or places where 'redundant' organisms predominate. One example is the long-haired rat, *Rattus villosissimus*, of central Australia. This distinctive rodent is often invisible at the landscape level for decades at a time, with small populations being clustered around desert oases. After drought-breaking rains its numbers erupt, and migratory hordes sweep across vast areas at speeds of $1-2 \, \text{km day}^{-1}$; dominating all other small mammals at these times, its burrow systems are used by at least 17 other species of vertebrates, and the excavated soil alters the dynamics of both the soil seed bank and trajectory of plant succession (Dickman 2003a). Despite its usual low profile, the long-haired rat is clearly not 'redundant' in any ecological sense of the word.

Ecologically, losses of species can have several consequences, and these depend largely on what the species do. At least three kinds of roles can be distinguished (Kinzig et al. 2001).

In the first instance, species interact with each other directly and indirectly, and influence each other's population sizes, use of resources and evolution. Interactions are sometimes obvious, such as when predators limit populations of their prey, competitively superior species invade the ranges of subordinates, or species depend on each other for provision of resources, as with the fungal and algal partners in lichens. In these one-on-one interactions, loss of one species can either liberate or doom the other. In other situations, interaction pathways occur among suites of species and make the task of predicting the impacts of a single extinction more difficult (Dickman 2003b). Such indirect interactions include trophic cascades, apparent competition, keystone predation and many others (Fig. 1.1).

As one example, let us consider a special kind of trophic cascade termed 'mesopredator release'. Here, if a top predator suppresses the numbers of a smaller predator, it may indirectly benefit the smaller predator's prey. Loss of the top predator, for example, the coyote (*Canis latrans*), may release mesopredators such as house cats (*Felis catus*), which can then deplete populations of scrub-breeding birds and lizards (Crooks & Soulé 1999). Local extinctions are

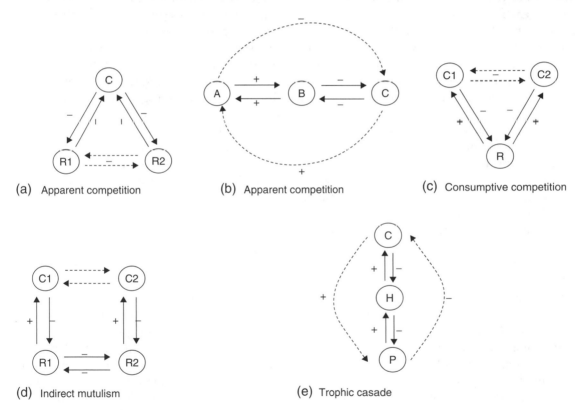

(a) Apparent competition

(b) Apparent competition

(c) Consumptive competition

(d) Indirect mutulism

(e) Trophic casade

Fig. 1.1 Examples of indirect interactions between species: (a) and (b) depict apparent competition; (c) consumptive competitio;, (d) indirect mutualism; and (e) trophic cascade. In (b), A, B and C represent species in the same trophic level; in (e), C represents top consumer or predator, H represents herbivore and P represents primary producer. In other interactions, C represents consumer species and R represents resource species. Direct effects between species are shown by solid arrows, indirect effects by broken arrows. Arrow heads show the species affected, and + and − show the direction of the effect. (Redrawn from Dickman, 2003b.)

sometimes followed by bewilderingly diverse effects, such as increases in plant damage following the loss of carnivores (Schmitz et al. 2000), increases in fish populations in the absence of feral horses (Levin et al. 2002) or, most famously, depressed reproductive success in flowering plants when house cats are absent (Darwin 1859). Such effects are often explicable within the framework of indirect interactions.

Second, some species affect their physical environment, and this in turn modulates the resources that are available to others. These species, termed 'ecosystem engineers' by Jones et al. (1994), often have dramatic and powerful effects on the environment, so the consequences of their extinction can be expected to be far-reaching. Two kinds of engineers were defined by Jones et al. (1994): 'autogenic engineers' change the environment via their own physical structures (e.g. corals that form reefs, trees that produce hollows), whereas 'allogenic engineers' change the environment by transforming materials from one state into another (e.g. burrowing animals such as the long-haired rat, noted above) (Fig. 1.2).

Let us consider two examples of the effects of losing engineer species. Firstly, the woylie (*Bettongia penicillata*) is a small (1 kg) marsupial that once occurred over most of southern Australia. Studies in the tiny current range of the species in the continent's far south-west show that individual woylies displace about 4.8 t of soil annually, and contribute significantly to infiltration of water, seed-bank dynamics and dispersal of hypogeal fungi (Garkaklis et al. 2004). Its disappearance from semi-arid habitats has increased rainfall run-off, and hence soil erosion, and appears to slow the establishment and growth of vascular plants. Secondly, the passenger pigeon once occurred in staggeringly large numbers (3–5 billion individuals) in eastern North America, but it was extirpated in the wild by 1900. Huge roosting and nesting aggregations of this species are suspected to have caused breakages of tree branches and limbs, which in turn increased fuel loads on the forest floor and influenced the frequency and intensity of fires (Ellsworth & McComb 2003). As passenger pigeons consumed vast numbers of red oak (*Quercus rubra*) acorns, the recent expansion of northern red oak forest and decline of white oak *Q. alba* may be further consequences of the pigeon's demise.

A third ecological role performed by species is the provision of 'ecosystem services'. These include fixation of energy and nutrients, cycling of water and minerals, formation of soil, transformation of gases and maintenance of climate. Early ecosystem-level studies suggested that particular species of plants are disproportionately important in fixing energy and matter (Waring 1989), so we may expect that loss of these species would compromise one or more ecosystem services (Ehrlich & Ehrlich 1992). Intriguingly, some research indicates that only a few species – perhaps a dozen – are needed to perform geochemical services, but key services such as primary productivity and uptake of CO_2 diminish with declining species richness (Naeem et al. 1995; Schläpfer et al. 2005).

Extinctions do not always have immediate or detectable effects, especially if the species lack strong engineering or keystone credentials. However, if losses are cumulative, ecosystem functioning may decline gradually until the system collapses. This scenario has been popularized as the 'rivet hypothesis' (Ehrlich & Ehrlich 1981), which likens species to rivets supporting an aircraft wing. Loss of just one or two rivets may increase only slightly the chance that the wing will fail, but catastrophe occurs when the next rivet is lost and the aircraft crashes. There is little, if any, evidence that collapse occurs catastrophically, but examples of progressive (and sometimes rapid) loss of ecological function abound.

Historical overexploitation of fisheries provides a good example. In many parts of the world, as readily exploited species of fish and shellfish have declined in catches, they have been replaced sequentially by ecologically similar species, thus delaying the onset of obvious system failure. However, as replacement species have themselves become progressively overfished (the last 'rivets'), coastal ecosystems have

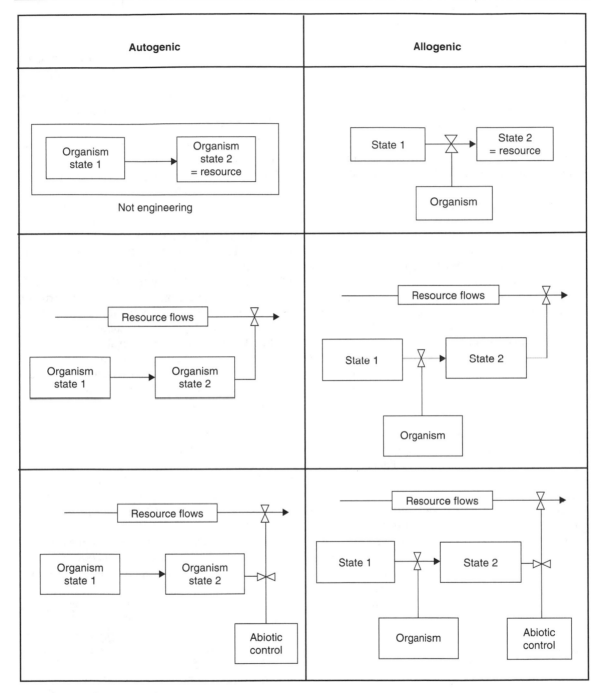

Fig. 1.2 Examples of ecosystem engineering by autogenic and allogenic engineer species. The point of modulation, where an engineer species alters the resource flow that is available to others, is represented by opposing arrow heads. (Redrawn from Jones et al. 1994.)

collapsed on a global scale (Jackson et al. 2001). In this situation, recovery of the original ecosystem is difficult or impossible to achieve, and degraded systems may then persist in alternative states for long periods (Suding et al. 2004).

A true conservationist is a man who knows that the world is not given by his fathers but borrowed from his children.

(John James Audubon 1785–1851.)

References

Australian Terrestrial Biodiversity Assessment (2002) http://audit.ea.gov.au/ANRA/ vegetation/docs/biodiversity

Bennett, P.M. & Owens, I.P.F. (1997) Variation in extinction risk among birds: chance or evolutionary predisposition? *Proceedings of the Royal Society of London Series B – Biological Sciences* **264**: 401–8

Bowman, D.M.J.S. (1993) Biodiversity: much more than biological inventory. *Biodiversity Letters* **1**: 163.

Bryant, D., Burke, L., McManus, J. & Spalding, M. (1998) *Reefs at Risk: a Map-based Indicator of Potential Threats to the World's Coral Reefs*. World Resources Institute, Washington, DC; International Center for Living Aquatic Resource Management, Manila; and United Nations Environment Programme–World Conservation Monitoring Centre, Cambridge.

Burgess, B.B. (2001) *Fate of the Wild: the Endangered Species Act and the Future of Biodiversity*. University of Georgia Press, Athens, Georgia.

Cardillo, M. (2003) Biological determinants of extinction risk: why are smaller species less vulnerable? *Animal Conservation* **6**: 63–9.

Cardillo, M., Purvis, A., Sechrest, W., Gittleman, J.L., Bielby, J. & Mace, G.M. (2004) Human population density and extinction risk in the World's carnivores. *Public Library of Science – Bology* **2**: 909–14

Cardillo, M., Mace, G.M., Jones, K.E., et al. (2005) Multiple causes of high extinction risk in large mammal species. *Science* **309**: 1239–41.

Crooks, K.R. & Soulé, M.E. (1999) Mesopredator release and avifaunal extinctions in a fragmented system. *Nature* **400**: 563–6.

Darwin, C. (1859) *On the Origin of Species by Means of Natural Selection or the Preservation of Favoured Races in the Struggle for Life*. J. Murray, London.

Diamond, J.M. (1989) Overview of recent extinctions. In *Conservation for the Twenty-first Century* (Eds D. Western & M.C. Pearl), pp. 37–41. Oxford University Press, Oxford.

Dickman, C.R. (2003a) Positive effects of rodents on biota in arid Australian systems. In *Rats, Mice and People: Rodent Biology and Management* (Eds G.R.

Singleton, L.A. Hinds, C.J. Krebs & D.M. Spratt), pp. 69–74. Australian Centre for International Agricultural Research, Canberra.

Dickman, C.R. (2003b) Species interactions: indirect effects. In *Ecology: an Australian Perspective* (Eds P. Attiwill and B. Wilson), pp. 158–70. Oxford University Press, Melbourne.

Ehrlich, P.R. & Ehrlich, A.H. (1981) *Extinction: the Causes and Consequences of the Disappearance of Species*. Random House, New York.

Ehrlich, P.R. & Ehrlich, A.H. (1992) The value of biodiversity. *Ambio* **21**: 219–26.

Ellsworth, J.W. & McComb, B.C. (2003) Potential effects of passenger pigeon flocks on the structure and composition of presettlement forests of eastern North America. *Conservation Biology* **17**: 1548–58.

Faith, D.P. (1992) Conservation evaluation and phylogenetic diversity. *Biological Conservation* **61**: 1–10.

Fisher, D.O., Blomberg, S.P. & Owens, I.P.F. (2003) Extrinsic versus intrinsic factors in the decline and extinction of Australian marsupials. *Proceedings of the Royal Society of London Series B – Biological Sciences* **270**, 1801–8.

Flannery, T.F. (1994) *The Future Eaters*. Reed Books, Sydney.

Garkaklis, M.J., Bradley, J.S. & Wooller, R.D. (2004) Digging and soil turnover by a mycophagous marsupial. *Journal of Arid Environments* **56**: 569–78.

Gaston, K.J. (1996) Species richness: measure and measurement. In *Biodiversity: a Biology of Numbers and Difference* (Ed. K.J. Gaston), pp. 77–113. Blackwell Science, Oxford.

Heywood, V.H. (Ed.) 1995. *Global Biodiversity Assessment*. Cambridge University Press, Cambridge.

Hughes, J.B., Daily, G.C. & Ehrlich, P.R. (1997) Population diversity: its extent and extinction. *Science* **280**: 689–92.

IUCN. (2004) *2004 IUCN Red List of Threatened Species*. Species Survival Commission, International Union for the Conservation of Nature and Natural Resources, Gland, Switzerland.

Jackson, J.B.C., Kirby, M.X., Berger, W.H., et al. (2001) Historical overfishing and the recent collapse of coastal ecosystems. *Science* **293**: 629–38.

Johnson, C.N., Delean, S. & Balmford, A. (2002) Phylogeny and the selectivity of extinction in Australian marsupials. *Animal Conservation* **5**: 135–42

Jones, C.G., Lawton, J.H. & Shachak, M. (1994) Organisms as ecosystem engineers. *Oikos* **69**: 373–86.

Kinzig, A.P., Pacala, S.W. & Tilman, D. (Eds) (2001) *The Functional Consequences of Biodiversity: Empirical Progress and Theoretical Extensions*. Princeton University Press, Princeton, New Jersey.

Levin, P.S., Ellis, J., Petrik, R. & Hay, M.E. (2002) Indirect effects of feral horses on estuarine communities. *Conservation Biology* **16**: 1364–71.

Main, A.R. (1978) Ecophysiology: towards an understanding of Late Pleistocene marsupial extinction. In *Biology and Quaternary Environments* (Eds D.Walker & J.C. Guppy), pp. 169–83. Australian Academy of Science, Canberra.

Manire, C.A. & Gruber, S.H. (1990) Many sharks may be headed towards extinction. *Conservation Biology* **4**: 10–11.

Manne, L.L., Brooks, T.M., & Pimm, S.L. (1999) Relative risk of extinction of passerine birds on continents and islands. *Nature* **399**: 258–61.

Martin, P.S. (1984) Prehistoric overkill: the global model. In *Quaternary Extinctions* (Eds P.S. Martin & R.G. Klein), pp. 354–403. University of Arizona Press, Tucson, Arizona.

May, R.M. (1990) Taxonomy as destiny. *Nature* **347**: 129–30.

May, R.M. (2000) The dimensions of life on Earth. In *Nature and Human Society – the Quest for a Sustainable World* (Eds P.H. Raven & T. Williams), pp. 30–45. National Academic Press, Washington, DC.

McKinney, M.L. (1997) Extinction vulnerability and selectivity: combining ecological and paleontological views. *Annual Review of Ecology and Systematics* **28**: 495–516.

Miller, G.H., Magee, J.W., Johnson, B.J., et al. (1999) Pleistocene extinction of *Genyornis newtoni*: human impact on Australian megafauna. *Science* **283**: 205–8

Miller, R.R., Williams, J.D. & Williams, J.E. (1989) Extinctions of North American fishes during the past century. *Fisheries* **14**: 22–38.

Myers, N. (1988): Threatened biotas: 'hotspots' in tropical forests. *The Environmentalist*, **8**: 1–20.

Myers, N. (1990): The biodiversity challenge: expanded hotspots analysis. *The Environmentalist* **10**: 243.

Myers, N, R., Mittermeier, A., Mittermeier, C.G., da Fonseca, G.A.B. & Kent, J. (2000): Biodiversity Hotspots for Conservation Priorities. *Nature* **403**: 853–8

Naeem, S., Thompson, L.J., Lawler, S.P., Lawton, J.H. & Woodfin, R.M. (1995) Empirical evidence that declining biodiversity may alter the performance of terrestrial ecosystems. *Philosophical Transactions of the Royal Society of London, Series B* **347**: 249–62.

Norse, E.A. & Carlton, J.T. (2003)World Wide Web buzz about biodiversity. *Conservation Biology* **17**(6): 1475–6.

Oldfield, M.L. (1984) *The Value of Conserving Genetic Resources*. Sinauer Associates, Sunderland, MA.

Owens, I.P.F. & Bennett, P.M. (2000) Ecological basis of extinction risk in birds: Habitat loss versus human persecution and introduced predators. *Proceedings of the National Academy of Sciences of the United States of America*, **97**, 12144–8.

OTA (1987) *Technologies to Maintain Biological Diversity*. US Congress Office of Technology Assessment, US Government Printing Office, Washington, DC.

Pimm, S. L. (1991) *The Balance of Nature? Ecological Issues in the Conservation of Species and Communities*. University of Chicago Press, Chicago, IL, pp. 30–45.

Pimm, S. L. (2000) Biodiversity is us. *Oikos* **90**: 3–6.

Pimm, S.L. (2001) *The World According to Pimm: a Scientist Audits the Earth*. McGraw Hill, New York.

Pimm, S.L., Russell, G.J., Gittlemn, J.L. & Brooks, T.M. (1995) The future of biodiversity. *Science* **269**: 347–50.

Purvis, A. & Hector, A. (2000) Getting the measure of biodiversity. *Nature* **405**: 212–19.

Purvis, A., Agapow, P.-M., Gittleman, J.L., & Mace, G.M. (2000a) Nonrandom extinction and the loss of evolutionary history. *Science* **288**: 328–30.

Purvis, A., Gittleman, J.L., Cowlishaw, G. & Mace, G.M. (2000b) Predicting extinction risk in declining species. *Proceedings of the Royal Society of London, Series B* **267**: 1947–52

Reich, P.B., Cavender-Bares, J., Craine, J., Walters, M.B., Oleksyn, J., Westoby, M. & Wright, I.J. (2003) The evolution of plant functional variation: traits, spectra, and strategies. *International Journal of Plant Sciences* **164**: S143–64.

Ricketts, T.H., Dinerstein, E., Olson, D.M., et al. (1999) *Terrestrial Ecoregions of North America: a Conservation Assessment*. Island Press, Washington, DC, 485 pp.

Roberts, J. M., Harvey, S. M., Lamont, P. A. & Gage, J. D. (2000) Seabed photography, environmental assessment and evidence for deep-water trawling on the continental margin west of the Hebrides. *Hydrobiologia* **44**, 173–83.

Roberts, R.G., Flannery, T.F., Ayliffe, L.K., et al. (2001) New ages for the last Australian megafauna: continent-wide extinction about 46,000 years ago. *Science* **292**: 1888–92.

Russell, G. J., T. M. Brooks, M. L. McKinney & C. G. &erson (1998) Taxonomic selectivity in bird and mammal extinctions: present and future. *Conservation Biology* **12**: 1365–76.

SCBD (2005) *Handbook of the Convention on Biological Diversity Including its Cartagena Protocol on Biosafety*, 3rd edn. Secretariat of the Convention on Biological Diversity, Montreal.

Schläpfer, F., Pfisterer, A.B. & Schmid, B. (2005) Non-random species extinction and plant production: implications for ecosystem functioning. *Journal of Applied Ecology* **42**: 13–24.

Schmitz, O.J., Hambäck, P.A. & Beckerman, A.P. (2000) Trophic cascades in terrestrial systems: a review of the effects of carnivore removals on plants. *American Naturalist* **155**: 141–53.

Soulé, M.E. (1991) Conservation tactics for a constant crisis. *Science* **253**: 744–749.

Stork, N.E. (1998) Insect diversity: facts, fiction and speculation. *Biological Journal of the Linnean Society* **35**: 321–37.

Suding, K.N., Gross, K.L. & Houseman, G.R. (2004) Alternative states and positive feedbacks in restoration ecology. *Trends in Ecology and Evolution* **19**: 46–53.

Walker, B.H. (1992) Biodiversity and ecological redundancy. *Conservation Biology* **6**: 18–23.

Waring, R.H. (1989) Ecosystems: fluxes of matter and energy. In *Ecological Concepts: the Contribution of Ecology to an Understanding of the Natural World* (Ed. J.M. Cherrett), pp. 17–41. Blackwell Scientific Publications, Oxford.

Waters, T. (1992) Sympathy for the devil. *Discover* **13**: 62

Watling, L. & Norse, E.A. (1998): Disturbance of the seabed by mobile fishing fear: a comparison to forest clearcutting. *Conservation Biology*, **12**, 1180–97

WCMC (1992) *Global biodiversity: Status of the Earth's Living Resources*. World Conservation Monitoring Centre, Cambridge.

Williams, J. D., Warren Jr., M.L., Cummings, K.S., Harris, J.L. & Neves, R.J. (1992) Conservation status of freshwater mussels of the United States and Canada. *Fisheries*, **18**, 6–22.

Williams, P.H., Gaston, K.J. & Humphries, C.J. (1994) Do conservationists and molecular biologists value differences between organisms in the same way? *Biodiversity Letters* **2**: 67–78.

Woinarski, J.C.Z., Milne, D.J. & Wanganeen, G. (2001) Changes in mammal populations in relatively intact landscapes of Kakadu National Park, Northern Territory, Australia. *Austral Ecology* **26**: 360–70.

Wroe, S., Field, J., Fullagar, R. & Jermiin, L.S. (2004) Megafaunal extinction in the late Quaternary and the global overkill hypothesis. *Alcheringa* **28**: 291–331.

2

Prioritizing choices in conservation

Georgina M. Mace, Hugh P. Possingham and Nigel Leader-Williams

The last word in ignorance is the man who says of an animal or plant: 'What good is it?' If the land mechanism as a whole is good, then every part is good, whether we understand it or not. If the biota, in the course of aeons, has built something we like but do not understand, then who but a fool would discard seemingly useless parts? To keep every cog and wheel is the first precaution of intelligent tinkering.
(Aldo Leopold, *Round River*, Oxford University Press, New York, 1993, pp. 145–6.)

Introduction

We are in the midst of a mass extinction in which at least 10%, and may be as much as 50%, of the world's biodiversity may disappear over the next few hundred years. Conservation practitioners face the dilemma that the cost of maintaining global biodiversity far exceeds the available financial and human resources. Estimates suggest that in the late twentieth century only US$6 billion per year was spent globally on protecting biodiversity (James et al. 1999), even though an estimated US$33 trillion per year of direct and indirect benefits were derived from ecosystem services provided by biodiversity, implying an asset worth US$330 trillion (Costanza et al. 1997). Together these crude estimates suggest that there could be a 500-fold underinvestment in conserving the world's biodiversity. However, even if these estimates are wildly wrong, the imbalance of funding is seriously inconsistent with best business prac-

tice in other sectors. In business, many companies spend about 10% of the value of their capital assets each year on maintaining those assets, although the figure varies depending on the type of asset. For example, 30% might be spent for computers compared with 5% for buildings: contrast that with 0.02% for biodiversity! Furthermore, the scale of underinvestment in biodiversity may be exaggerated by the effects of poor governance, sometimes even corruption, on achieving success in conservation (Smith et al. 2003). Given such problems, conservation scientists and nongovernment organizations (NGOs) supporting international conservation efforts are beginning to develop systems to more effectively target investment in biodiversity conservation (Johnson 1995; Kershaw et al. 1995; Olson & Dinerstein 1998; Myers et al. 2000; Possingham et al. 2001; Wilson et al. in press).

One fundamental resource allocation question facing conservation scientists and practitioners is whether conservation goals are best met by managing single species as opposed to whole ecosystems (Simberloff 1998). Efforts in conservation priority setting have historically concentrated on ecosystem-based priorities – determining where and when to acquire protected areas (Ferrier et al. 2000; Margules & Pressey 2000; Pressey & Taffs 2001; Meir et al. 2004). There has been comparatively little work on the question of how to allocate conservation effort between species. Despite the tension between ecosystem-based and species-based conservation, we believe there is merit in considering the issue of resource allocation between species because:

1. a 'fuzzy' idea such as ecosystem management holds little appeal for the general public, who prefer to grasp simpler messages conveyed by charismatic species such as tigers (Leader-Williams & Dublin 2000);
2. data on species, whether through direct counts of indicator species (Heywood 1995), or through assessments of threat (Butchart et al. 2005), provide some of the most readily available, repeatable and explicit monitoring and analytic systems with which to assess the success or otherwise of conservation efforts (Balmford et al. 2005).
3. in practice, almost regardless of their ultimate goal, conservation bodies often end up directing conservation actions to species and species communities (see e.g. figure 1 of Redford et al. 2003), probably because these are tangible and manageable components of ecosystems.

The topic of setting priorities for conservation is immense, so here we restrict ourselves to different methods for setting priorities between species. We explore the issues that a systematic approach should consider, and we show how simple scoring systems may lead to unintended consequences. We also recommend an explicit discussion of attributes of the species that make them desirable targets for conservation effort. Using a case study, we show how different perspectives will affect the outcome, and so as an alternative we present a method based on economic optimization. Ultimately, any decisions about 'what to save first' should in-

clude judgments that cannot be made by scientists or managers alone. Involving wider societal and political decision-making processes is vital to gain local support for, and ensure the ultimate success of, all conservation planning.

Single species approaches

Species-based conservation management approaches have, until fairly recently, concentrated on a single species, such as keystone species, umbrella species, indicator species or flagship species (see Leader-Williams & Dublin 2000). Keystone and umbrella species differ in the importance of their ecological role in an ecosystem:

1. **keystone species** have a disproportionate effect on their ecosystem, due to their size or activity, and any change in their population will have correspondingly large effects on their ecosystem (e.g. the sole fruit disperser of many species of tree);
2. **umbrella species** have such demanding habitat and/or area requirements that, if we can conserve enough land to ensure their viability, the viability of smaller and more abundant species is almost guaranteed.

In contrast, 'flagship species' encompass purely strategic objectives:

3. **flagship species** are chosen strategically to raise public awareness or financial support for conservation action.

Furthermore, definitions for indicator species can encompass both ecological and strategic roles:

4. **indicator species** are intended either to represent community composition or to reflect environmental change. With respect to the latter, indicator species must respond to the particular environmental change of con-

cern and demonstrate that change when monitored.

One species may be a priority species for more than one reason, depending on the situation or context in which the term is used. However, the terms 'keystone' and 'umbrella' are likely to remain more of a fixed characteristic or property of that species. In contrast, the term 'flagship' and, possibly to a lesser extent, 'indicator' may be more context-specific.

Promoting the conservation of a specific focal species may help to identify potential areas for conservation that satisfy the needs of other species and species assemblages (Leader-Williams & Dublin 2000). For example, the umbrella species concept (Simberloff 1998) can represent an efficient first step to protect other species. In addition, minimizing the number of species that must be monitored once a protected area has been created will reduce the time and money that must be devoted to its maintenance (Berger 1997).

Alternatively, conservation managers and international NGOs may choose to focus on the most charismatic 'flagship' species, which stimulate public support for conservation action, and that in turn may have spin-off benefits for other species. For example, use of the giant panda (*Ailuropoda melanoleuca*) as a logo by the World Wildlife Fund (WWF) has been widely accepted (Dietz et al. 1994) as a successful mechanism for conserving many other species across a wide variety of taxonomic groups. Furthermore, other mammalian and avian 'flagships' have been used to promote the conservation of large natural ecosystems (i.e. Mittermeier 1986; Goldspink et al. 1998; Downer 1996; Johnsingh & Joshua 1994; Western 1987; Dietz et al. 1994).

Nevertheless, the context of what may constitute a charismatic species can differ widely across stakeholders. For example, the tiger (*Panthera tigris*) is among the most popular flagship species in developed countries, but those in developing countries who suffer loss of life and livelihood because of tigers or other large predators have a different view (Leader-Williams & Dublin 2000). Such dissonance is best avoided by promoting locally supported flagship species (Entwistle 2000). For example, the discovery of a new species of an uncharismatic, but virus-resistant, wild maize, with its possible utilitarian value for human food production, highlighted the conservation value of the Mexican mountains in which it was found (Iltis 1988). This increase in local public awareness led to the establishment of a protected area that conserves parrots and jaguars (*Panthera onca*), orchids and ocelots (*Leopardus pardalis*), species that many consider charismatic. Hence this species of wild maize served as a strategically astute local flagship species. Another way of promoting local flagships is to prioritize those species that bring significant and obvious local benefits (Goodwin & Leader-Williams 2000), such as the Komodo dragon, *Varanus komodoensis* (Walpole & Leader-Williams 2001), which generates tourism. Similarly, species that can be hunted for sport, such as the African elephants (*Loxodonta africana*), may contribute directly to community conservation programmes (Bond 1994).

Several questions can arise from promoting conservation through single species (Simberloff 1998). One of these is how should individual species be prioritized? The common response is to begin with species that are most at risk of extinction, the critically endangered species. Many countries and agencies take this approach. However, there may be no known management for some of these species, and if there is, it may be risky and/or expensive. This can lead to a large share of limited conservation resources being expended with negligible or uncertain benefit (Possingham et al. 2002). On the other hand, when taking an ecosystem approach, managers might choose to focus on the keystone species that play the most significant role in the ecosystem. Unfortunately in many ecosystems we do not know the identity of keystone species. Often, after intensive study, they turn out to be invertebrates or fungi (Paine 1995), groups that are unlikely to attract public or government support unless ways can be found to make them locally relevant.

Another problem with single species conservation arises when the management of one focal species is detrimental to the management of another focal species. For example, in the Everglades of Florida, management plans for two charismatic, federally listed birds are in conflict. One species, the Everglades snail kite (*Rostrhamus sociabilis plumbeus*), has been reduced to some 600 individuals by wetland degradation and agricultural and residential development. It feeds almost exclusively on freshwater snails of the genus *Pomacea* and requires high water levels, which increase snail production. The snail kite is thus an extreme habitat specialist (Ehrlich et al. 1992). The other species, the wood stork *Mycteria americana,* has been reduced to about 10,000 pairs by swamp drainage, habitat modification and altered water regimes. Ironically, the US Fish and Wildlife Service opposed a proposal by the Everglades National Park to modify water flow to improve stork habitat on the grounds that the change would be detrimental to the kite (Ehrlich et al. 1992) (an added thought-provoking detail is that both species are common in South America).

Another issue is that few studies have been carried out to assess the effectiveness of one focal species in adequately protecting viable populations of other species (Caro et al. 2004). For example, the umbrella-species concept is often applied in management yet rarely tested. The grizzly bear (*Ursus arctos*) has been recognized as an umbrella species but, had a proposed conservation plan for the grizzly bear in Idaho been implemented, taxa such as reptiles would have been underrepresented (Noss et al. 1996). Similarly, in a smaller scale study, the areas where flagship species, such as jaguar, tapir (*Tapirus terrestris*) and white-lipped peccary (*Tayassu pecari*), were most commonly seen did not coincide with areas of vertebrate species richness or abundance (Caro et al. 2004). Although these results may not hold true for all other protected areas based around flagship spe-

cies, it does highlight the need for more field-based studies to determine the most appropriate approach for conserving the most biodiversity. As a result of problems associated with single species management, focus has been turning towards multiple species approaches.

Multispecies approaches

Methods based on several focal species, or protecting a specific habitat type, might be a more appropriate means of prioritizing protected areas (Lambeck 1997; Fleishman et al. 2000; Sanderson et al. 2002b). A frequent criticism of setting conservation priorities based on a single focal species is that it is improbable that the requirements of one species would encapsulate those of all other species (Noss et al. 1996; Basset et al. 2000; Hess & King 2002; Lindenmayer et al. 2002). Hence, there is a need for multispecies strategies to broaden the coverage of the protective umbrella (e.g. Miller et al. 1999; Fleishman et al. 2000, 2001; Carroll et al. 2001).

Among the different multispecies approaches, Lambeck's (1997) 'focal species' approach seems the most promising because it provides a systematic procedure for selecting several focal species (Lambeck 1997; Watson et al. 2001; Bani et al. 2002; Brooker 2002; Hess & King 2002). In Lambeck's (1997) innovative approach, a suite of focal species are identified and used to define the spatial, compositional and functional attributes that must be present in a landscape. The process involves identifying the main threats to biodiversity and selecting the species that is most sensitive to each threat. The requirements of this small and manageable suite of focal species guide conservation actions. The approach was extended by Sanderson et al. (2002a), who proposed the 'landscape species approach'. They defined landscape species by their 'use of large, ecologically diverse areas and their impacts on the structure and function of natural ecosystems . . . their requirements in time and space make them particularly susceptible to human alteration and use of wild landscapes'. Because landscape species require large, wild areas, they could potentially serve an umbrella function (*sensu* Caro & O'Doherty 1999) – meeting their needs would provide substantial protection for the species with which they co-occur. Like other focal-species approaches, this method of setting priorities carries inherent biases (Lindenmayer et al. 2002), and may be constrained by incomplete or inconsistent data.

Ecosystem and habitat-based approaches

Some conservation scientists believe that setting conservation priorities at the scale of ecosystems and habitats is more appropriate for developing countries with limited resources for conservation, inadequate information about single species and pressing threats such as habitat destruction. Logically, how much effort we place in conserving a particular ecosystem should take into account factors such as: how threatened it is, how well represented that ecosystem is in that country's protected area network, the number of species restricted to that ecosystem (endemic species), the cost of conserving the ecosystem and the likelihood that conservation actions will work. One can debate the relative importance of each of these factors – for example, some consider the the number of endemic species is paramount, whereas others prefer the notion of 'equal representation' whereby a fixed percentage of every habitat type is conserved.

The main goals of an ecosystem approach are to:

1. maintain viable populations of all native species *in situ*;
2. represent, within protected areas, all native ecosystem types across their natural range of variation;
3. maintain evolutionary and ecological processes;

4. manage over periods of time long enough to maintain the evolutionary potential of species;
5. accommodate human use and occupancy within these constraints (Grumbine 1994).

Although the financial efficiencies inherent in managing an ecosystem rather than several single species are attractive, this approach is also not without its problems. First, compared with a species, ecosystem boundaries are harder to define, so determining the location, size, connectivity and spacing of protected areas to conserve the full range of ecosystems, and variation within those ecosystems, is more difficult (Possingham et al. 2005). Second, individual species are more interesting to people and will attract greater emotional and financial investments than ecosystems. Third, although ecological services are provided by ecosystems, individual species often play pivotal roles in the provision of these services, particularly for direct uses such as tourism or harvesting. Finally, the main problem faced by managers wishing to implement an ecosystem approach is the lack of data available on how ecosystems function. This manifests itself in confusion about how much of each ecosystem needs to be conserved to protect biodiversity adequately in a region. In contrast, for the better known single species, the issue of adequacy can be dealt with using population viability analysis and/or harvesting models (Beissinger & Westphal 1998; this volume, Chapter 15).

Systematic conservation planning

Systematic conservation planning (or gap-analysis in the USA: Scott et al. 1993) focuses on locating and designing protected areas that comprehensively represent the biodiversity of each region. Without a systematic approach, protected area networks have the tendency to occur in economically unproductive areas (Leader-Williams et al. 1990), leaving many habitats or ecosystems with little or no protection (Pressey 1994). The systematic conservation planning approach can be divided into six stages (Margules & Pressey 2000).

1. Compile biodiversity data in the region of concern. This includes collating existing data, along with collecting new data if necessary, and if time and funds permit. Where biodiversity data, such as habitat maps and species distributions, are limited more readily available biophysical data may be used that reflect variation in biodiversity, such as mean annual rainfall or soil type.
2. Identify conservation goals for the region, including setting conservation targets for species and habitats, and principles for protected area design, such as maximizing connectivity and minimizing the edge-to-area ratio.
3. Review existing conservation areas, including determining the extent to which they already meet quantitative targets, and mitigate threats.
4. Select additional conservation areas in the region using systematic conservation planning software.
5. Implement conservation action, including decisions on the most appropriate form of management to be applied.
6. Maintain the required values of the conservation areas. This includes setting conservation goals for each area, and monitoring key indicators that will reflect the success of management (see below).

Ultimately, conservation planning is riddled with uncertainty, so managers must learn to deal explicitly with uncertainty in ways that minimize the chances for major mistakes (Margules & Pressey 2000; Araújo & Williams 2000, Wilson et al 2005), and be prepared to modify their management goals appropriately through adaptive management.

Systematic conservation planning can complement species-based approaches because it focuses on removing the threat of development and it compliments a long tradition of species recovery plans that concentrate on mitigating threats. The degree to which different countries use species-based planning as opposed to systematic conservation planning depends on historical, cultural and legislative influences. Even with systematic conservation planning, however, the better surveyed species or species groups often feature as the units for assessment. In other words, the conservation value of different areas is often assessed on the presence or conservation status of the species within it, simply because these are the best known elements of biodiversity. Systematic conservation planning approaches have become popular and widespread, partly because they are supported by several decision-support software packages (Possingham et al 2000, Pressey et al 1995, Williams et al 2000, Garson et al 2002).

Methods for setting conservation priorities of species

Prioritizing species, habitats and ecosystems by their perceived level of endangerment has become a standard practice in the field of conservation biology (Rabinowitz 1981; Master 1991; Mace & Collar 1995; Carter et al. 2000; Stein et al. 2000). The need for a priority-setting process is driven by limited conservation resources that necessitate choices among a subset of all possible species in any given geographical area, and distinct differences among species in their apparent vulnerability to extinction or need for conservation action. This need has led to the development of practical systems for categorizing and assessing the degree of vulnerability of various components of biodiversity, particularly vertebrates (e.g. Millsap et al. 1990; Mace & Lande 1991; Master 1991; Reed 1992; Stotz et al. 1996), and more recently ecoregions (Hoekstra et al. 2005).

Methods used for assessing the conservation status of species are varied but follow three general styles (Regan et al. 2004), rule-based, point scoring and qualitative judgement. Perhaps the best known system is that developed by the IUCN (International Union for the Conservation of Nature and Natural Resources) – The World Conservation Union – which uses a set of five quantitative rules with explicit thresholds to assign a risk of extinction (Mace & Lande 1991; IUCN 2001). Other methods adopt point-scoring approaches where points are assigned for a number of attributes and summed to indicate conservation priority (Millsap et al. 1990; Lunney et al. 1996; Carter et al. 2000). Other methods assess conservation status using qualitative criteria; judgements about a species' status are determined intuitively based on available information and expert opinion (Master 1991). One widely applied system is the biodiversity status-ranking system developed and used by the Natural Heritage Network and The Nature Conservancy (Master 1991; Morse 1993). This ranking system has been designed to evaluate the biological and conservation status of plant and animal species and within-species taxa, as well as of ecological communities.

Rule-based methods

Quantitative rule-based methods can be used to estimate the extinction risk of a species and thus contribute to determining priority areas for conservation action. For example, the IUCN Red List places species in one of the following categories: extinct (EX), extinct in the wild (EW), critically endangered (CR), endangered (EN), vulnerable (V), near threatened (NT) or least concern (LC), based on quantitative information for known life history, habitat requirements, abundance, distribution, threats and any specified management options of that species, and in a data deficient (DD) category if there are insufficient data to make an assessment (IUCN 2001). The IUCN system is based around five criteria (A to E) which reflect different ways in which a species might qualify for any of the threat categories (CR, EN, VU). A species is placed in a category if it meets one or

more of the criteria – for example because there are less than 250 mature individuals of the Norfolk Island green parrot (*Cyanoramphus cookii*) in the wild it is immediately listed as endangered under criterion D of the IUCN Red List protocol. A similar species can meet a higher category of threat if it meets alternative criteria. For example, the orange-bellied parrot (*Neophema chrysogaster*) also has less than 250 mature individuals but it is listed as critically endangered, under criterion C2b, because the population is also in decline and all the individuals are in a single subpopulation. One conceptual problem with rule-based methods is that a species that just missed out on being listed as, say, endangered on several criteria would be ranked as vulnerable, equal with a species that may have only just met the criteria for being vulnerable.

The rule-based methods have the advantage that they are completely explicit about what feature of the species led to it being listed as threatened. In the IUCN system, assessors have to list the criteria whereby the species qualified for a particular category of threat, and also have to provide documentation to support this information – usually in the form of scientific surveys or field reports that detail the information used. As a result, listings may be continually updated and improved as new data become available. Normally this will allow a new consensus among experts, but in the exceptional cases where this is not agreed, the IUCN have a petitions and appeals process to resolve matters. For example, in 2001 some of the listings of marine turtle species were disputed among experts. On this occasion, IUCN implemented their appeal procedure and provided a new assessment (http://www.iucn.org/themes/ssc/redlists/petitions.html). The wide use of the IUCN system also means that there is an ever increasing resource of best-practice documentation and guidelines, which aid consistent and comparable approaches by different species assessors (see *http://www.iucn.org/themes/ssc/redlists.htm*).

Point scoring method

The point scoring method for assigning conservation priority involves assigning a series of scores to each species based on different parameters relating to their ecology or conservation status, which together will determine their relative priority. One method of dealing with the scores is then to simply sum them to give an overall conservation priority, although this can be misleading. Beissinger et al. (2000) suggest that a categorical approach based on a combination of scores might be more accurate in determining overall conservation priority.

An example of a point scoring system is that developed by Partners in Flight (PIF) in 1995 in an effort to conserve non-game birds and their habitats throughout the USA (Carter et al. 2000). The PIF system involves assigning a series of scores to each species ranging from 1 (low priority) to 5 (high priority) for seven parameters that reflect different degrees of need for conservation attention. The scores are assigned within physiographical areas and the seven parameters are based on global and local information. Three of the parameters are strictly global and are assigned for the entire range of the bird: breeding distribution (BD), non-breeding distribution (ND) and relative abundance (AR). Other parameters are threats to breeding (TB), threats to non-breeding (TN), population trend (PT) and, locally, area importance (AI). The scores for each of these seven parameters are obtained independently (Carter et al. 2000). The PIF then uses a combination of approaches, including the summing of scores, to determine an overall conservation priority (Carter et al. 2000), with species that score highly on several parameters achieving high priority. Although this method of defining bird species of high conservation priority is thought to be reliable, like other methodologies, it is hindered by the lack of data on species distribution, abundance and populations trends, particularly in areas outside the USA to which many of these species migrate (Carter et al. 2000).

A problem with some point-scoring methods is that there is no explicit link to extinction risk, the weightings of each criteria, from 1 to 5 in the example above, are completely arbitrary, and there is an infinity of ways in which the scores could be combined: adding, multiplying, taking the product of the largest three values, and so forth. A related problem is that point-scoring methods can generate an artificially high ranking for a species when criteria are interrelated. For example, a system that prioritized species because they needed large home ranges, had slow reproductive rates and small litter sizes might end up allocating unreasonably high scores to any large-bodied species. All three of these traits are associated with relatively large body size, but they are not necessarily so much more vulnerable.

Conservation status ranks method

Status ranks are based primarily on objective factors relating to a species' rarity, population trends and threats. Four aspects of rarity are typically considered: the number of individuals, number of populations or occurrences, rarity of habitat, and size of geographic range. Ranking is based on an approximately logarithmic scale, ranging from 1 (critically imperiled) to 5 (demonstrably secure). Typically species with ranks from 1 to 3 would be considered of conservation concern and broadly overlap with species that might be considered for review under the Endangered Species Act or similar state or international statutes.

The NatureServe system (Master 1991) is one example of a system that uses status ranks. Developed initially by The Nature Conservancy (TNC) and applied throughout North America, the NatureServe system uses trained experts who evaluate quantitative data and make intuitive judgements about species vulnerability. The aim of the NatureServe system is to determine the relative susceptibility of a species or ecological community to extinction or extirpation. To achieve this, assessments consider both deterministic and stochastic processes that can lead to extinction. Deterministic factors include habitat destruction or alteration, non-indigenous predators, competitors, or parasites, over-harvesting and environmental shifts such as climate change. Stochastic factors include, environmental and demographic stochasticity, natural catastrophes and genetic effects (Shaffer 1981).

NatureServe assessments are performed on a basic unit called an element. An element can be any plant or animal species or infraspecific taxon (subspecies or variety), ecological community, or other non-taxonomic biological entity, such as a distinctive population (e.g. evolutionarily significant unit or distinct population segment, as defined by some agencies) or a consistently occurring mixed species aggregation of migratory species (e.g. shorebird migratory concentration area) (Regan et al. 2004). Defining elements in this way ensures that a broad spectrum of biodiversity and ecological processes are identified and targeted for conservation (Stein et al. 2000). This approach is believed to be an efficient and effective approach to capturing biodiversity in a network of reserves (e.g. Jenkins 1976, 1996). Assessment results in a numeric code or rank that reflects an element's relative degree of imperilment or risk of extinction at either the global, national or subnational scale (Master et al. 2000).

Back to basics – extinction risk versus setting priorities

The discussion above has reviewed methods for categorizing species according to their conservation priority. Running throughout is a tendency to equate conservation priority with extinction risk; yet these are clearly not the same thing (Mace & Lande 1991). Extinction risk is only one of a range of considerations that determine priorities for action or for conservation funding. The threat assessment is really an assessment of **urgency**, and an answer to the question of how quickly action needs to be taken. Hence, all other things being equal, the critically endangered species will be most likely to become extinct first if nothing is done. However, this is by no means the only consideration that should be used by a conservation planner. How then should extinction risk be used for priority-setting? It may be easier to make the analogy with a different system altogether. For example, the priority-setting systems used by Triage nurses in hospital emergency departments categorize people according to how urgently they need to be seen; those seen first are the ones that appear to have the most urgent and threatening symptoms. The symptoms can be very diverse, however, and some may turn out upon inspection and diagnosis to be less serious than might have been expected. Medical planning across the board would not use the triage system to allocate resources. The same is true for conservation planning. As with ill and injured people, our first sorting of cases should be according to urgency, and should also be precautionary (i.e. take more risks with listing species that are in fact not threatened than with failing to list those that really are). However, once the diagnosis is made, and the manager is reasonably sure that most critical cases are now known and diagnosed, a more systematic planning process should follow.

Variables other than risk

Now we consider a whole range of new variables other than risk. Table 2.1 shows a range of variables – grouped under headings of biological value (i.e. what biologists would consider), economic value, social and cultural value, urgency and practical issues. Under each of these headings are a range of attributes that might contribute to a species priority. The first three columns concern values, but the last two are rather different. Urgency is a measure that can be complicated to implement – i.e. high urgency may indicate that if nothing is done now, then it

Table 2.1 Classes and kinds of issues that are considered in priority-setting exercises for single-species recovery

Biological value	Economic value	Social and cultural value	Urgency	Practical issues
Degree of endemism	Cost of management or recovery	Scientific and educational benefits	Threat status = extinction risk	Feasibility and logistics
Relictual status	Direct economic benefits	Cultural status (e.g. ceremonial)	Time limitation, i.e. opportunities will be lost later	Recoverability, i.e. reversibility of threats, rate of species response
Evolutionary uniqueness	Indirect economic benefit	Political status (e.g. symbolic or emblematic)	Timeliness, i.e. likelihood of success varies with time	Popularity – will there be support from the community?
Collateral benefits to other species	Ecological services	Popularity		Responsibility, i.e. how much is this also someone else's responsibility?
Collateral costs to other species		Local or regional significance		Land tenure
Ecological uniqueness				Governmental/agency jurisdictions
Keystone species status				
Umbrella species status				

will be too late. This measure is not a value score that can easily be added to the others, and a moderate score has little meaning. Practical issues are also rather different, and will vary greatly in their nature and importance depending on the context. Some species that are considered urgent cases may be extremely impractical and/or costly to attend to. This set of considerations is probably not complete, but it does illustrate the point that there are more things to think about than extinction risk.

This initial classification by the value type is hard to manage in a priority-setting system. Therefore, in Table 2.2, we classify these into six criteria reflecting the nature of the attribute (importance, feasibility, biological benefits, economic benefits, urgency and chance of success). This classification has the advantage that the different questions are more or less independent of one another, and each addresses a question that public, policy-makers and scientists can all address, and for which they can provide at least relative scores.

Interestingly, the criteria that biologists commonly consider, and which form the basis of most formal decision-processes, fall under one heading (biological benefits). Yet in practice, the other five criteria (Table 2.2) also influence real decisions. Would it not, therefore, be preferable to incorporate these other criteria explicitly in the process of setting priorities?

Turning criterion-based ranks into priorities

A potential next step would be to add the scores from Table 2.2. By allocating a score of 1, 2 or 3 to each criterion and then adding the ranks, an overall priority could be calculated. We advise against this for several reasons. First, the different variables are not equal; we might for example wish to weight the biological issues more highly. Second, they are not additive: as mentioned earlier both urgency and chance of success are all or nothing decisions. For

Table 2.2 Criteria for setting prorities. The different kinds of considerations from Table 2.1 are classified into six criteria (rows), each of which can be qualitatively assessed for a particular species

Criterion	Explanation	Subcriteria	Scores
Importance	'Does anyone care?' A measure of how much support there is likely to be	Social and cultural importance (including charisma) Responsibility – how much of the species status depends on this project?	Important (I) Moderately important (M) Unimportant (U)
Feasibility	'How easy is this to achieve?' An assessment of the difficulty associated with this project	Logistical and political, source of funds, community attitudes Biological	Feasible (F) Moderately difficult (M) Difficult (D)
Benefits	'What good will it do?' A measure of how much good will result from the project.	Reduction in extinction risk, increase in population size, extent of occurrence Collateral biological benefits, to other species or processes	Highly beneficial (H) Moderately beneficial (M) Unclear benefits (U)
Costs	'What will it cost?' An assessment of the relative economic costs of the project (or gains). In this criterion there are both postive and negative aspects which have to be weighed against each other	Direct and indirect costs of project Direct and indirect social and economic costs and benefits that will flow from the project	Expensive Moderately costly Inexpensive
Urgency	'Can it be delayed?' A measure of whether the project is time-limited, or whether it can be delayed	Extinction risk, potential for loss of opportunity if delayed	Urgent Moderately urgent Less urgent
Chance of success	'Will it work?' An assessment of whether or not the project will work	Will it meet its specified objectives?	Achievable Uncertain Highly uncertain

example, if chance of success is nil we would not wish to invest in that species at all, so it would seem more logical to multiply other scores by the chance of success. Third, although we have sorted the issues into more-or-less independent categories, there still are associations between them. For example, the feasibility and chance of success are likely to be positively correlated, as are biological benefits and importance. Hence, simple scoring can lead to double-counting, which is not what was intended.

Multicriteria decision-analysis is one decision-making tool for choosing between priorities that rate differently for separate criteria. There are innumerable ways of carrying out a multicriteria analysis, and the process can be complex and may lead to ambiguous results. An expedient process at this stage is to invite a range of experts representing different perspectives to rate the priorities explicitly. For example, given the possible set of scores in Table 2.2, what set would they most wish to see in the top priorities versus those lower down? This sounds complicated but in practice we think it is feasible.

A good example of this approach was developed for UK birds by the Royal Society for the Protection of Birds (Avery et al. 1995). Three criteria were used: global threat, national decline rates and national responsibility, and each was rated high, medium or low. However, by simply adding these scores, globally endangered species that are stable, and for which the UK has medium responsibility, had the

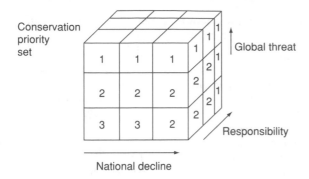

Fig. 2.1 The conservation cube. (From Avery et al. 1995.)

same priority as globally secure species exhibiting slow decline in the UK. This would not be most people's choice; whatever their status in the UK, a globally endangered species probably should be in the category of highest priority. Hence, Avery et al. (1995) set priorities using their conservation cube (Fig. 2.1). Here they evaluated each of the 27 possible circumstances into three categories for priority. In their system, any globally threatened species and any species declining at a high rate nationally are the highest priority.

This approach can be taken more generally using the six criteria in Table 2.2. By asking what would be the criteria associated with top priority species, it is possible to assemble a profile. For example, whereas a species conservation 'idealist' might choose to ignore importance, feasibility, economic benefits and chance of success, and to focus just on the most urgent and most threatened forms, a more 'political' approach would be to maximize import-

ance and economic benefits and minimize risk of failure. Hence the two profiles would look quite different (Fig. 2.2). Figure 2.2 illustrates the different approaches – see how you would score the criteria in Table 2.2 to make your own set!

Here we are effectively creating a complex rule set that maps any species into one of three categories without adding or multiplying the scores for different criteria. The method suffers from its somewhat arbitrary nature. Below we suggest that optimal allocation of funds between species can be achieved more rigorously if we place the problem within an explicit framework in which we can apply decision theory.

A decision theory approach – optimal allocation

A major problem with using scores or ranks for threatened species to determine funding and action priorities is that these methods were not designed for that task – they were designed to determine the relative level of threat to a suite of species (Possingham et al. 2002). Hence they cannot provide the solution to the problem of optimal resource allocation between species – this problem should be formulated then solved properly (Possingham et al. 2001).

Optimal allocation is one simple and attractive approach to prioritization that could inform decisions about how to allocate resources between species. It requires information about

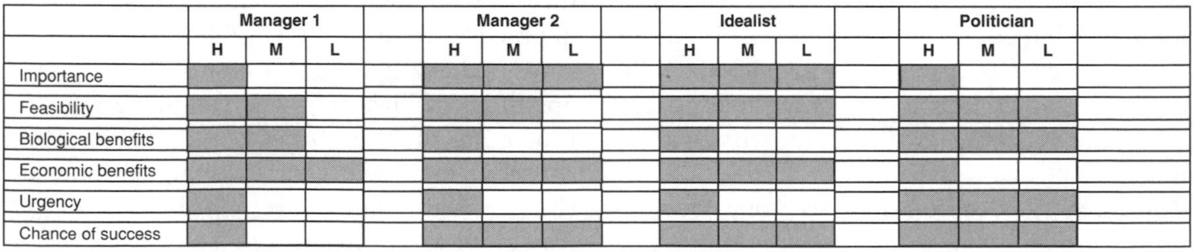

	Manager 1				Manager 2				Idealist				Politician			
	H	M	L		H	M	L		H	M	L		H	M	L	
Importance																
Feasibility																
Biological benefits																
Economic benefits																
Urgency																
Chance of success																

Fig. 2.2 Priority sets for four different people. The blocked out cells indicate the conditions under which assessors would choose to include species in their priority set, according to how they scored on the variables in Table 2.2 as H, high; M, medium; L, low.

the relationship between the resources allocated to the species and the reduction in probability of extinction. Here we use expert opinion and/or population models to estimate the relationship between percentage recovery (measured, for example, in terms of probability of not becoming extinct) and the funds allocated to that species.

For poorly known taxa the curves showing this relationship would very much be a reflection of expert opinion, garnered by asking questions about how much it might cost to give a particular species a 90% chance of not becoming extinct in the long term. Given a set amount of money for a set period in the conservation budget, the optimal allocation of

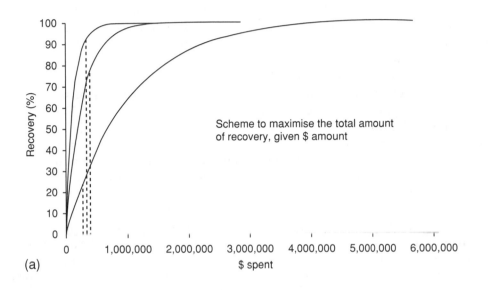

(a)

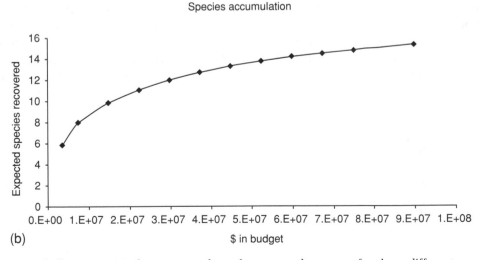

(b)

Fig. 2.3 Optimal allocation. (a) Three curves show the expected recovery for three different species given certain amounts of investment. If the manager has a specified budget (in this case $1 million), the optimal allocation among species that achieves the greatest total amount of recovery will result if funds are allocated as shown by the vertical dotted lines (see Possingham et al. 2002). (b) Increasing investment leads to gradually increasing numbers of species recovered.

funds can be determined between species. This occurs when the rate of gain of recovery for each species is equal, such that there is no advantage in shifting resources from one species to another (see Fig. 2.3 and Possingham et al. 2002). The implicit objective is to maximize the mean number of recovered species given a fixed budget and assuming all species are of equal 'value'.

Using the set of species plotted in Fig. 2.3, we estimate the costs of recovery, and then find the optimal allocation of funds per species. The species accumulation curve shows the total expected number of species that can be recovered given a conservation budget. The algorithm will tend first to select species that show large recovery for relatively low costs. Slow responders will be conserved later. Given an annual budget basis, the more intractable conservation problems may never be funded because the selection process will always favour allocation of resources to the species that provide the greatest gains for the smallest costs (the low-hanging fruit).

So how would these two approaches: criteria-based prioritization and optimal allocation of resources differ in practice? Obviously there is no general answer to this, other than a priori we do not expect them to be the same. The outcome of a small case study, based on real species and the expertise of two real conservation managers is shown in Fig. 2.4.

When species are rated highly by the criteria the two approaches give similar results, but at low criterion scores there can be much variability. Perhaps the only general conclusion here is that inevitably the optimal allocation approach will favour some species that, on the basis of the criteria, would not be given high priority. In practice, sensible management could use both approaches – the criteria to select high-priority sets and the financial algorithm to then maximize the benefits from the finite resources available to conservation.

Conclusions

Priority setting needs to consider a range of variables, and although this undoubtedly occurs, it is not always transparent. Although much effort has gone into biologically based systems, in practice other societal value judgments are often included. We suggest that, if conservation goals are to be achieved, it is vital to be explicit about what these are, and to decide upon them in an open and consultative manner before choices are made.

Different people and organizations, and different sectors in society, will make different choices in their value judgments. Approaches to understanding these choices are important so we can interpret the differences in setting priorities.

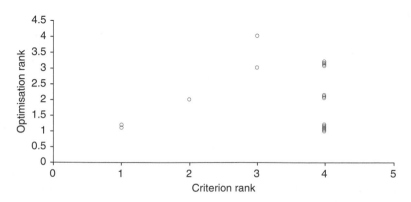

Fig. 2.4 Comparison of priority ranks for18 species using the criteria-based method versus optimal allocation of funds.

We recommend using more than one method to set priorities, and the comparison can be informative. We also recommend that decisions about resource allocation be formulated more explicitly in terms of objectives, constraints and costs.

For if one link in nature's chain might be lost, another might be lost, until the whole of things will vanish by piecemeal.
(Thomas Jefferson (1743–1826) in Charles Miller, *Jefferson and Nature*, 1993.)

References

Araújo, M.B. & Williams, P.H. (2000) Selecting areas for species persistence using occurrence data. *Biological Conservation* **96**: 331–45.

Avery, M., Gibbons, D.W., Porter, R., Tew, T., Tucker, G. & Williams, G. (1995) Revising the British Red Data List for birds: the biological basis of UK conservation priorities. *Ibis* **137**(S1): 232–9.

Balmford, A., Bennun, L., ten Brink, B., et al. (2005) Science and the Convention on Biological Diversity's 2010 target. *Science* **307**: 212–13.

Bani, L., Baietto, M., Bottoni, L. & Massa, R. (2002) The use of focal species in designing a habitat network for a lowland area of Lombardy, Italy. *Conservation Biology* 16: 826–31.

Basset, Y., Novotny, V., Miller, S.E. & Pyle, R. (2000) Quantifying biodiversity: experience with parataxonomists and digital photography in Papua New Guinea and Guyana. *Bioscience* **50**(10): 899–908.

Beissinger, S.R. & Westphal, M.I. (1998) On the use of demographic models of population viability in endangered species management. *Journal of Wildlife Management* **62**: 821–41.

Beissinger, S.R., Reed, J.M., Wunderle, Jr, J.M., Robinson, S.K. & Finch, D.M. (2000) Report of the AOU Conservation Committee on the Partners in Flight species prioritization plan. *Auk* **117**: 549–61.

Berger, J. (1997) Population constraints associated with the use of black rhinos as an umbrella species for desert herbivores. *Conservation Biology* **11**: 69–78.

Bond, I. (1994) Importance of elephant sport hunting to CAMPFIRE revenue in Zimbabwe. *TRAFFIC Bulletin* **14**: 117–19.

Brooker L. (2002) The application of focal species knowledge to landscape design in agricultural lands using the ecological neighbourhood as a template. *Landscape and Urban Planning* **60**: 185–210.

Butchart, S.H.M., Stattersfield, A.J., Bennun, L.A., et al. (2005) Measuring global trends in the status of biodiversity: Red List Indices for birds. *Public Library of Science Biology* **2**: 2294–304.

Carroll, C., Noss, R.F. & Paquet, P.C. (2001) Carnivores as focal species for conservation planning in the Rocky Mountain region. *Ecological Applications* **11**: 961–80.

Caro, T. & Doherty, G. (1999) On the use of surrogate species in conservation. *Conservation Biology* **13**: 805–14.

Caro, T., Engilis, A., Fitzherbert, E. & Gardner, T. (2004) Preliminary assessment of the flagship species concept at a small scale. *Animal Conservation* **7**: 63–70.

Carter, M.F., Hunter, W.C., Pashley, D.N. & Rosenberg, K.V. (2000) Setting conservation priorities for landbirds in the United States: the Partners in Flight Approach. *Auk* **117**: 541–8.

Costanza, R., d'Arge, R., de Groot, R., et al. (1997) The value of the world's ecosystem services and natural capital. *Nature* **387**: 253–60.

Dietz, J.M., Dietz, L.A. & Nagagata, E.Y. (1994) The effective use of flagship species for conservation of biodiversity: the example of lion tamarins in Brazil. In *Creative Conservation: Interactive Management of Wild and Captive Animals* (Eds P.J.S. Olney, G.M. Mace & A.T.C. Feistner), pp. 32–49. Chapman & Hall, London.

Downer, C. (1996) The mountain tapir, endangered 'flagship' species of the high Andes. *Oryx* **30**: 45–58.

Ehrlich, P.R., Dobkin, D.S. & Wheye, D. (1992) *Birds in Jeopardy: the Imperiled and Extinct Birds of the*

United States and Canada, including Hawaii and Puerto Rico. Stanford University Press, Stanford, CA.

Entwistle, A. (2000) Flagships for the future? *Oryx* **34**: 239.

Ferrier, S., Pressey, R.L. & Barrett, T.W. (2000) A new predictor of the irreplaceability of areas for achieving a conservation goal, its application to real-world planning, and a research agenda for further refinement. *Biological Conservation* **93**: 303–25.

Fleishman, E., Murphy, D.D. & Brussard, P.F. (2000) A new method for selection of umbrella species for conservation planning. *Ecological Applications* **10**: 569–79.

Fleishman, E., Blair, R.B. & Murphy, D.D. (2001) Empirical validation of a method for umbrella species selection. *Ecological Applications* **11**: 1489–501.

Garson, J., Aggarwal, A. & Sarkar, S. (2002) Birds as surrogates for biodiveristy: an analysis of a data set from southern Québec. *Journal of Biosciences* **27**:347–60.

Goldspink, C.R., Holland, R.K., Sweet, G. & Stjernsredt, R. (1998) A note on the distribution of the puku, Kobus vardoni, in Kasanka National Park, Zambia. *African Journal of Ecology* **36**: 23–33.

Goodwin, H.J. & Leader-Williams, N. (2000) Tourism and protected areas: distorting conservation priorities towards charismatic megafauna? In *Priorities for the Conservation of Mammalian Diversity: Has the Panda had its Day?* (Eds A. Entwistle & N. Dunstone), pp. 257–75. Cambridge University Press, Cambridge.

Grumbine, R.E. (1994) What is ecosystem management? *Conservation Biology* **8**: 27–38.

Hess, G.R. & King, T.J. (2002) Plannng for wildlife in a suburbanizing landscape I: selecting focal species using a Delphi survey approach. *Landscape and Urban Planning* **58**: 25–40.

Heywood, V.H. (1995) *Global Biodiversity and Assessment.* Cambridge University Press, New York.

Hoekstra, J.M., Boucher, T.M., Ricketts, T.H. & Roberts, C. (2005) Confronting a biome crisis: global disparities of habitat loss and protection. *Ecology Letters* **8**: 23–9.

Iltis, H.H. (1988) Serendipity in the exploration of biodiversity: what good are weedy tomatoes? In *Biodiversity* (Ed. E.O. Wilson), pp. 98–105. National Academy Press, Washington, DC.

IUCN (2001) *IUCN Red List Categories: Version 3.1.* Species Survival Commission, International Union for the Conservation of Nature and Natural Resources, Gland, Switzerland.

James, A.N., Gaston, K.J. & Balmford, A. (1999) Balancing the earth's accounts. *Nature* **401**: 323–4.

Jenkins, R.E. (1976) Maintenance of natural diversity: approach and recommendations. In *Transactions of the 41st North American Wildlife and Natural Resources Conference* 21–25 March, Washington, DC (Ed. K. Sabol), pp. 441–51.

Jenkins, R.E. (1996) Natural Heritage Data Centre Network: managing information for managing biodiversity. In *Biodiversity in Managed Landscapes: Theory and Practice* (Eds R.C. Szaro & D.W. Johnston), pp. 176–92. Oxford University Press, New York.

Johnson, N.C. (1995) *Biodiversity in the Balance: Approaches to Setting Geographic Conservation Priorities.* Biodiversity Support Program, World Wildlife Fund, Washington, DC.

Johnsingh, A.J.T. & Joshua, J. (1994) Conserving Rajaji and Corbett National Parks: the elephant as a flagship species. *Oryx* **28**: 135–40.

Kershaw, M., Mace, G.M. & Williams, P.H. (1995) Threatened status, rarity and diversity as alternative selection measures for protected areas: a test using Afrotropical antelopes. *Conservation Biology* **9**: 324–34.

Lambeck, R.J. (1997) Focal species: a multi-species umbrella for nature conservation. *Conservation Biology* **11**: 849–56.

Leader-Williams, N. & Dublin, H.T. (2000) Charismatic megafauna as 'flagship species'. In *Priorities for the Conservation of Mammalian Diversity: has the Panda had its Day?* (Eds A. Entwistle & N. Dunstone), pp. 53–81. Cambridge University Press, Cambridge.

Leader-Williams, N., Harrision, J. & Green, M.J.B. (1990) Designing protected areas to conserve natural resources. *Science Progress* **74**: 189–204.

Lindenmayer, D.B., Manning, A.D., Smith, P.L., et al. (2002) The focal-species approach and landscape restoration: a critique. *Conservation Biology* **16**: 338–45.

Lunney, D., Curtin, A., Ayers, D., Cogger, H.G. & Dickman, C.R. (1996) An ecological approach to identifying the endangered fauna of New South Wales. *Pacific Conservation Biology* **2**: 212.

Mace, G.M. & Collar, N.J. (1995) Extinction risk assessment for birds through quantitative criteria. *Ibis* **137**: 240–46.

Mace, G.M. & Lande, R. (1991) Assessing extinction threats: toward a reevaluation of IUCN threatened species categories. *Conservation Biology* **5**: 148–57.

Margules, C.R. & Pressey, R.L. (2000) Systematic conservation planning. *Nature* **405**: 243–53.

Master, L.L. (1991) Assessing threats and setting conservation priorities. *Conservation Biology* **5**: 559–63.

Master, L.L., Stein, B.A., Kutner, L.S. & Hammerson, G. (2000) Vanishing assets: conservation status of US species. In *Precious Heritage. The Status of Biodiversity in the United States* (Eds B.A. Stein, L.S. Kutner & J.S. Adams), 93–118. Oxford University Press, New York.

Meir, E., Andelman, S. & Possingham, H.P. (2004) Does conservation planning matter in a dynamic and uncertain world? *Ecology Letters* **7**: 615–22.

Miller, B., Reading, R., Strittholt, J., et al. (1999) Using focal species in the design of nature reserve networks. *Wild Earth* **8**: 81–92.

Millsap, B.A., Gore, J.A., Runde, D.E. & Cerulean, S.I. (1990) Setting priorities for conservation of fish and wildlife species in Florida. *Wildlife Monographs* **111**: 1–57.

Mittermeier, R.A. (1986) Primate conservation priorities in the Neotropical region. In *Primates: The Road to Self-sustaining Populations* (Ed. K. Benirschke), pp. 221–40. Springer-Verlag, New York.

Morse, L.E. (1993) Standard and alternative taxonomic data in the multi-institutional Natural Heritage Data Center Network. In *Designs for a Global Plant Species Information System* (Eds. F.A Bisby, G.F. Russell & R. Pankhurst), pp. 127–56. Oxford University Press, Oxford.

Myers, N., Mittermeier, R.A., Mittermeier, C.G., da Fonseca, G.A.B. & Kent, J. (2000) Biodiversity hotspots for conservation priorities. *Nature* **403**: 853–8.

Noss R.F., Quigley, H.B., Hornocker, M.G., Merrill, T. & Paquet, P.C. (1996) Conservation biology and carnivore conservation in the Rocky Mountains. *Conservation Biology* **10**: 949–63.

Olson, D.M. & Dinerstein, E. (1998) The global 200: a representation approach to conserving the Earth's most biologically valuable ecoregions. *Conservation Biology* **12**: 502–15.

Paine, R.T. (1995) A conversation on refining the concept of keystone species. *Conservation Biology* **9**: 962–64.

Possingham, H., Ball, I. & Andelman, S. (2000) Mathematical methods for identifying representative reserve networks. In *Quantitative Methods for Conservation Biology* (Eds S. Ferson & M. Burgman), pp. 291–305. Springer-Verlag, New York.

Possingham, H.P., Andelman, S.J., Noon, B.R., Trombulak, S. & Pulliam, H.R. (2001) Making smart conservation decisions. In *Research Priorities for Conservation Biology* (Eds G. Orians & M. Soulé), pp. 225–44. Island Press, Covelo, California.

Possingham, H.P., Andelman, S.J., Burgman, M.A., Mendellin, R.A., Master, L.L. & Keith, D.A. (2002) Limits to the use of threatened species lists. *Trends in Ecology and Evolution* **17**: 501–7.

Possingham, H.P., Franklin, J., Wilson, K. & Regan, T.J. (2005) The roles of spatial heterogeneity and ecological processes in conservation planning. In *Ecosystem Function in Heterogeneous Landscapes* (Eds G.M. Lovett, C.G. Jones, M.G. Turner & K.C. Weathers), pp. 389–406. Springer-Verlag, New York.

Pressey, R.L. (1994) *Ad hoc* reservations: forward or backward steps in developing representative reserve systems? *Conservation Biology* **8**: 662–8.

Pressey, R.L. & Taffs, K.H. (2001) Sampling of land types by protected areas: three measures of effectiveness applied to western New South Wales. *Biological Conservation* **101**: 105–17.

Pressey, R.L., Ferrier, S., Hutchinson, C.D., Sivertsen, D.P. & Manion, G. (1995) Planning for negotiation: using an interactive geographic information system to explore alternative protected area networks. In *Nature Conservation 4: the Role of Networks* (Eds D.A. Saunders, J.L. Craig & E.M. Mattiske), pp. 23–33. Surrey Beatty and Sons, Sydney.

Rabinowitz, D. (1981) Seven forms of rarity. In *The Biological Aspects of Rare Plant Conservation* (Ed. H. Synge), pp. 205–17. Wiley, Chichester.

Redford, K.H., Coppolillo, P., Sanderson, E.W., et al. (2003) Mapping the conservation landscape. *Conservation Biology* **17**: 116–31

Reed, J.M. (1992) A system for ranking conservation priorities for neotropical migrant birds based on relative susceptibility to extinction. In *Ecology and Conservation of Neotropical Migrant Landbirds* (Eds J.M.H. Hagan III & D.W. Johnston), pp. 524–36. Smithsonian Institution Press, Washington, DC.

Regan, T.J., L.L. Master & G.A. Hammerson. (2004) Capturing expert knowledge for threatened species assessments: a case study using Natureserve conservation status ranks. *Acta Oecologica* **26**: 95–107.

Sanderson, E.W., Redford, K.H., Chetkiewicz, C.L.B., et al. (2002a) Planning to save a species: the jaguar as a model. *Conservation Biology* **16**: 58–72.

Sanderson, E.W., Redford, K.H., Vedder, A., Ward, S.E. & Coppolillo, P.B. (2002b) A conceptual model for conservation planning based on landscape species requirements. *Landscape and Urban Planning* **58**: 41–56.

Scott, J.M., Davis, F., Csuti, B., et al. (1993) Gap analysis: a geographic approach to protection of biological diversity. *Wildlife Monographs* **123**: 1–41.

Shaffer, M.L. (1981) Minimum population sizes for species conservation. *BioScience*, 31, 131-134

Simberloff, D. (1998) Flagships, umbrellas, and keystones: is single-species management passe' in the landscape era. *Biological Conservation* **83**: 247–57.

Smith, R.J., Muir, R.D.J., Walpole, M.J., Balmford, A. & Leader-Williams, N. (2003) Governance and the loss of biodiversity. *Nature* **426**: 67–70.

Stein, B.A., Kutner, L.S. & Adams, J.S. (2000) *Precious Heritage. The Status of Biodiversity in the United States*. Oxford University Press, New York.

Stotz, D.F., Fitzpatrick, J.W., Parker, III, T.A. & Moskovits, D.K. (1996) *Neotropical Birds: Ecology and Conservation*. University of Chicago Press, Chicago, IL

Walpole, M.J. & Leader-Williams, N. (2002) Ecotourism and flagship species in conservation. *Biodiversity and Conservation* **11**: 543–547.

Watson, J., Freudenberger, D. & Paull, D. (2001) An assessment of the focal species approach for conserving birds in varigated landscapes in southeastern Australia. *Conservation Biology* **15**: 1364–73.

Western, D. (1987) Africa's elephants and rhinos: flagships in crisis. *Trends in Ecology and Evolution* **2**: 343–6.

Williams, P.H., Burgess, N.D. & Rahbek, C. (2000) Flagship species, ecological complementarity and conserving the diversity of mammals and birds in sub-Saharan Africa. *Animal Conservation* **3**: 249–60.

Wilson, K.A., Westphal, M.I., Possingham, H.P. & Elith, J. (2005) Sensitivity of conservation planning to different approaches to using predicted species distribution data. *Biological Conservation* **122**: 99–112.

Wilson, K.A., McBride, M., Bode, M. & Possingham, H.P. (In press) Prioritising the allocation of conservation resources between biodiversity hotspots. *Nature*.

3

What is biodiversity worth? Economics as a problem and a solution

David Pearce*, Susanna Hecht and Frank Vorhies

In spite of the cost of living, it's still popular.
(Laurence J. Peter, USA educator and writer, 1919–88.)

Introduction – what is the problem?

The entire history of human 'civilization' is one of converting unutilized land to human-oriented uses. Much of this conversion has occurred in very recent history – since AD 1700 – and virtually all of it has involved converting woodlands to croplands (Vitousek et al. 1997). Between 1700 and 1980, 1.2 billion ha of agricultural land was gained at the expense of a roughly equal amount of forest (Richards 1990). Conversion on this scale involves loss of species, and very probably species diversity (there is a general relationship between geographical area and the number of species it supports (MacArthur & Wilson 1968)). The problem is further exacerbated because land conversion fragments the remaining ecosystems, and land-use changes frequently produce pollution. (This literature has exploded in recent years. Journals such as *Conservation Biology* and *Environmental Conservation*, plus a score of edited volumes, such as Laurance & Bierragaard (1999), have taken this as a central focus in conservation studies.)

Spectacular extinctions, like that of the passenger pigeon (*Ectopistes migratorius*) in the early part of the twentieth century, provide obvious cautionary tales about human excess. Once so numerous that their flocks would blacken the sky, unbridled human predation in the late nineteenth century wiped them from the planet. Creatures on the brink of extinction, such as Brazil's golden lion marmoset (*Callithrix chrysoleuca*) whose habitats in the Atlantic forest of Brazil have been reduced to perhaps 5% of their previous terrain (Laurance & Bierregaard 1999), are indicators of the plights of the countless far less glamorous organisms.

There is widespread alarm at the rates of species loss, although estimates are themselves very uncertain (Laurance & Bierregaard 1999; Myers et al. 2000). Most of the alarm is expressed by life scientists and by local forest-/resource-dependent communities that have witnessed these changes.

This alarm has given rise to some of the movements concerned with maintaining local biodiverse systems such as tropical woodlands. The rubber tappers of Brazil are perhaps the most iconic of these movements. In Acre,

* Regretfully David died in September 2005 while the book was in production.

Brazil, tappers stopped ranchers from clearing their traditional lands and developed a kind of forest tenure known as extractive reserves where access rights for extraction are maintained (and can even be sold), but the land itself remains outside of the market, and so cannot get sucked into speculative processes so characteristic of the western Amazon. These efforts to maintain both livelihoods and tree resources are quite widespread in the tropics (other examples are given by Peet & Watts (2003) and Hecht & Cockburn (1989)).

That most nations have now ratified the 1992 Convention on Biological Diversity – with the notable exception of the USA – is a sign that there is global political concern about biodiversity loss. Nevertheless, how effective is the Convention at conserving biodiversity? The Convention's financial agency, the Global Environment Facility (GEF), has only had an annual authorized budget for biodiversity conservation and international waters improvement, plus other leveraged funds, of about $350 million per year throughout the 1990s (D.W. Pearce, *Environment and Economic Development: the Economics of Sustainable Development*, unpublished manuscript 2005).

If we add to this funding by government agencies and conservation NGOs, then perhaps about $1 billion is directly spent annually on conserving the world's biodiversity through such conventional means as establishing and managing protected areas. If this is an indicator of the world's substantive commitment to biodiversity conservation, it is clearly trivial and hardly seems commensurate with the warnings and concerns of ecologists and others. In short, there is a 'commitment deficit', a huge gap between the scale of the problem as seen by many, and the will of politicians and citizens and politicians to tackle it.

Economists tend to be divided between those who regard loss rates as potentially catastrophic and those who find it hard to believe that much of economic value resides in diversity *per se*. The nature of economic value is not what many believe it to be, a problem arising from the popular confusion of terms such as 'commercial', 'financial' and 'economic'. To be sure, part of economics does deal with the exchange of marketed commodities and services ('commerce'), and it deals with the flow of funds in an economy and how those funds affect commercial activity ('finance'), but economics in its broadest sense is the study of the use and allocation of resources and need not necessarily have anything to do with financial flows.

The goal usually chosen by economists for the allocation of resources is 'utilitarianism' – the ethical theory first proposed by Jeremy Bentham and James Mill that argues that action should be directed toward achieving the greatest happiness for the greatest number of people. Today the part of economics that is relevant to global policy making and to issues such as environmental conservation is called 'welfare economics'. Here what matters is the welfare or wellbeing of people, and what underpins that wellbeing. We all know that some of those determinants relate to the provision of non-monetary services and goods, such as peace and quiet, clean air, access to the countryside, appreciation of wildlife and so on. What economics does, then, is to look for, and try to measure the wellbeing that people get from seemingly intangible but extremely real flows of goods and services provided by ecological systems.

What does biodiversity do for us?

What exactly is the problem of biodiversity loss? The truth is that no-one can be sure. Life scientists are still finding out what species do in ecosystems, and what they do may or may not be relevant to what humans want. There are two levels of ignorance: (i) how species affect ecosystem dynamics; and (ii) the broad anthropocentric value ascribed to those functions.

Given the lack of understanding of biodiversity and its relevance to humans, economics looks for and tries to measure the values or benefits that people derive from seemingly elusive environmental goods and services. Economics also deals with conflicts of interest or 'trade-offs'. Clearly commercial and financial interests can conflict with the non-monetary aspects of wellbeing. The conversion of natural areas to agriculture has brought benefits in terms of food production, but it has come at a cost of lost species and changes in broader ecosystem functions. Part of the problem is that the benefits of agriculture are commercial and tangible, whereas the benefits of species diversity and ecosystem integrity do not show up in the form of financial flows. Economics also tries to understand the dynamics of environmental externalities, i.e. the costs that accrue to economic production that are not calculated in the price of the product, but which may have significant environmental and social consequences.

Interestingly, today, there is a growing interest in biodiversity within the private sector (Vorhies, 2002). Several multinational companies – for example, British American Tobacco, BP, Lafarge, HSBC, Rio Tinto and Unilever, to name only a few – are beginning to look seriously at the biodiversity aspects of their operations. Profit-driven businesses, however, are still relatively poor at handling problems with competing interests, high levels of uncertainty and an array of externalities. Nevertheless, companies – especially in sectors such as mining, oil and gas, and agriculture – are beginning to realize that biodiversity issues can have both direct and indirect impacts on a company's financial performance.

Companies that are looking seriously at biodiversity are, in essence, adopting an economist's approach to conservation issues. First, they are trying to identify what biodiversity does for the company and its stakeholders. Second, they are trying to evaluate conservation benefits in terms of the value of real flows of corporate resources so that they are comparable

with corporate activities that destroy the environmental assets. In so doing, some companies are beginning to recognize a 'business case' for biodiversity conservation.

Ecological ignorance and economic ignorance

Nevertheless, the economics approach faces two serious challenges. First, we may not know what the benefits of conservation are. As Roughgarden (1995, p. 149) says:

> 'We have no ecological engineers to say how an ecosystem will change if we choose to eliminate its species, one by one. That is not a question we have been trying to answer – we have been trying to find out what the species actually in the ecosystem are doing. So there will be a time lag as information is developed on how particular systems will function if some of the components are eliminated.'

Second, to Roughgarden's challenge of ecological ignorance, we have to add the challenge of economic ignorance. It is only very recently that environmental economists have begun to investigate what ecosystems and species do for human beings and what the economic value of those services actually is. The conservation problem, then, is how to behave in face of this compounded ignorance.

One reason, therefore, for the political 'commitment deficit' could be that we are all being asked to pay out, or change our behaviour, for the conservation of something whose benefits are uncertain. Humans are not naturally given to very precautionary behaviour, as the very limited international efforts to reduce global warming testify (Pearce 2003). If radical efforts are to be made to conserve biodiversity, all scientists, including economists, have to do a much better job of explaining why biodiversity needs to be conserved.

The limits to action

There are, of course, limits to controlling biodiversity loss. Failure to acknowledge limits can lead to false expectations, and false expectations can lead to complacency. No doubt, for example, there are many who believe that because we now have an international agreement on biodiversity conservation, the problem is under control. Others may believe that all we need to do is to get the USA to ratify the biodiversity convention and then all will be solved.

Unfortunately, nearly all international environmental agreements often confirm little more than a 'business as usual' level of conservation, i.e. the agreements do not generate additional conservation compared with what would have happened anyway (Barrett, 2003). Also, as noted above, the earmarked financial commitment is largely divorced from the scale of the problem.

There are several dimensions to the problem of limits, but the most notable is population growth. Habitat conversion is closely linked to global expansion in human numbers. Two thousand years ago there were roughly 250 million people on our planet. Currently the number is around 6.5 billion. Some projections suggest that in 50 years time, there will be 50% more people, regardless of the population-reducing measures being taken now. This phenomenon is known as 'demographic inertia'. It is inconceivable that this rate of population growth can be accommodated without further land conversion and hence without further biodiversity loss. Biodiversity loss will continue even if consumption patterns remain unchanged or decrease and production efficiency increases.

We must recognize the unsustainable resource use in North America and Europe. We also must also be aware of the burgeoning consumption demand of developing economies, particularly the giant economies of China and India. Asia absorbs an increasingly larger portion of market share for global timber, and are important drivers for soybean and livestock expansion into tropical South American woodlands (Hecht et al. in press), as well as fossil fuels. Faced with the scale of the economic forces driving biodiversity loss, its protection becomes a task of slowing the rate of loss, rather than only trying to conserve what we currently have.

Why are economic incentives relevant?

People who convert land respond to economic incentives as well as other incentives. If rates of biodiversity loss are to be slowed, economic incentives must play a central role in policy measures. Although the traditional model of parks and protected areas – 'fence and forget' – can still have relevance, they fail to address the economic drivers of change. With growing populations and increasing consumer demands, habitat protection can lead to increasing conflicts. In Indonesia, for example, one-third of all forest sector land conflict was due to boundary changes and restricted forest access.

A very pragmatic reason for taking economics seriously is that it drives most land conversions, and it is welfare economics that has to inform the policies needed to slow the rate of landuse change. No-one argues that **only** the 'dismal science' matters. What economists say is that economics matters, it may well matter a bit more than other factors and it has a critical role to play in reducing the loss of biodiversity.

There are cases where local and regional communities have shown the capacity to manage biodiverse areas because they see benefits from doing so. These benefits can include goods – forest products of various kinds – and environmental services. Such perceived benefits generate economic incentives to conserve. In El Salvador, for example, community groups have developed forest management programmes, and are part of watershed councils at regional levels and now seek payment for environmental services (Hecht et al. in press). In the Amazon pilot project, communities have organized zoning of local land use to mediate and control forest use while supporting local livelihoods (Padoch 1999).

Does everything have to have a price?

Many find the economic arguments suspect. Simply because people **are** motivated by economic factors, does not mean that policy **should** be similarly motivated. But what is the alternative? There may be an appeal to 'morality' other than the utilitarian basis of economics, which is to allocate resources to meet the wants and desires of people. What is usually argued is that there is some 'higher' value residing in Nature which humans have at least an obligation, and perhaps a duty to serve – an environmental ethic.

The problem with this higher morality is that it frequently fails to provide recipes for **practical** action. In other words, arguments for an environmental ethic may be 'correct' in some sense, but not necessarily useful. Such higher moral views may not be widely shared by ordinary people or, even if shared, may be inconsistent with other personal motivations and desires. Again, economists are not saying that morality is unimportant, but they are questioning whether moral arguments can persuade the relevant parties to slow rates of habitat conversion.

Appeals to morality and to questions of externalities have largely defined regulatory policies, although with little sign that they have reduced the problem of biodiversity loss. This is especially so when the actions needed are either global or have to be directed at sovereign nations facing desperate developmental needs and rapid population growth, or where the state itself has been run by kleptocrats actively bent on plundering the nations' natural resources, such as Indonesia or Nigeria (Curran et al. 2004). In such cases, legal frameworks for

environmental regulation may exist but they are generally ignored.

Also, and very largely ignored in moral debates, environmentally ethical behaviour comes at a cost. In the developed world this might involve changing consumption habits in favour of 'green' products, and forgoing the returns to some kinds of land development. In the developing world, on the other hand, pro-biodiversity programmes can affect livelihoods profoundly, especially for the half the world living at incomes less than $2 a day. Obliging or persuading people to do something they otherwise would not have done imposes a cost on those people, in the same way that designating a forest as a national protected area may deprive them of their livelihoods by essentially criminalizing activities such as firewood collection, grazing and periodic hunting (Neuman 1999). Changing access rights can trigger significant conflict and affect the lives of many people.

Whatever view is taken on moral arguments, there is sufficient cause for concern to orient attention towards finding solutions that address the motivations of human beings to convert land and destroy biodiversity. This is the case for the economic approach.

What do we know about the value of biodiversity?

If economics is to help conservation, it has to do so in two stages. First, we need to demonstrate that biodiversity is economically important. Second, we must explore incentive mechanisms for capturing this economic value so that it persuades those who convert land to think and act differently. These two stages are sometimes called 'demonstration and capture'.

The economic values or benefits of biodiversity include four general components:

1. its contribution to ecosystem functions;
2. commercial and use values (timber, forest products);

3. non-use values;
4. its contribution to ecosystem resilience.

Ecosystem functions include watershed regulation, nutrient cyling and microclimate mediation, the provision of global services such as climate regulation and carbon sequestration, and evolutionary processes. These functions are mediated through the component organisms of an ecosystem – its biodiversity of plants, animals and microbes, all of which are interlinked in complex ways. This network of interacting organisms contrasts with the emphasis that modern economics places on particular elements of an ecosystem – its commodities: forest trees, non-timber products, animals, etc., that are conceptualized and economically valued independently of their larger ecosystem connections.

Commercial and use values involve the harvesting use and marketing of particular biodiversity commodities, such as timber, bush meat and medicinal plants.

Non-use values reflect people's willingness to pay for biodiversity conservation regardless of the uses made of biodiversity. Imagine someone who has never seen a tropical forest and, for one reason or another, can never visit one. Independently of any value they obtain, e.g. from climate change mitigation, they may nonetheless be willing to pay to conserve the forest. This is sometimes called an 'existence value'. Motivations for non-use value vary – some notion of 'stewardship', some notion of Nature's right to exist, a concern to leave an asset for future generations, aesthetics, and so on. It would be hard to explain donations to conservation societies if those donations were not partly based on non-use, environmental values.

Ecosystem resilience values derive from aggregated diversity – i.e. from the aggregated value of genetic diversity within species, species diversity **and** ecosystem diversity. Ecosystem resilience seems to be largely a function of **biological diversity**, even though the nature of connections between biodiversity and stability are a source of some controversy.

Resilience focuses on the dynamic capacities of the ecosystem, by measuring the degree of shock or stress that the system can absorb before moving from one state to another. These processes of change are marked by **discontinuities** and potential **irreversibilities** and may not be 'linear'. A modest change for instance, may result in some dramatic effect, whereas some major disruptions may have little effect on highly resilient systems. Diversity, it is argued, stimulates resilience perhaps because the functions of individual threatened or lost species can be replaced by other species in the same landscape system. The smaller the array of species the less chance there is of this substitution process taking place.

From an economic standpoint, the challenge is to identify and measure the value of ecosystem resilience. Unfortunately, neither is easy. Identifying how close a system might be to collapse is extremely difficult, yet one would expect willingness to pay to avoid that disaster to be related in some way to the chances that the collapse will occur. If the probabilities are known, the value sought is then the premium that would be paid to conserve resilience. Thus, one could argue that the entire cost of managing non-resilient systems is the 'economic willingness' to support of a vulnerable, unstable ecosystem, because these costs would be avoided if more diverse and therefore more resilient systems are adopted.

What is better known today about the economic values of biodiversity relates to the direct uses of ecosystems and somewhat to non-use values. Hardly anything is known about resilience values. Table 3.1 lists some 'representative' values for forest ecosystems that have been the subject of most applied environmental economics research.

Although clearly subject to many uncertainties, Table 3.1 does suggest that, of the 'known' values, carbon storage dominates. This impression, however, may be an artefact of the enormous efforts that have gone into carbon studies. It is, nevertheless, an interesting finding for

Table 3.1 Some representative economic values ($ ha^{-1} yr^{-1} unless stated otherwise) for forest ecosystems

Forest good or service	Tropical forests	Temperate forests
Timber:		
conventional logging	440	
sustainable logging	30–266	−4000 to +700 (NPV)
Fuelwood	40	–
Non-timber forest products	0–100	Small
Genetic information (drugs only)	0–3000	–
Recreation	2–470 (general; forests near towns) 1000 (unique forests)	80
Watershed benefits	15–850	−10 to +50
Climate benefits	360–2200 (GPV)†	90–400 (afforestation)
Contributions to resilience and insurance	?	?
Non-use values	2–12 4400 (unique areas)	2–45
Willingness to pay a premium for sustainable forest management	5–15% of timber prices?	5–15% of timber prices?

*NPV (net present value) = sum of all future benefits minus costs, discounted to allow for a lower value of $1 being placed on future gains and losses.
†GPV, gross present value = as above including costs.
(Source: Pearce & Pearce 2000; Pearce 2001)

policy purposes because it reveals that integrating forest and climate policies is crucial and mutually beneficial. Forests can be major beneficiaries of climate control regimes. The emphasis on environmental/forest **services** such as carbon sequestration provides a policy-relevant way for valuing biodiversity.

Capturing biodiversity values

Indentifying the economic values of biodiversity is just one part of the economic argument. The other part observes that only some of these values have markets. Markets in non-use value have evolved, and are inherent in the costs of maintaining parks as well as in mechanisms such as debt-for-nature swaps (where international debt is 'paid' by conserving land) and with the Global Environment Facility. As for use values in forests, for example, there are markets in timber, fuelwood, bush meat and other non-timber products. Carbon is also 'monetized' in the sense that carbon is traded, either under voluntary schemes or, increasingly, under the Kyoto Protocol mechanisms.

Similar markets for a variety of ecosystem goods and services can evolve. There are few markets in watershed regulation values, although Costa Rica's Forest Law paves the way for their recognition. Here forest owners can receive direct payment for conserving their forests, with the payment schedule being determined roughly in accordance with economic valuation studies of watershed benefits.

Although many payments for environmental services involve state or international transfers to communities or landowners, or state ownership of lands around reservoirs, there are other contexts where direct payment for maintaining biodiversity might be possible. In Costa Rica, when pollinator habitat was maintained around coffee plantations, productivity increased by around 20% and returns increased by $40,000 above 'normal'. As domestic bee populations had collapsed due to parasites, pollination was carried out by wild bees. A direct local market here could, in principle, evolve for owners in coffee landscapes to preserve significant areas of old growth forest.

Biodiversity benefits based on genetic information have been much debated. In the early years enthusiasm about the potential pharmaceutical returns to biodiversity was unconstrained as the economic value of 'cures' for AIDS, cancers, etc., were routinely invoked. Today there is far more caution. Some analysts consider the drug value of forests to be very small indeed, a few dollars per hectare (e.g. Simpson et al. 1996) and thus are unlikely to influence forest conservation. Other analyses (e.g. Rausser & Small 2000) suggest the values could be significant, maybe several thousand dollars per hectare, in megadiverse areas such as India or China.

Although this divergence of estimates was originally thought to be due to differences in modelling the way the search for genetic information is conducted (whether randomly or using prior knowledge), it now seems that the divergence in estimates is most likely due to the selection of parameter values in the models. Only the passage of time can improve on this debate.

What is known is that 'bioprospecting' contracts between pharmaceutical companies and forested countries have not resulted in major financial flows. This reflects a range of issues. Indigenous communities were reluctant to assist schemes that deprived them of land or resources with no patent rights or protections on their native intellectual property. Biologists worried that markets developed for 'cures' might cause local depletion or extinction. The market for rhino horn and Amazon dolphin (*Inia geoffrensis*) penises (a kind of Viagra *avant la lettre*) has indeed driven these species to the brink of oblivion. Also, industries that depend on a chemical derived from a wild plant may, if possible, simply create laboratory templates and avoid the need for the habitats of the original organism. For example, the original source of Tamaxofen, a breast cancer treatment drug, was the Pacific yew (*Taxus brevifolia* Nutt),

and now the drug is entirely synthetic. Other wild plants such as rubber may simply be domesticated.

Ignorance as an argument for conservation

The uncertainty about how biodiversity functions and what it does for us is a reason for more conservation not less. We may gain information about the value of biodiversity by delaying habitat conversion. In economics this concept is known as 'quasi-option value' (QOV). (In the finance literature, it is known as 'real option value'. Unfortunately, there is a quite different notion of option value in the environmental economics literature, which is unrelated to quasi-option value.)

Quasi-option value measures the value of information gained by postponing a decision that has irreversible consequences. For example, conversion of primary tropical forest land to agriculture tends to be irreversible. Although forests do grow back, the biodiversity profile of the new secondary forest may be different. Arguments for delay are, of course, arguments for at least temporary conservation and, if the resulting information supports non-conversion, for longer-term conservation.

Very few studies exist that give insights into the size of quasi-option value. There are some cases where QOV is used to support a regional development strategy. For example, the state of Acre in Brazil pays rubber producers a surcharge to ensure that tapping (and its generally positive impacts on forests) remains economically viable so that they continue to extract forest products rather than convert lands to pasture. This payment can be seen as a type of QOV supporting local forest practices while other kinds of markets (fair trade, tourism, certified forestry) develop.

Other studies disagree. Bulte et al. (2002) argued that once uncertainty and irreversibility are included, Costa Rica has 'too little' tropical forest cover. However, one outcome of the Bulte's study was that QOV turns out to be considerably less important for forest conserve-or-convert decisions than introducing the notion of a rising relative valuation of forests. The idea here is simple but is often not incorporated in economic appraisals of conservation. As people get richer their willingness to pay for environmental goods is likely to grow somewhat faster than the general price level: there is a 'relative price effect'. Bulte et al. (2002) note that what matters for the decision to convert forests are (i) the value of this relative price effect, (ii) the presence of global environmental benefits and (iii) the extent to which markets can be developed in those benefits. They conjecture that QOV remains unimportant relative to these other considerations.

Whether quasi option value can be marshalled 'in defence' of conservation therefore may be an open question. However, one final argument emerges from the context of uncertainty and irreversibility – the precautionary principle. This principle gives the benefit of the doubt to conservation.

The obvious bias in current decision-making is that 'development' is presumed best unless conservation can be shown to have very high values. The 'safe minimum standards' approach to precaution inverts the process and declares conservation to be best unless the 'development' benefits are shown to be substantial. How large do 'substantial' benefits have to be is left unanswered. We do know, however, that a great deal of habitat conversion takes place for markedly little benefit to the converter, while often incurring substantial costs to local communities, the biodiversity and even planetary processes. There are many examples of road developments throughout the tropics that caused immense socio-biotic disruption for minimal economic gain (Hecht & Cockburn 1989; Peet & Watts 2003).

If compensation mechanisms could be found to stop the land conversion, the suggestion is

that comparatively modest payments could save a great deal of biodiversity. This kind of thinking lies behind more recent movements aimed at simply paying owners, or *de facto* owners, not to clear forest. The solution is simple, but it is contingent on payments being forthcoming and on their size being modest.

That such a system might work is suggested by the impact of remittances in some places in Central America. In El Salvador, civil war exiled one-sixth of its population who continue to send monies to support rural relatives. Two-thirds of the foreign exchange comes in the form of remittances that go directly to many rural households. Remittances and wages have permitted households to purchase rather than produce basic grains, and this in turn has been instrumental in landscape recuperation by taking pressures for subsistence clearing off more marginal land (Hecht et al. in press). In addition, the elimination of livestock subsidies undermined the expansion of pasture, a historically dominant activity stimulating forest conversion. Today more than 60% of the terrain has woodland cover.

Another approach is to develop 'protected productive areas' in which biodiversity conservation forms an integral part of a productive process. Such an approach is developing in sectors that have a significant presence on the landscape, such as oil and gas pipelines or mining operations. Rio Tinto Zinc, for example, now has a published biodiversity strategy and is committed to integrating biodiversity conservation into its mining operations.

Community-based development programmes such as the Amazon pilot project and the southern Mexico organic farm and fair trade movements incorporate large areas of traditional populations into productive protection by giving growers a better price for growing their coffee in biodiversity friendly ways. Finally, there are production systems that themselves incorporate rich patterns of biodiversity. Complex rubber and fruit groves in Asia are systems that engage the biodiversity and human occupation of the landscape, and thus integrate economic, domestic and environmental concerns. Compared with monospecific production systems, these diverse landscapes embrace substantial biodiversity, and such approaches will clearly have to be part of any planetary future (Bray et al. 2003; Vandemeer & Perfecto 2003).

Conclusions

Arguing for biodiversity conservation on moral grounds is fraught with difficulty. This does not mean it should not take place, simply that it may have little substantive influence in policy arenas. The economic approach has a morality of its own, but it is based on the notion of economic incentives, of sending signals to those who make the decisions to destroy diversity that they may want to think and act differently. If the benefits of conservation are perceived as being higher than the benefits of conversion, there should be inducement to maintain habitat or, at least, use it sustainably. Even here, expectations must be realistic: there are limits to action set by population growth, distribution of resources and the problems of corrupt governments.

The first step of the economic approach is to estimate, even roughly, the economic values of biodiversity. Such a process helps to identify what needs to be targeted. For example, the available literature suggests that carbon storage is a major economic benefit of forest biomass. The second step is to find ways of developing markets for the biodiversity benefits that currently have no markets. Promising developments have occurred, from payments to forest owners to maintain forests because of the environmental services they provide, to global 'deals' whereby payments are made to reflect non-use values.

A third step is to strengthen the business case for biodiversity within the strategic planning of economic institutions. If biodiversity conservation is a critical issue for the global community, then biodiversity conservation needs to transcend the arenas of government policy-makers and NGO activists.

A fourth step might be to support biodiversity maintenance and recovery through 'development from below'. Remittances are now a significant part of the global economy, and they could be applied to actively support programmes to enhance forest and biodiversity resurgence. This process is described (cf. Hecht et al. in press), but still remains understudied.

How much biodiversity such economic measures can save remains, of course, an open question. The chances are that it will be more than other approaches can deliver.

There is no higher priority for conservation biologists than to improve their understanding of economics.
(D. W. Orr in *Conservation Biology*, 2005, p.1318.)

References

Barrett, S. (2003) *Environment and Statecraft: the Strategy of Environmental Treaty-making*. Oxford University Press, Oxford.

Bray, D. Merino, L. Negreros, P. Segura, G. Torres, JM & Vester. H. (2003) Mexico's community-managed forests as a global model for sustainable landscapes. *Conservation Biology* **17**(3): 672. doi: 10.1046/j.1523-1739.2003.01639.x.

Bulte, E., van Soest, D., van Kooten, C. & Schipper, R. (2002) Forest conservation in Costa Rica when non-use benefits are uncertain and rising. *American Journal of Agricultural Economics* **84**(1): 150–60.

Curran, L.M., Trigg, S.N., McDonald, A.K., et al. (2004) Lowland forest loss in protected areas of Indonesian Borneo. *Science* **303**: 1000–3.

Hecht, S.B & Cockburn, A. (1990) *Fate of the Forest*. Harper Collins, New York.

Hecht, S.B., Kandel, S., Rosa, H., Gomes, I. & Cuellar, N.(In press) The political ecology of forest resurgence in El Salvador. *World Development*.

Laurance, W. & Bierregaard, R. (Eds) (1999) *Tropical Forest Remnants*. University of Ghicago Press, Chicago.

MacArthur, R. & Wilson, E. (1968) *The Theory of Island Biogeography*. Princeton University Press, Princeton, NJ.

Myers, N., Mittermeier, R., Mittermeier, C., Fonseca, G. de & Kent, J. (2000) Biodiveristy hotspots for conservation priorities. *Nature* **403**: 853–8.

Neuman, R. (1999) *Imposing Wilderness*. University of California Press, Berkeley, CA.

Padoch, C. (ed.) (1999) *Varzea: Diversity, Development, and Conservation of Amazonia's Whitewater Floodplains*. New York Botanical Garden Press, Bronx, NY.

Pearce, D.W. (2001) The economic value of forest ecosystems. *Ecosystem Health* **7**(4): 284–96.

Pearce, D.W. (2003) Will global warming be controlled? Reflections on the irresolution of humankind. In *Challenges to the World Economy: Festschrift for Horst Siebert* (Eds R. Pethig & M. Rauscher), pp. 367–82. Springer-Verlag, Berlin.

Pearce, D.W. & Pearce, C. (2000) *The Value of Forest Ecosystems*. Convention on Biological Diversity, Montreal. www.biodiv.org/doc/publications/cbd-ts-04.pdf.

Peet, R. & Watts, M. (2003) *Liberation Ecologies*. Routledge, New York.

Posey, D.A. (1999) *Cultural and Spiritual Values of Biodiversity*. IT Publications, London.

Rausser, G. & Small, A. (2000) Valuing research leads: bioprospecting and the conservation of genetic resources. *Journal of Political Economy* **108**(1): 173–206.

Richards, J. (1990) Land transformation. In *The Earth as Transformed by Human Action*. (Eds B. Turner, H. Clark., R. Kates., J. Richards., J. Mathews & W. Meyer), pp. 163–78. Cambridge University Press, Cambridge.

Roughgarden, J. (1995) Can economics protect biodiversity? In *The Economics and Ecology of Biodiversity Decline: the Forces Driving Global Change* (Eds T. Swanson), pp. 149–55. Cambridge University Press, Cambridge:

Simpson, D., Sedjo, R. & Reid, J. (1996) Valuing biodiversity for use in pharmaceutical research. *Journal of Political Economy* **104** (1): 163–85.

Vandermeer, J. & Perfecto, I. (1997) The agroecosystem: the need for the conservationists lens. *Conservation Biology* **11**: 1–3.

Vitousek, P., Moony, K., Lubichectko, J. & Melillo, J. (1997) Human domination of Earth's ecosystems. *Science* **277**: 494–9.

Vorhies, F. (project director) (2002) *Business and Biodiveristy: the Handbook for Corporate Action*. Earthwatch Institute (Europe); Union for the Conservation of Nature and Natural Resources, Gland, Switzerland; and the World Business Council for Sustainable Development.

4

Impacts of modern molecular genetic techniques on conservation biology

Eli Geffen, Gordon Luikart and Robin S. Waples

I am the family face;
Flesh perishes, I live on,
Projecting trait and trace
Through time to times anon,
And leaping from place to place
Over oblivion.

(Thomas Hardy, 'Heridity' in *Moments of Vision*, 1917.)

Introduction

Conservation biology strives to conserve biodiversity and biological processes in ecosystems, of which genetic variation is a key component. Genetic variation is the underlying foundation of higher levels of biodiversity (e.g. populations and species). Without genetic variation, populations could not evolve and adapt to future environmental changes. Because DNA (deoxyribonucleic acid) is fundamental to all biological systems, the practice of conservation often requires genetic studies. Beyond the measurement and conservation of genetic variation *per se*, the uses of molecular genetic techniques in conservation biology include:

1. identification of individuals, species, populations and conservation units;

2. detection of hybrid zones and admixed populations;
3. quantification of dispersal and gene flow;
4. estimation of current and historical population size;
5. assessment of parentage, relatedness, reproductive success, mating systems and social organization.

Molecular markers also assist forensic detection of illegally killed and trafficked plants and animals or their body parts. Finally, markers that are under selection (and thus influence fitness) can identify locally adapted populations that could have special value for conservation.

Two developments in molecular biology have had unprecedented significance for conservation biology: the PCR (polymerase chain reaction) process and the discovery of microsatellites. Since its development in 1985, PCR has transformed the life sciences, including conservation biology, due to the ease (and still declining cost) with which it generates millions of copies of any DNA fragment from minuscule quantities. The PCR technique has allowed the non-destructive study of living specimens and their long-dead ancestors. A surge of mitochondrial DNA (mtDNA) sequence studies on phylogeny, hybridization and gene flow among populations ensued, including some based on fragments of museum skins and specimens preserved in ethanol (Brown & Brown 1994). For example, ancient bones of the Laysan duck (*Anas laysanensis*) were identified by mtDNA analysis from lava tubes on the main Hawaiian islands, where they apparently had become extinct (Cooper et al. 1996). These data justified reintroduction and suggested that many island endemics may be relics of former cosmopolitan species (Wayne et al. 1999).

Microsatellites consist of a length of DNA in which sequences of one to four nucleotides are repeated many times (e.g. $[AC]_n$, where $n = 5$ to 50 repeats). The number of repeats defines an allele at a locus. Microsatellites are typically highly variable, often with > 10 alleles per locus in a population. They are widely dispersed in eukaryotic genomes and inherited in a Mendelian fashion. They can be amplified by PCR from only tiny amounts (one-to-several molecules) of DNA and thus can be salvaged from partially degraded DNA, such as in museum skins, dried faeces or fossil bones. Because of these features, microsatellites have become the most widely used molecular genetic marker. Numerous other PCR-based molecular markers and analysis systems exist, including SNPs (single nucleotide polymorphisms), and direct sequencing of PCR products (see Sunnucks (2000), Morin et al. (2004) and (Schlotterer 2004) for reviews).

Genetics is a key component of many aspects of conservation biology. From the design of reserves to the management of breeding programmes, molecular techniques are indispensable and are increasingly being used to address questions of conservation relevance. Molecular biology is undoubtedly the fastest evolving field

of science. Conservation biologists can make use of these emerging techniques, which are rapidly transforming the field to one that is more molecular oriented. Conservation biology is an inexact science because new crises emerge every day and in most cases solutions are but extrapolations from related cases. Molecular biology is helping to change that trend by allowing conservation biologists to quickly scan a wide range of individual and population characteristics at a given site. Genetic data are most useful in conjunction with more traditional data, such as demographics, life history, distribution, etc. Rapid gain of detailed information on a population at risk may allow better understanding of the system at hand, and more sound recommendations for the decision makers.

Systematics and hybridization

Defining a species can be vital to its legal protection – for example, under the Convention on International Trade in Endangered Species (CITES) agreement, which regulates trade in endangered species, or the Endangered Species Act (ESA) in the USA. The ESA efforts to restore the red wolf, *Canis rufus*, to its native North American range began 25 years ago, founded on the belief that the red wolf was a distinct species. More recent genetic analyses from captive individuals and museum skins (Wayne & Jenks 1991; Roy et al. 1996), however, found no unique genetic characters in the red wolf and suggested a close genetic relationship to the coyote, *Canis latrans*. Reich et al. (1999) hypothesized that red wolves arose as a result of hybridization between grey wolves, *Canis lupus*, and coyotes during the past 2500 years, thus calling into question their conservation status under the ESA. Although this conclusion has been disputed (Wilson et al. 2000), the red wolf genetic studies have highlighted the issues of how to determine what constitutes a valid unit for conservation purposes and what conservation value should be afforded to hybrids.

What molecular tools are available for deducing the systematic status of an animal? One promising source of information about evolutionary relationships among species and populations is the circular, 16,000 base pair segment of DNA contained in mitochondria. The genes in mtDNA are well defined, and numerous universal primers, which target particular DNA segments in specific genes and operate effectively across a wide range of taxa, are available commercially. As each cell contains many more copies of mtDNA than nuclear DNA, mtDNA is easier to extract from minute, degraded samples. In most organisms, mtDNA is maternally inherited, so only one sequence copy can be extracted (as opposed to two for nuclear genes – one from each parental chromosome). Disadvantages of mtDNA include:

1. it represents the evolution only of maternal DNA and provides no direct information about genetic contributions of males;
2. it is inherited as a unit, so represents only a single marker and phylogenies based on it can be less robust that those based on nuclear DNA, for which it is typically possible to assay a large number of independent markers.

A frustrating paradox is that nuclear genes have the potential to provide more robust phylogenies but have been used less commonly, in part because many require specific primers and cloning before sequencing is possible. Nuclear genes that have been used commonly in systematic studies include those associated with the male-inherited Y chromosome genes in mammals (e.g. Lundrigan et al. 2002; Makova & Li 2002) and the highly variable major histocompatibility complex (MHC) region (Holmes & Ellis, 1999). As an alternative to sequence data, phylogenetic reconstruction can be achieved from short interspersed nucleotide elements (SINEs; Shedlock & Okada, 2000). Short interspersed nucleotide elements are dispersed throughout eukaryotic genomes in great numbers. Because an insertion (i.e. a small DNA segment that was inserted into the sequence of a gene) is an essentially irreversible event, the sequence of the insertions can be traced through a lineage to infer common ancestry among taxa. Short interspersed nucleotide elements have been used to infer phylogeny of African mammals, primates and reptiles, among other taxa (e.g. Nikaido et al. 2003). The abundance of molecular data has promoted development of several new statistical methods for phylogenetic reconstruction that have been discussed elsewhere (Felsenstein 1981, 2003; Hendy 1993; Hillis et al. 1996; Larget & Simon 1999).

An example of the use of phylogenetic reconstruction in conservation is the taxonomic status of endangered subspecies of the leopard (*Panthera pardus*). The leopard has an extensive geographical distribution, and in many regions it is quite common. However, some subspecies are extremely rare (e.g. the Arabian, *P. p. nimer*, and the Amur, *P. p. orientalis*, subspecies). Uphyrkina et al. (2001) used phylogenetic reconstruction to determine whether these rare subspecies of leopard are genetically unique (Fig. 4.1). These results can be used as guidelines for management of this species. For example, a highly phylogenetically distinct subspecies might have high conservation value and merit separate management (e.g. without interbreeding with other subspecies, as can occur in zoos and reintroduction programmes).

Non-invasive sampling and population size estimation

It is difficult to monitor or evaluate the population status of many threatened and endangered species because they live at low densities, roam over large areas, inhabit regions that are difficult to work in or have an elusive life style. Furthermore, many of these species are large (e.g. marine mammals), dangerous (e.g. carnivores) or secretive (e.g. nocturnal marsupials), meaning that trapping individuals for the purpose of

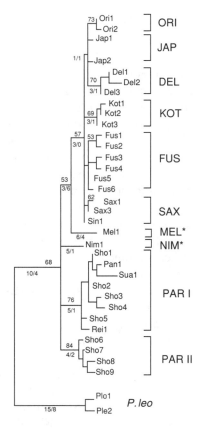

Fig. 4.1 Phylogenetic relationships among the leopard mtDNA haplotypes from combined NADH-5 (611 bp) and control region (CR, 116 bp) mtDNA. (Modified from Uphyrkina et al. 2001.) Lion (*Panthera leo*) samples were used as outgroup species. Maximum parsimony tree is shown. Numbers above branches represent bootstrap support (100 replicates); only those with > 50% are shown. Numbers below show number of steps/ number of homoplasies. ORI stands for *Panthera pardus orientalis*, SAX for *P. p. saxicolor*, MEL for *P. p. melas*, KOT for *P. p. kotiya*, FUS for *P. p. fusca*, DEL for *P. p. delacouri*, JAP for *P. p. japonensis*, NIM for *P. p. nimer*, and PAR I and II for the two African clusters. Both ORI and NIM have a distinct position on the species phylogenetic tree, indicating their genetic uniqueness.

tagging is complex and expensive, even if the necessary permits can be secured. Scats, pellets, hair, feathers, egg shells, sloughed skin, urine and other body fluid secretions contain minute amounts of DNA that can be amplified by PCR. Hair was collected by hair-traps from black bears (Boersen et al. 2003) and from sleeping nests of chimpanzees (Morin et al. 1994), and a systematic survey for kit fox scats was carried out using trained dogs (Smith et al. 2001). Consequently, an array of molecular (Bellmain et al. 2004) and statistical (Valière et al. 2002) methods are being developed to monitor animal populations with-

out the need to handle, or even observe, the subjects. For example, hair or scats collected from brown bears (*Ursus arctos*) (Bellmain & Taberlet 2004) and scats collected from coyotes (Kohn et al. 1999) have been used to estimate population size (abundance) and to track individual movements and home ranges.

A prerequisite for such techniques is that samples are correctly identified, often using species-specific DNA sequences. Sequences of any of the mtDNA genes (e.g. cytochrome *b*) are often sufficient to allow distinction between scats (or other material) from several species at a study site (e.g. black bear (*Ursus americanus*) versus brown bear (*Ursus arctos*), or wolf versus red fox (*Vulpes vulpes*)). The DNA extracted from each scat (or other material) can be subjected to microsatellite analysis, which can identify different individuals based on their unique multilocus genotypes (their DNA 'fingerprint'). An interesting complication could arise if, for example, a brown bear eats a black bear and the black bear's DNA shows up in the faecal sample, or if one wolf urinates on another wolf's faeces; in these cases, individual identification will be difficult because DNA from more than one individual would be amplified.

From a smear of faeces or a pinch of hair follicles, the molecular detective can identify the sex (e.g. using sex-linked genes such as ZFX/ZFY, which are carried by either sex chromosome; Lucchini et al. 2002), reproductive status and parasite load of the subjects (Kohn & Wayne 1997; Fedriani & Kohn 2001). Further sleuthing can provide estimates of population abundance, based on a variation of the standard ecological practice of mark-recapture. In this case, however, the 'marks' are the naturally occurring DNA fingerprints of individuals, and recaptures are detections of a DNA finger print more than once among non-invasive samples (e.g. faecal samples). Abundance can be estimated as the asymptote of a curve plotting the cumulative number of unique genotypes (*y* axis) as a function of the number of new samples collected (Kohn et al. 1999; Banks et al. 2003). Failure to find new genotypes in additional samples suggests that most of the population has already been sampled.

Although non-invasive techniques are powerful, they (i) can be difficult to develop, (ii) require a pilot study to validate reliability, and (iii) usually require repeated genotyping of each locus on each sample to avoid genotyping errors (and thus cost more in time and money than when using fresh tissue samples). Low concentrations or partially degraded DNA can lead to genotyping errors during PCR amplification. For example, low DNA concentration in a sample occasionally causes an amplification of only one allele in a heterozygote (termed allelic drop out), an error that can yield a false homozygous genotype, leading to biased estimates, especially when a small number of individuals are involved (Taberlet et al. 1999). New DNA extraction protocols and software to detect and control for scoring and other errors are being developed. In coming years, those efforts, along with systematic use of rigorous laboratory and scoring protocols, automation of protocols, and error rate quantification and reporting (Broquet & Petit 2004) should help overcome many of these methodological problems.

Genetic diversity within populations

Why do species become extinct? This is one of the most debated questions in conservation biology (Caughley 1994). Deterministic forces, such as unrelenting harvest or incremental losses of habitat, obviously can place species at high risk. When populations become small, however, random events become relatively more important and may play a major role in many extinctions. For example, in a population of 1000, if males and females are equally likely, the sex ratio will seldom deviate far from 1:1, ensuring sufficient females to produce the next generation. In a population of 10, however, two or fewer will be females 5% of the time, just by chance. Large variations in birth rates, age structure and other demographic processes also occur by chance in small populations. For the same reason, small populations are prone to lose genetic diversity because the rate of genetic drift (random fluctuations in allele frequency) increases and alleles become extinct by chance faster than they are being generated by mutation. Loss of diversity constrains long-term evolution, because genetic variation is the raw material for natural selection to act upon. More diverse populations are better able to accommodate environmental variation and the outbreaks of disease.

On shorter time frames, loss of diversity reduces fitness primarily due to the expression of deleterious, recessive alleles. In large populations, selection keeps such alleles at a low frequency, so they usually occur in heterozygotes, where their deficiencies are masked by a copy of 'normal' alleles. In small populations, deleterious alleles can drift to high frequencies just by chance and become expressed in homozygotes, thus reducing fitness of the population through inbreeding depression. Populations that decline rapidly in size are said to suffer a genetic bottleneck, so termed because only alleles that make it through the bottleneck will survive in the population.

Which random processes, demographic or genetic, pose greater risks to small populations? This also has been one of the most hotly debated topics in conservation biology. For many years it was commonly believed that demographic stochasticity was more likely to cause extinctions (e.g. Lande 1993), the argument being that populations were likely to become extinct through random fluctuations in size before cumulative losses of genetic diversity became severe enough to seriously reduce fitness. However, recent studies demonstrate that genetic factors quite often play an important role in the extinction process (Spielman et al. 2004).

A number of empirical studies have found a correlation between reduced heterozygosity (and other measures of genetic diversity) and lowered individual fitness (Reed & Frankham 2003); more homozygous (inbred) individuals often have lower survival and fecundity. For example, the energetic cost of burrowing, a trait essential to survival in the pocket gopher (*Thomomys* spp.), was significantly lower in populations with higher genetic variability (Hildner & Soulé 2004). Many small populations of endangered species are restricted to isolated patches in the wild, or even housed in captivity as part of breeding programmes. Such populations have no immigration, a natural process that counteracts the fixation of deleterious alleles and loss of heterozygosity by importing novel alleles from other populations.

Saccheri et al. (1996) dramatically illustrated the importance of this natural process in a butterfly metapopulation in Finland. Subpopulations with low levels of heterozygosity had a significantly higher subsequent probability of extinction (after controlling for environmental and demographic extinction risks). This was probably the result of inbreeding depression that affected larval survival, adult longevity and egg hatching rate. It appears that the other populations were rescued from this fate by receiving sufficient immigration, bringing novel alleles into the population.

Conservation biologists have drawn on fundamental principles of population genetics to develop the concept of **genetic rescue**, which occurs when immigrants make a positive contribution to fitness over and above the demographic effects of simply adding more individuals. This rescue effect is most likely to occur if the recipient population is small, isolated and suffering from inbreeding depression. Under these circumstances, genetically divergent immigrants can import new alleles into the population to counteract the tendency for erosion of genetic diversity and to mask deleterious alleles responsible for inbreeding depression. The Finnish butterfly laboratory study (Saccheri et al. 1996) illustrates how this process can function in the wild. Vila et al (2003) showed that a single breeding immigrant into a severely bottlenecked and geographically isolated Scandinavian population of grey wolf could recover genetic diversity.

Animal breeders have long practiced a form of genetic rescue by periodically injecting 'new blood' into their broodlines. Direct interventions to effect genetic rescue of natural populations of conservation interest is an exciting new development with some apparent successes. For example, by the 1980s numerous developmental and reproductive abnormalities indicated that the endangered Florida panther, *Felis concolor coryi*, was suffering from inbreeding depression. Population genetic models indicated that a brief episode of high gene flow (using animals from Texas), followed by subsequent generations of low gene flow, could genetically restore the population by reducing the frequency of deleterious alleles without substantially reducing the frequency of alleles responsible for local adaptation (Hedrick 1995). Although long-term results will not be known for several generations, preliminary data suggest that this strategy may be working (Hedrick 2001).

However, such interventions are risky with no guarantee of success. In fact, it is quite possible that genetic rescue attempts could reduce fitness rather than increase it. Just as matings between genetically similar individuals can lead to inbreeding depression, interbreeding of genetically divergent individuals can lead to outbreeding depression, either through dilution of locally adapted genes or disruption of gene complexes that function effectively together (Lynch 1991). Furthermore, a host of behavioural, ecological and demographic factors (e.g. unintentional importation of exotic diseases; McCallum & Dobson 2002) can influence the consequences of human manipulated migration. Therefore, although the concept of genetic rescue may seem elegantly simple and empirical examples document its potential benefits, developing testable models to predict when genetic rescue may seem likely to succeed (and fail) is a major challenge for the future (Tallmon et al. 2004). Rescue is most likely to occur (without outbreeding depression) when gene flow is being restored into inbred populations that only **recently** became small and isolated such that little time existed for adaptive differentiation to develop.

Gene flow among populations

Not only is nature patchy, but habitat fragmentation is accelerating as roads, agriculture, logging and other developments divide continuous habitats into isolated patches, disrupting immigration as well as reducing population sizes (Hanski & Gaggiotti 2004). In many species, local populations are connected by dispersal into larger metapopulations, and these connectivities can be essential to the long-term persistence of the metapopulation as a whole, for both demographic and genetic reasons. Estimating the rate and pattern of migration among patches is thus vitally important for the conservation biologist.

Genetic markers are well suited to the study of gene flow, or movement of genes among populations, because they integrate information about migration or isolation over evolutionary time frames. Genetic markers thus can provide information not only about contemporary migration, but also historical patterns of connectivity. For example, the African wild dog, *Lycaon pictus*, is among the most endangered canid species. Girman et al. (2001) showed that although populations cluster into two genetic units (eastern and southern), the admixture zone spans much of the current geographical range of the wild dog. The authors concluded that the Selous population in Tanzania is an appropriate source of individuals for reintroduction into Masai Mara and Serengeti, where wild dogs declined precipitously in recent years. This example illustrates that population genetic analysis is not a theoretical exercise but an important tool for developing translocation plans, long-term management programmes and reserve design (Palumbi 2003).

Genetic analysis of population structure commonly comprises three main stages:

1. identification and enumeration of populations;
2. analysis of relationships among populations;
3. evaluation of patterns of differentiation as a function of geographical distance.

The first step, determining how many populations exist, is a necessary precursor to many subsequent types of analyses. In some cases, candidate populations are easy to infer from the discontinuous geographical distribution of individuals, and standard statistical methods can be used to test the null hypothesis that all samples belong to a single random mating population. In other cases, distributions may be continuous or overlapping, making it difficult to collect meaningful samples for statistical tests. In this situation, clustering methods (Pritchard et al. 2000; Manel et al. 2005) can be used to estimate the number of gene pools present in a mixed sample and assign individuals to specific gene pools. Pritchard et al. (2000) used this approach to show that at least three populations of the endangered Taita thrush, *Turdus helleri*, occur in Kenya. This method can be powerful if strong genetic differences exist among populations, but its general applicability is still being evaluated.

Once populations are identified, it is important to examine their genetic relationships to gain insights into patterns of migration. The first step is typically calculation of a genetic distance between pairs of populations. A commonly used measure is the fixation index (F_{st}), which measures the fraction of the total variation in allele frequency that is found between populations. The F_{st} is inversely related to the number of migrants (N_m) per generation between the populations of interest. Allele or haplotype frequencies are used to calculate F_{st} or related genetic distances, some specific to microsatellites. A matrix of pairwise genetic distances can be visualized as a tree network connecting all populations or as a two- or three-dimensional plot (e.g. fig. 1 in Girman et al. 2001). Analysis of molecular variance (AMOVA; Excoffier et al. 1992) is a procedure that allows the overall genetic variance to be partitioned into components of interest, such as geographical subdivisions or temporal replicates. In the Australian green turtle (*Chelonia mydas*), AMOVA was used to show that about 99% of the genetic variation in microsatellite loci was contained within rookeries. In contrast, only 22.5% of the genetic variation in mtDNA haplotypes occurred within rookeries, whereas 77.5% was partitioned among regions and none among rookeries within regions (FitzSimmons et al. 1997). The combined genetic and tagging evidence allowed the authors to conclude that the observed genetic subdivision is due to migration of turtles from the south Great Barrier Reef through the courtship area of the north Great Barrier Reef population.

Another topic of interest is the role of geographical distance in shaping the observed genetic structure. Understanding the relationship between geographical and genetic distance is important for any conservation plan. When this association is high, geographical distance can be a meaningful barrier to dispersal and care should be taken to conserve populations that are close enough together to permit sufficient genetic exchange. If an association between geographical and genetic distance is not found, it may indicate that few barriers to dispersal exist even at large spatial scales, but it could also mean that the populations are isolated and historical factors have shaped the present-day structure.

How are individuals associated with populations? Typically, we assign individuals based on the collection site. However, this approach risks misclassifying migrant individuals. Applying an assignment test (Paetkau et al. 1995) – a powerful statistical tool that 'assigns' each individual to the most likely population of origin based on its multilocus genotype – has the potential to provide information for a broad range of questions of conservation relevance (Manel et al. 2005). For example, assignment tests and related analyses have been used to document male-biased dispersal in the whitefooted mouse, *Peromyscus leucopus* (Mossman & Wasser 1999); to show that treating wolverines (*Gulo gulo*) from Montana as a single population is not a sound conservation strategy, even though they have high apparent dispersal capability (Cegelski et al. 2003); to highlight risks of fragmentation due to overharvest of the kelp *Laminaria digitata* in the English Channel (because gene flow from adjacent, continuous strands is generally more important than distant transport by currents; Billot et al. 2003); and to evaluate introgression of coyote genes into the red wolf (Miller et al. 2003). A precursor of assignment tests known as genetic stock identification (Pella & Milner 1987; Brown et al. 1999) has been used for many years to help manage mixed-stock fisheries of Pacific salmon and other commercial species to avoid unsupportable harvest of depressed wild populations. For example, real-time (24-h turnaround) genetic analysis of samples from a Chinook salmon (*Oncorhynchus tshawytscha*) fishery in the Lower Columbia River has helped managers determine when endangered populations from the upper Columbia and Snake River basins enter the fishery, at which point the fishery can be closed (Shaklee et al. 1999).

Effective population size

The effective population size (N_e) is one of the most important parameters in conservation genetics because it influences the rate of loss of genetic variation, the rate of inbreeding (mating between relatives) and the efficiency of selection in eliminating deleterious alleles and maintaining adaptive ones. A rough approximation of N_e is the number of breeding individuals in a population that leave offspring that survive to reproductive age. The effective population size is defined more technically as the size of the ideal population that loses genetic variation at the same rate as the population being studied. In an 'ideal' population, population size is constant, sex ratio is equal and variation in reproductive success among individuals is random. All of these provisions are typically violated in real populations, with the result that $N_e/N < 0.5$, and sometimes a great deal less (Frankham 1995).

Several recent studies of marine species have estimated N_e to be three to six orders of magnitude lower than N. For example, Hauser et al. (2002) used variation at seven microsatellite loci to estimate N_e in the New Zealand snapper, *Pagrus auratus*, using two independent molecular-based methods. Scale samples were collected beginning in 1950 around the time a commercial fishery started to harvest the Tas-

man Bay population. Genetic variation (allelic richness and heterozygosity) was much lower in 1998 than in the samples from 1950, a result that would not be expected in large populations. Allele frequency changes over this period were also typical of those found in small populations. The effective size estimates consistent with these observed genetic changes were 46 and 176, respectively, for the heterozygosity loss and temporal change methods. In contrast, the census size was estimated in the mid-1980s to be 3.3 million fish. Hedgecock (1994) proposed a hypothesis to explain this phenomenon in marine species with very high fecundity and very high mortality of eggs and larvae: most families produce no offspring that survive to reproduce, and the next generation is derived from progeny of a very few families that are 'sweepstakes' winners in the reproductive lottery. This hypothesis and the empirical estimates of tiny N_e/N ratios remain controversial, but they demonstrate that even large populations can be at risk of losing genetic variation, and that monitoring of genetic variation and N_e can be useful, even when the census size is large.

The effective population size, N_e, can be calculated from demographic data, such as lifetime variance in reproductive success, but these data are difficult to obtain for most species. Furthermore, demographic methods often overestimate N_e because they do not include all factors causing N_e to be less than N. For these reasons, methods for estimating and monitoring N_e based on molecular markers were developed and have made an important contribution to conservation (reviewed by Schwartz et al. 1999).

Genetic bottlenecks

A population bottleneck, or rapid reduction in N_e, generates characteristic genetic signatures that can be detected with realistic samples (e.g. c.30 individuals scored for 10–20 molecular markers). One signature is a deficit of rare alleles (frequency < 0.10), which develops in small, declining populations. In a large stable population, most alleles occur at low frequency (Fig. 4.2; Luikart et al. 1998). During a bottleneck, rare alleles are lost first, leading to an apparent excess of alleles at moderate frequency. Populations in which a large fraction of alleles are at intermediate frequencies thus are likely to have recently experienced a bottleneck. Another signature, detectable using microsatellite data, depends on the ratio of the number of alleles to the range in allele sizes (Garza & Williamson 2001). During a bottleneck, the number of alleles declines faster than the range, leading to a low ratio. Yet another signature is an excess of heterozygosity (i.e. Hardy Weinberg expected H_e) compared with the theoretical equilibrium gene diversity expected for a large, stable population (Luikart & Cornuet 1998). All these kinds of information (allele length, allele frequencies and heterozygosity excess) are used in the Bayesian approach for detecting bottlenecks developed by Beaumont (1999). Thus the Beaumont approach should, in theory, be the most powerful; however, its performance and reliability has not been thoroughly evaluated.

The signature of a bottleneck event is an alarm call for those who monitor populations of an endangered species. For example, if a strong bottleneck is detected, it would be prudent to initiate monitoring of the genetic and demographic status of the population – and perhaps take action such as translocations (as in the example of the Florida panther, above). In extreme cases, such as the African cheetah (*Acinonyx jubatus*), the effects of an apparently ancient bottleneck event (approximately 10,000 years ago) are still observed today in the form of very low genetic variability on a continental scale (Menotti-Raymond & O'Brien 1993).

Detecting selection and local adaptation

Most studies in conservation genetics have used markers assumed to be neutral (i.e. not associ-

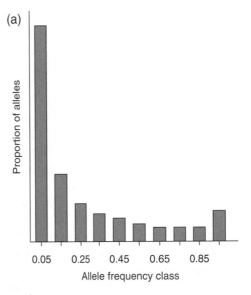

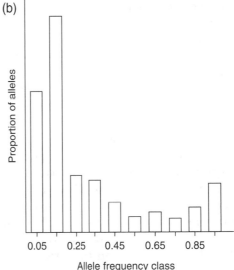

Fig. 4.2 Genetic signature of a population bottleneck: a mode-shift in the distribution of allele frequencies. Large stable populations (i.e. populations near mutation-drift equilibrium) have a large proportion of alleles at low frequency (a). Why? Because new mutations are rare (occurring as a single copy), and new alleles usually fluctuate at low frequency until they are lost via random genetic drift. However, a bottleneck causes rapid loss of rare alleles and generates a deficit of alleles at low frequency (frequency < 0.10) (b). This shifts the mode of the distribution from the low-frequency class (0.0–0.1) to an intermediate frequency class (e.g. 0.1–0.2). Bottlenecks can be thought of as strong sampling events where rare alleles are lost.

ated with fitness), largely because population genetics theory and models are best developed for neutral alleles. For example, methods for estimating levels of gene flow, effective population size, bottlenecks, mating system characteristics (F_{is}) and some methods for inferring phylogenies, all assume that markers are neutral. Applications not requiring neutral markers include parentage and relatedness estimation.

Although in general the assumption of neutrality for molecular markers may be reasonable, with the increasing number of markers in a typical data set, it is likely that some will be under selection. In addition, more and more studies are using markers located in genes (e.g. SNPs, Morin et al. 2004), making selection effects or 'signatures' more likely. Markers in genes (e.g. introns) are more likely to be affected by selection than most markers that are seldom near genes and thus unlikely to be under selection directly or through linkage to a selected gene. Fortunately, new statistical tests now make it feasible to identify loci under selection. Two important uses are (i) excluding selected loci for applications in which neutrality must be assumed, and (ii) using selected loci to help identify locally adapted populations with special value or concern for conservation.

Loci under selection should be excluded from inferences about population demography and evolutionary history, because selection can bias inferences – even if only one out of many markers is under selection (Landry et al. 2002; Luikart et al. 2003; Storz et al. 2003). For example, Allendorf & Seeb (2000) studied 36 markers from four populations of salmon, and found one locus with extremely high F_{st} (0.71) relative to the other loci: mean F_{st} with and without the outlier locus was 0.20 and 0.09, respectively (Fig. 4.3). In this example, one strong outlier locus more than doubled the estimation of population differentiation. The locus was probably under selection because such a high F_{st} is extremely unlikely by chance alone, at a neutral locus. Fortunately, several computer programs are now freely available to allow tests for outliers and to help differentiate between selected and neutral loci.

Once markers under selection are identified, they might be used in conservation to help design translocation programmes. For example, if two populations are candidates as sources for translocation into a small or declining population, the source with the least genetic differentiation at selected loci (fewest F_{st} outliers), relative to the declining population, might be used preferentially. This is true especially if the source with few F_{st} outliers also has the most similar environment or habitat compared with that of the recipient population. These views expand further the concept of 'genetic rescue', and in the future may become guidelines in translocation programmes.

Molecular markers, if confirmed as adaptive, also may be used to prioritize or rank populations for conservation importance. For example, a population containing a high proportion of adaptive and unique alleles might be of higher conservation value than another population with fewer such alleles. Adaptive markers (and other adaptive characters) could be integrated along with neutral markers (and other non-genetic data) when prioritizing populations for conservation (Fig. 4.4). Unfortunately, prioritizing preservation of one population based on a sample of adaptive genes could actually reduce diversity across the rest of a species' gene pool. This could jeopardize the adaptive potential of a species to future environmental changes (Luikart et al. 2003). For example, if we prioritize conservation of one population based on a few divergent adaptive genes unique to that population, we might lose adaptive genes in other populations that would improve the species persistence in future environments. Further difficulties arise in predicting which genes will be adaptive in future environments. Thus, although the use of adaptive gene markers for prioritizing populations is desirable, it can be difficult and risky to apply effectively. More research is needed to assess the usefulness of adaptive markers in conservation.

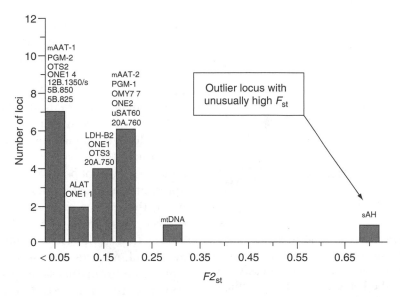

Fig. 4.3 Distribution of F_{st} (differentiation among populations) for 21 molecular marker loci (including mtDNA, microsatellites and allozymes and RAPDs (random amplified polymorphic DNA)) from sockeye salmon, *Oncorhynchus nerka*, populations from Cook Inlet, Alaska. (Modified from Allendorf & Seeb 2000.) $F2_{st}$ differs from the classic 'F_{st}' in that all minor alleles (at low frequency) are pooled to make only two allele classes, before computing F_{st}. This allows a less biased comparison between multi-allelic microsatellites and allozymes, which are mostly bi-allelic. Note that mtDNA is expected to have a relatively high F_{st} because the mtDNA effective population size (N_e) is small and thus drift is strong. N_e is small because only females (half the population) transmit mtDNA, and because mtDNA is haploid (half the number of chromosomes compared with nuclear genes). This makes the mtDNA effective population size only one-quarter the effective size of the nuclear genes (assuming a 50:50 sex ratio).

A powerful tool for evaluating selection in specific genes is the relative frequency of DNA substitutions at sites that do and do not result in changes in the amino acid sequence of proteins. Because of redundancies in the genetic code, some mutations are synonymous (S, no change in amino acid sequence, hence are considered neutral), while others are non-synonymous (N, result in a change). As most mutations are deleterious, N mutations are generally much rarer because they are quickly eliminated. In an analysis of DNA sequence data for the transferrin gene (important in binding and sequestering iron), Ford (2001) found very high N/S ratios in salmon but not other vertebrates – indicating strong, positive selection for transferrin variation in Pacific salmon. The positively selected sites occur primarily on the outside of the mol-

ecule in regions subject to binding by bacterial proteins. One possible explanation for this result is an 'arms race' for access to iron between pathogenic bacteria (for which iron is often a limiting nutrient) and the host salmon, which must continually change the structure of transferrin to keep ahead of bacterial mutations.

Populations often differ in many phenotypic and life-history traits. How does one decide under what circumstances these differences are important to conserve, and if so, which traits are the most valuable? Has a particular trait evolved many times within the species (hence it might be regenerated again in the future if lost) or only once? Joint analysis of genetic and life-history data can provide a powerful means to help set conservation priorities. Waples et al. (2004) examined chinook

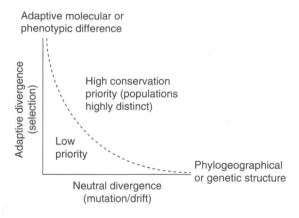

Adaptive molecular or phenotypic difference

Fig. 4.4 Adaptive markers could be treated separately from (but integrated with) neutral markers when prioritizing populations for conservation. Populations with the highest diversity and uniqueness for both adaptive and neutral markers would receive highest priority for conservation. Other non-genetic information (life history, morphology, environment) should also be integrated when ever possible for prioritizing or identifying populations for conservation. (From Luikart et al. 2003.)

salmon (*O. tshawytscha*) populations from California to British Columbia and mapped life-history variation on a tree depicting genetic relationships. They focused on differences in run timing (the season adults enter fresh water to begin their spawning migration), which commonly is used by managers to define management and conservation units. In coastal basins and the Lower Columbia River, populations with different run timing co-occur in many distinct genetic lineages, providing strong evidence for repeated, parallel evolution of run-timing differences. In these regions, genetic differences between populations from the same river basin but having different run-timing are typically small enough that they can be explained by fairly recent divergence (within about 100 years) or, more likely, low levels of ongoing gene flow. A very different pattern, however, was found in the interior Columbia and Snake River Basins. In this region east of the Cascade Mountains, all spring-run populations are strongly differentiated

from all fall-run populations, to the extent that they are behaving largely as separate biological species even where they overlap in distribution. Furthermore, the interior spring-run populations have a unique suite of tightly correlated life-history traits that perhaps has evolved only once within the species. The results helped to identify evolutionarily significant units (ESUs; Waples 1991) of Chinook salmon – groups of populations that collectively represent major components of genetic diversity of the species as a whole and which are believed to be on largely independent evolutionary trajectories. In coastal areas and the lower Columbia River, spring- and fall-run populations from the same geographical area are part of the same ESU, but in the interior of Columbia spring- and fall-run populations are in separate ESUs. Over half the ESUs of Chinook salmon are now protected as threatened or endangered 'species' under the Endangered Species Act of the USA.

Forensic genetics and conservation

Recent molecular techniques allow forensic scientists to extract DNA from tiny remains at a crime scene and relate it to an offender. Conservation biologists have used related methodology to trace the source of whale meat sold at the Japanese markets (Cipriano & Palumbi 1999) and to monitor illegal ivory trade (Comstock et al. 2003). The power of forensic science is especially important in the marine environment and in remote wilderness areas where poaching of threatened species is otherwise difficult to detect (Avise 1998). In future years, molecular approaches for species and population identifications may become standard procedure with law enforcement agencies. Mitochondrial genes are useful for species identification using large databases available online (e.g. NCBI Genbank; DNA Surveillance, Ross et al. 2003). Microsatellites and other highly variable markers can be used to identify the source population of poached specimens by comparison with a reference database using the assignment test (Manel et al. 2002).

Assignment tests and microsatellite genotyping were used to detect fraud in a fishing tournament in Finland. A fisherman claimed to have caught an excessively large salmon (*Salmo salar*) in the local Saimaa Lake. Officials doubted that the salmon was of local origin and had a genetics laboratory genotype nine micosatellite loci in the fish as well as in samples from Saimaa Lake and nearby fishing areas. The alleles in the 'winning' fish were so uncommon in the Saimaa Lake sample that its multilocus genotype was extremely unlikely (probability $<1/10,000$) to have originated in the lake (Primmer et al. 2000); instead, its genotype was much more likely in other populations that were subjected to fishing. When confronted with this information, the fisherman confessed to purchasing the big fish at a distant fish market.

The future

Non-invasive and forensic techniques will become standard tools for the conservation biologist in coming years. Forensic and biodiversity inventorying studies could benefit from emerging ambitious projects to 'barcode' (i.e. to sequence a single mtDNA gene) all species on the planet (e.g. Hebert et al. 2004). It will be important, however, to couple these emerging molecular techniques with more traditional morphological analyses of vouched specimens to confidently match genotypes with actual species. Such information and mobile PCR or genotyping machines could allow rapid (on site) identification of species from tiny tissue samples. Combined with the availability of GPS technology, much of the information that required years of tedious fieldwork will be obtained via the Internet and at the laboratory bench.

Rapid identification of genes expressed in a variety of organisms has been achieved by the systematic sequencing of cDNA libraries. Specific transcripts, generally known as expressed sequence tags (ESTs), are prepared from different tissues or developmental stages of a single organism. The ESTs can be used to construct catalogues of tissue-specific or stage-specific genes. Such libraries constructed for endangered or keystone species may help monitor environment-related stress and developmental disorders in these populations. Similarly, single nucleotide polymorphisms (SNPs) recognized for key genes (coding and non-coding regions of the genome) in a target species can predict, for example, the resistance of a population to specific diseases (i.e. having or lacking gene-mediated resistance) and the need for vaccination. For example, Liu & Lamont (2003) scanned a chicken population for susceptibility to *Salmonella* using key SNPs. We anticipate that similar applications would be developed rapidly for conservation purposes.

Another exciting tool is micro-array technology (Gibson 2002; Pfunder et al. 2004), which opens up new perspectives for biodiversity monitoring. A single DNA micro-array contains many thousands of genetically based characteristics (cDNA or oligonucleotides) on one microscopic glass slide (termed 'genome chip'). This technology promises to monitor the whole genome on a single chip so that the researcher can have a better picture of the interactions among thousands of genes simultaneously. A 'Mammalia Chip', for example, could include redundant diagnostic markers to unambiguously identify all European mammal species (Pfunder et al. 2004). Such application could serve as a forensic tool for poaching control or for scanning scats and hair samples (Davison et al. 2002). Micro-arrays were designed originally to measure gene expression (e.g. production of mRNA), but now also exist for measuring DNA sequence variation (e.g. genotyping hundreds of loci simultaneously). A chip designed for a specific endangered species can detect expression changes in multiple genes. Understanding adaptive phenotypic variations in a species is most important for conservation purposes because these expression changes are intimately connected to fitness. Scanning by micro-array analysis a portion of a population could reveal whether individuals are behaviourally or environmentally stressed and the reasons causing it, the sex and reproductive state of individuals, parasite load, ability of individuals to accommodate various selection pressures, etc. (Gibson 2002). Combining non-invasive methodology with micro-array technology is a powerful tool that in the future could provide a complete profile of a population from a single sampling trip or as a means to monitor populations over a long period in unprecedented detail.

The first rule of intelligent tinkering is to keep the parts
(Aldo Leopold, *Round River*, Oxford University Press, New York, 1993, p. 146.)

References

Allendorf, F.W & Seeb, L.W. (2000) Concordance of genetic divergence among sockeye salmon populations at allozyme, nuclear DNA, and mitochondrial DNA markers. *Evolution* **54**: 640–51.

Avise, J.C. (1998) Conservation genetics in the marine realm. *Journal of Heredity* **89**: 377–82.

Banks, S.C., Hoyle, S.D., Horsup, A., Sunnucks, P. & Taylor, A.C. (2003) Demographic monitoring of an entire species (the northern hairy-nosed wombat, *Lasiorhinus krefftii*) by genetic analysis of non-invasively collected material. *Animal Conservation* **6**: 101–7.

Beaumont, M.A. (1999) Detecting population expansion and decline using microsatellites. *Genetics* **153**: 2013–29.

Bellmain, E. & Taberlet, P. (2004) Improved noninvasive genotyping method: application to brown bear (*Ursus arctos*) faeces. *Molecular Ecology Notes* **4**: 519–22.

Bellmain, E., Swenson, J.E., Tallmon, D., Brunberg, S. & Taberlet, P. (2004) Estimating population size of elusive animals with DNA from hunter-collected feces: four methods for brown bears. *Conservation Biology* **19**: 150–61.

Billot, C., Engel, C., Rousvoal, S., Kloareg, B & Valero, M. (2003) Current patterns, habitat discontinuities and population genetic structure: the case of the kelp *Laminaria digitata* in the English Channel. *Marine Ecology Progress Series* **253**: 111–21.

Boersen, M.R., Clark, J.D. & King, T.L. (2003) Estimating black bear population density and genetic diversity at Tensas River, Louisiana using microsatellite DNA markers. *Wildlife Society Bulletin* **31**: 197–207.

Broquet, T. & Petit, E. (2004) Quantifying genotyping errors in noninvasive population genetics. *Molecular Ecology* **13**: 3601–8.

Brown, B., Smouse, P., Epifanio, J. & Kobak, C. (1999) Mitochondrial DNA mixed stock analysis of American shad: coastal harvests are dynamic and variable. *Transactions of the American Fisheries Society* **128**: 977–94.

Brown, T.A. & Brown, K.A. (1994) Ancient DNA – using molecular-biology to explore the past. *Bioessays* **16**: 719–26.

Caughley, G. (1994) Directions in conservation biology. *Journal of Animal Ecology* **63**: 215–44.

Cegelski, C., Waits, L. & Anderson, N. (2003) Assessing population structure and gene flow in Montana wolverines (*Gulo gulo*) using assignment-based approaches. *Molecular Ecology* **12**: 2907–18.

Cipriano, F. & Palumbi, S.R. (1999) Genetic tracking of a protected whale. *Nature* **397**: 307–8.

Comstock, K.E., Ostrander, E.A. & Wasser, S.K. (2003) Amplifying nuclear and mitochondrial DNA from African elephant ivory: a tool for monitoring the ivory trade. *Conservation Biology* **17**: 1840–3.

Cooper, A., Rhymer, J., James, H.F., et al. (1996) Ancient DNA and island endemics. *Nature* **381**: 484.

Davison, A., Birks, J.D.S., Brookes, R.C., Braithwaite, T.C. & Messenger, J.E. (2002) On the origin of faeces: morphological versus molecular methods for surveying rare carnivores from their scats. *Journal of Zoology* **257**: 141–3.

Excoffier, L., Smouse, P. & Quattro, J. (1992) Analysis of molecular variance inferred from metric distances among DNA haplotypes: application to human mitochondrial DNA restriction data. *Genetics* **131**: 479–91.

Fedriani, J.M. & Kohn, M.H. (2001) Genotyping faeces links individuals to their diet. *Ecology Letters* **4**: 477–83.

Felsenstein, J. (1981) Evolutionary trees from DNA sequences: a maximum likelihood approach. *Journal of Molecular Evolution* **17**: 368–76.

Felsenstein, J. (2003) *Inferring Phylogenies*. Sinauer Associates, Sunderland, MA.

FitzSimmons, N.N., Moritz, C., Limpus, C.J., Pope, L. & Prince, R. (1997) Geographic structure of mitochondrial and nuclear gene polymorphisms in Australian green turtle populations and male-biased gene flow. *Genetics* **147**: 1843–54.

Ford, M.J. (2001) Molecular evolution of transferrin: evidence for positive selection in salmonids. *Molecular Biology and Evolution* **18**: 639–47.

Frankham, R. (1995) Effective population-size adult-population size ratios in wildlife – a review. *Genetical Research* **66**: 95–107.

Garza, J.C. & Williamson, E.G. (2001) Detection of reduction in population size using data from microsatellite loci. *Molecular Ecology* **10**: 305–18.

Gibson, G. (2002) Microarrays in ecology and evolution: a preview. *Molecular Ecology* **11**: 17–24.

Girman, D.J., Vila, C., Geffen, E., et al. (2001) Patterns of population subdivision, gene flow, and genetic variability in the African wild dog (*Lycaon pictus*). *Molecular Ecology* **10**: 1702–23.

Hanski, I.A. & Gaggiotti, O.E. (2004) *Ecology Genetics and Evolution of Metapopulations: Standard Methods for Inventory and Monitoring*. Academic Press, San Diego, CA.

Hauser, L., Adcock, G.J., Smith, P.J., Ramirez, J.H.B. & Carvalho, G.R. (2002) Loss of microsatellite diversity and low effective population size in an overexploited population of New Zealand snapper (*Pagrus auratus*). *Proceedings of the National Academy of Sciences USA* **99**: 11742–7.

Hebert, P.D.N., Stoeckle, M.Y., Zemlak, T.S. & Francis, C.M. (2004) Identification of birds through DNA barcodes. *Public Library of Science* **2**: 1657–63.

Hedgecock, D. (1994) Temporal and spatial genetic-structure of marine animal populations in the California current. *California Cooperative Oceanic Fisheries Investigations Reports* **35**: 73–81.

Hedrick, P. (1995) Gene flow and genetic restoration: the Florida panther as a case study. *Conservation Biology* 9: 996–1007.

Hedrick, P. (2001) Conservation genetics: where are we now? *Trends in Ecology and Evolution* **16**: 629–36.

Hendy, M.D. (1993) Spectral-analysis of phylogenetic data. *Journal of Classification* **10**: 5–24.

Hildner, K.K. & Soulé, M.E. (2004) Relationship between the energetic cost of burrowing and genetic variability among populations of the pocket gopher, *T. bottae*: does physiological fitness correlate with genetic variability? *Journal of Experimental Biology* **207**: 2221–7.

Hillis, D.M., Moritz, C. & Mable, B.K. (1996) *Molecular Systematics*, 2nd edn. Sinauer Associates, Sunderland, MA.

Holmes, E.C. & Ellis, S.A. (1999) Evolutionary history of MHC class I genes in the mammalian order Perissodactyla. *Journal of Molecular Evolution* **49**: 316–24.

Kohn, M.H. & Wayne, R.K. (1997) Facts from feces revisited. *Trends in Ecology and Evolution* **12**: 223–7.

Kohn, M.H., York, E.C., Kamradt, D.A., et al. (1999) Estimating population size by genotyping faeces. *Proceedings of the Royal Society of London Series B – Biological Sciences* **266**: 657–63.

Lande, R. (1993) Risks of population extinction from demographic and environmental stochasticity and random catastrophes. *American Naturalist* **142**: 911–27.

Landry, P.A., Koskinen, M.T. & Primmer, C.R. (2002) Deriving evolutionary relationships among populations using microsatellites and (delta mu)(2): all loci are equal, but some are more equal than others. *Genetics* **161**: 1339–47.

Larget, B. & Simon, D.L. (1999) Markov chain Monte Carlo algorithms for the Bayesian analysis of phylogenetic trees. *Molecular Biology and Evolution* **16**: 750–9.

Liu, W. & Lamont, S.J. (2003) Candidate gene approach: potentional association of caspase-1, inhibitor of apoptosis protein-1, and Prosaposin gene polymorphisms with response to *Salmonella enteritidis* challenge or vaccination in young chicks. *Animal Biotechnology* **14**: 61–76.

Lucchini, V., Fabbri, E., Marucco, F., Ricci, S., Boitani, L. & Randi, E. (2002) Noninvasive molecular tracking of colonizing wolf (*Canis lupus*) packs in the western Italian Alps. *Molecular Ecology* **11**: 857–68.

Luikart, G. & Cornuet, J.M. (1998) Empirical evaluation of a test for identifying recently bottlenecked populations from allele frequency data. *Conservation Biology* **12**: 228–37.

Luikart, G., Sherwin, W.B., Steele, B.M. & Allendorf, F.W. (1998) Usefulness of molecular markers for detecting population bottlenecks via monitoring genetic change. *Molecular Ecology* **7**: 963–74.

Luikart, G., England, P.R., Tallmon, D., Jordan, S., Taberlet, P. (2003) The power and promise of population genomics: from genotyping to genome typing. *Nature Reviews Genetics* **4**: 981–94.

Lundrigan, B.L., Jansa, S.A. & Tucker, P.K. (2002) Phylogenetic relationships in the genus *Mus*, based on paternally, maternally, and biparentally inherited characters. *Systematic Biology* **51**: 410–31.

Lynch, M. (1991) The genetic interpretation of inbreeding depression and outbreeding depression. *Evolution* **45**: 622–9.

Makova, K.D. & Li, W.H. (2002) Strong male-driven evolution of DNA sequences in humans and apes. *Nature* **416**: 624–6.

Manel, S., Berthier, P. & Luikart, G. (2002) Detecting wildlife poaching: Identifying the origin of individuals with Bayesian assignment tests and multilocus genotypes. *Conservation Biology* **16**: 650–9.

Manel, S., Gaggiotti, O. & Waples, R.S. (2005) Assignment tests: matching the biological questions with appropriate methods. *Trends in Ecology and Evolution* **20**: 136–42.

McCallum, H. & Dobson, A. (2002) Disease, habitat fragmentation, and conservation. *Proceedings of the Royal Society of London, Series B* **269**: 2041–9.

Menotti-Raymond, M. & O'Brien, S.J. (1993) Dating the genetic bottleneck of the African cheetah. *Proceedings of the National Academy of Sciences USA* **90**: 3172–6.

Miller, C.R., Adams, J.R. & Waits, L.P. (2003) Pedigree-based assignment tests for reversing coyote (*Canis latrans*) introgression into the wild red wolf (*Canis rufus*) population. *Molecular Ecology* **12**: 3287–301.

Morin, P.A., Wallis, J., Moore, J.J. & Woodruff, D.S. (1994) Paternity exclusion in a community of wild chimpanzees using hypervariable simple sequence repeats. *Molecular Ecology* **3**: 469–77.

Morin, P.A., Luikart, G. & Wayne, R.K. (2004) SNPs in ecology, evolution and conservation. *Trends in Ecology and Evolution* **19**: 208–16.

Mossman, C.A. & Waser, P.M. (1999) Genetic detection of sex-biased dispersal. *Molecular Ecology* **8**: 1063–7.

Nikaido, M., Nishihara, H., Hukumoto, Y. & Okada, N. (2003) Ancient SINEs from African endemic mammals. *Molecular Biology and Evolution* **20**: 522–7.

Paetkau, D., Calvert, W., Stirling, I. & Strobeck, C. (1995) Microsatellite analysis of population structure in Canadian polar bears. *Molecular Ecology* **4**: 347–54.

Palumbi, S.R. (2003) Population genetics, demographic connectivity, and the design of marine reserves. *Ecological Applications* **13**(Supplement): S146–58.

Pella, J. & Milner, G.B. (1987) Use of genetic marks in stock composition analysis. In: *Population Genetics and Fishery Management* (Eds N. Ryman & F. Utter), pp. 247–76. University of Washington Press, Seattle, WA.

Pfunder, M., Holzgang, O. & Frey, J.E. (2004) Development of microarray-based diagnostics of voles and shrews for use in biodiversity monitoring studies, and evaluation of mitochondrial cytochrome oxidase I vs. cytochrome b as genetic markers. *Molecular Ecology* **13**: 1277–86.

Primmer, C.R., Koskinen, M.T. & Piironen, J. (2000) The one that did not get away: individual assignment using microsatellite data detects a case of fishing competition fraud. *Proceedings of the Royal Society of London, Series B – Biological Sciences* **267**: 1699–704.

Pritchard, J.K., Stephens, M. & Donnelly, P. (2000) Inference of population structure using multilocus genotype data. *Genetics* **155**: 945–59.

Reed, D.H. & Frankham, R. (2003) Correlation between fitness and genetic diversity. *Conservation Biology* **17**: 230–7.

Reich, D.E., Wayne, R.K. & Goldstein, D.B. (1999) Genetic evidence for a recent origin by hybridization of red wolves. *Molecular Ecology* **8**: 139–44.

Ross, H.A., Lento, G.M., Dalebout, M.L., et al. (2003) DNA surveillance: web-based molecular identification of whales, dolphins and porpoises. *Journal of Heredity* **94**: 111–14.

Roy, M.S., Geffen, E., Smith, D. & Wayne, R.K. (1996) Molecular genetics of pre-1940 red wolves. *Conservation Biology* **10**: 1413–24.

Saccheri, I.J., Brakefield, P.M. & Nichols, R.A. (1996) Severe inbreeding depression and rapid fitness rebound in the butterfly *Bicyclus anynana* (Satyridae). *Evolution* **50**: 2000–13.

Schlotterer, C. (2004) The evolution of molecular markers – just a matter of fashion? *Nature Reviews Genetics* **5**: 63–9.

Schwartz, M.K., Tallmon, D. & Luikal, G. (1999) Genetic estimators of population size: many methods, much potential, unknown utility. *Animal Conservation* **2**: 321–3.

Shaklee, J.B., Beacham, T.D., Seeb, L. & White, B.A. (1999) Managing fisheries using genetic data: case studies from four species of Pacific salmon. *Fisheries Research* **43**: 45–78.

Shedlock, A.M. & Okada, N. (2000) SINE insertions: powerful tools for molecular systematics. *Bioessays* **22**: 148–60.

Smith, D.A., Ralls, K., Davenport, B., Adams, B. & Maldonado, J.E. (2001) Canine assistants for conservationists. *Science* **291**: 435.

Spielman, D., Brook, B.W. & Frankham, R. (2004) Most species are not driven to extinction before genetic factors impact them. *Proceedings of the National Academy of Sciences USA* **101**: 15261–4.

Storz, J.F. & Nachman, M.W. (2003) Natural selection on protein polymorphism in the rodent genus *Peromyscus*: evidence from interlocus contrasts. *Evolution* **57**: 2628–35.

Sunnucks, P. (2000) Efficient genetic markers for population biology. *Trends in Ecology and Evolution* **15**: 199–203.

Taberlet, P., Waits, L.P. & Luikart, G. (1999) Noninvasive genetic sampling: look before you leap. *Trends in Ecology and Evolution* **14**: 323–7.

Tallmon, D.A., Luikart, G. & Waples, R.S. (2004) The alluring simplicity and complex reality of genetic rescue. *Trends in Ecology and Evolution* **19**: 489–96.

Uphyrkina, O., Johnson, W.E., Quigley, H., et al. (2001) Phylogenetics, genome diversity and origin of modern leopard, *Panthera pardus*. *Molecular Ecology* **10**: 2617–33.

Valière, N., Berthier, P., Mouchiroud, D. & Pontier, D. (2002) Gemini: software for testing the effects of genotyping errors and multitubes approach for individual identification. *Molecular Ecology Notes* **2**: 83–6.

Vila, C., Sundqvist, A.K., Flagstad, O., et al. (2003) Rescue of a severely bottlenecked wolf (*Canis lupus*) population by a single immigrant. *Proceedings of the Royal Society of London Series B – Biological Sciences* **270**: 91–7.

Waples, R.S. (1991) Pacific salmon, *Oncorhynchus* spp., and the definition of 'species' under the Endangered Species Act. *Marine Fisheries Reviews* **53**: 11–22.

Waples, R.S., Teel, D.J., Myers, J. & Marshall, A. (2004) Life history divergence in chinook salmon: historic contingency and parallel evolution. *Evolution* **58**: 386–403.

Wayne, R.K. & Jenks, S.M. (1991) Mitochondrial DNA analysis implying extensive hybridization of the endangered red wolf *Canis rufus*. *Nature* **351**: 565–8.

Wayne, R.K., Leonard, J. & Cooper, A. (1999) Ancient DNA: full of sound and fury. *Annual Review of Ecology and Systematics* **30**: 457–77.

Wilson, P.J., Grewal, S., Lawford, I.D., et al. (2000) DNA profiles of the eastern Canadian wolf and the red wolf provide evidence for a common evolutionary history independent of the gray wolf. *Canadian Journal of Zoology* **78**: 2156–66.

5

The role of metapopulations in conservation

H. Resit Akçakaya, Gus Mills
and C. Patrick Doncaster

Nothing in the world is single;
All things, by a law divine,
In one another's being mingle.
(Percy Bysshe Shelley (1792–1822), 'Love's Philosophy')

Introduction

Wherever wildlife management concerns the movement of individuals across structured habitat, its scale of operations will encompass metapopulation dynamics. The goal of this essay is to review the potential applications of metapopulation concepts and models in reserve design and conservation management. Our perspective is forward-looking. We show how some key problems of where to direct conservation effort and how to manage populations can be addressed in the context of regional habitat structure and the survival and renewal of habitat patches. We also mention several cases of successful metapopulation management and point out practical problems (for example, see Box 5.1)

We emphasize:

1. that the viability of a population may depend on surrounding populations, in which case metapopulation processes influence or determine reserve design and management options;

2. that understanding the dynamic processes of populations requires models, which make assumptions that need validating;
3. that the principle limitation of metapopulation models is their single-species focus.

Conservation strategies clearly depend on the particular social, economic and ecological circumstances of each region, and concepts such as the metapopulation can seem irrelevant to practical concerns. We aim to show, nevertheless, that an understanding of metapopulation dynamics can be vital to asking pertinent questions and seeking potential solutions. The conceptual framework of metapopulation dynamics tells us what information is needed in order to build case-specific models relevant to any of a wide range of issues. These issues include: the potential disadvantages of habitat corridors, or hidden benefits of sink habitat; the optimal schedule for translocations or reintroductions; the relative merits of reducing local extinctions against increasing colonizations; the optimum distribution of habitat improvement; and the advantages of increasing life spans of ephemeral habitats.

Concepts

We define a **metapopulation** as a set of discrete populations of the same species, in the same general geographical area, that may exchange individuals through migration, dispersal, or human-mediated movement (based on a very similar definition by Hanski & Simberloff 1997). Older, more restrictive definitions of metapopulation (e.g. Hanski & Gilpin 1991) reflect particular approaches to modelling, for example, by requiring that populations have independent (uncorrelated) fluctuations, are all equally connected by dispersal (Levins' 'island–island' model), or that one population is much larger and less vulnerable than the others (MacArthur and Wilson's 'mainland–island' model). Most criticisms of the metapopulation concept (e.g. Dennis et al. 2003) arise from shortcomings of these more restrictive definitions (Baguette & Mennechez 2004). Over the past decade, the trend in metapopulation concepts has moved from abstract models toward real-world applications. Our more general definition has only two requirements: (i) populations are geographically discrete; (ii) mixing of individuals between populations is less than that within them – otherwise the regional assemblage of local populations may be more aptly described as a single panmictic population. Within these limits, the definition encompasses all levels of variation between populations in colonization rates (including the extreme of 'source–sink' systems, detailed later in this essay) and in extinction rates (including synchronous extinctions, detailed later in this essay). We emphasize that a metapopulation is a dynamic system of linked populations, as opposed to simply a patchy habitat, and many of its demographic processes are visible only through the filter of models.

Although the focus of this essay is on species conservation in habitat fragmented by human activities, metapopulations occur in a variety of forms without any human intervention. Many species depend on habitat patches created by natural disturbances such as fires. Other examples of natural metapopulations include species inhabiting discrete water bodies such as ponds and lakes; despite the physical isolation of freshwater habitats, their populations of aquatic plants and invertebrates may be widely interconnected by birds inadvertently transporting propagules between them (Figuerola & Green 2002), and their populations of amphibians are often interconnected by seasonal dispersal through the landscape. Amongst mammals the Ethiopian wolf (*Canis simensis*) is naturally confined to rodent-rich alpine meadows, but is threatened with extinction by the intervening terrain between plateaux becoming too hostile to allow safe passage (Macdonald & Sillero 2004). Mountain sheep (*Ovis canadensis*) populations in southern California inhabit mountain 'islands' in a desert (Fig. 5.1); this species cannot live for long in the desert, but it can migrate through it (Bleich et al. 1990).

A **sink** is a population with deaths exceeding births and extinction only averted by immigrants exceeding emigrants. Conversely, a **source** is a population with a net outflux of individuals. The identification of sources and sinks is complicated by temporal and spatial variability, and density dependence in demography and dispersal (detailed later in this essay).

Habitat corridors are more-or-less linear strips of habitat with a designed or incidental function of increasing dispersal among populations. We focus specifically on human-modified habitat, additional to natural linear features (such as riparian habitat) that may already link populations. Corridors such as field margins supplement hedgerows which were planted to meet needs not directly related to conservation, but which are increasingly nurtured for their conservation value. Corridors may provide a continuous stretch of habitat between populations, or discontinuous patches that improve connectivity in 'stepping-stone' fashion. A corridor for movement in one direction may simultaneously act as a barrier in the

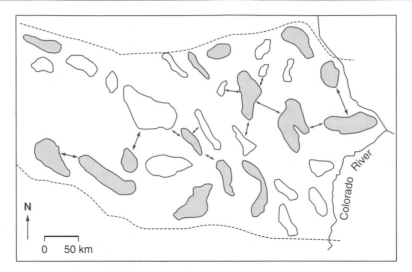

Fig. 5.1 Populations of mountain sheep (*Ovis canadensis*) in southern California. Shaded areas indicate mountain ranges with resident populations, arrows indicate documented intermountain movements; the dotted lines show fenced highways. (After Bleich et al. 1990; reprinted from Akçakaya et al. 1999 with permission from Applied Biomathematics.)

perpendicular direction (such as road verge: Rondinini & Doncaster 2002).

Issues and options

Does conservation need metapopulation concepts?

Animals and plants may occupy metapopulations wherever landscapes are either naturally heterogeneous, or fragmented as a result of human activities such as habitat loss to urbanization, agriculture and transportation routes. Metapopulations are thus relevant to the conservation of any patchy or fragmented habitat. They are also relevant to the conservation of a single population if its dynamics depend on those of neighbouring populations.

One misunderstanding is that the use of the metapopulation concept in conservation requires or implies the conservation or management of species as multiple populations. In some cases, maintaining more than one popu-

lation does increase the persistence of the species as a whole, but this is neither universal, nor a necessary result of using a metapopulation approach. Thus, what conservation needs is not necessarily metapopulations *per se*, but the metapopulation approach and concepts, which permit assessment of the persistence of a species that happens to exist in a metapopulation, either naturally or due to habitat loss and fragmentation. The metapopulation concept is important because species that exist in a metapopulation face particular issues related to environmental impacts, and have conservation options that can be evaluated more completely, or only, in a metapopulation context. These are discussed in the next two sections.

Environmental impacts in a metapopulation context

Metapopulations can be affected by impacts on their entirety or on the individual components. Impacts studied at the regional level include roads and other dispersal barriers that decrease connectivity of populations, and habitat fragmentation that divides a homogeneous population into several smaller populations. The effects of such factors on the overall viability of the species involve interactions among populations (e.g. dispersal and recolonization), and as such they can be assessed or studied only in a metapopulation context.

Impacts such as hunting or fishing may reduce reproduction or survival of individuals in particular populations. For example, hunting pressure or fishing mortality may differ between neighbouring populations, and failure to incorporate the variation into quotas may result in overexploitation, even if the regional harvest is set at a conservative (precautionary) level (Smedbol & Stephenson 2004). An overall harvest level set for a metapopulation may even lead to a series of local extinctions (or a serial collapse of stocks), if most hunters (fishermen) focus on the same few populations with easiest access. After these are locally extinct, the focus shifts to remaining populations with the easiest access. Thus, many local extinctions can occur serially, although the overall (regional) harvest quota is precautionary and is never exceeded. Dynamics of these sorts may have contributed to the collapse of the Newfoundland cod fishery in 1992 with the loss of 40,000 jobs and no recovery in sight.

Conservation and management in a metapopulation context

Conservation options for species that exist in metapopulations include those that aim to increase the size or persistence of individual populations, as well as those that aim to benefit the metapopulation.

The conservation options at the single population level include habitat protection or improvement, regulation of harvest, reduction of predation and removal of exotic species. Even these measures that target individual populations may need to be evaluated in a metapopulation context, because the presence of other populations may change the relative effectiveness of alternative options. An example of this is the effectiveness of reducing seed predation for *Grevillea caleyi*, an endangered understory shrub of Australian eucalypt forest. The few remaining populations of this species are found within a small area at the interface between urban development and remnant native vegetation, and are threatened by habitat destruction, adverse fire regimes and very high seed predation (Auld & Scott 1997). Seed predators include weevils in the canopy and native mammals at the soil surface. Seed germination is triggered by fires, which also kill existing plants. Thus, the frequency and intensity of fires are important components of the species' ecology. A study focusing on a single small population (Regan et al. 2003) concluded that predation reduction improved the chances of long-term persistence of small populations substantially. However, a metapopulation study (Regan & Auld 2004) concluded that management of fires is crucial for the long-term persistence of *G. caleyi* populations, and that predation management was rather ineffective by itself. The reason for this difference is that the number of seeds entering the seed bank after predation is extremely low for a single small population, and there is a substantial risk that all seeds will be depleted in the seed bank due to viability loss and germination. Reducing predation rates for a small population would therefore substantially reduce its risk of extinction. For the metapopulation, however, its seed bank is large enough to always contain available seeds, and a reduction in predation rates does not have a substantial effect on its risk of extinction. At the metapopulation level it is more important to ensure adequate seed production, regular germination and plant survival in years when there are no fire events (Regan & Auld 2004). Thus, for the regional persistence of *G. caleyi* fire management appears to be a much more important strategy, a conclusion that was not as apparent when only a single population was considered, even though both actions – fire management and

predation control – can target a single population or all populations in the metapopulation.

The conservation options at a metapopulation level include reserve design, reintroduction and translocation, dispersal corridors and
management actions geared to local population
dynamics (such as sources and sinks). We
discuss these below.

RESERVE DESIGN

Reserve design is a complex topic that almost
always involves multiple species, as well as
social, political and economic constraints. Here
we focus on only one aspect: directing conservation effort at a subset of the populations of a
target species, in order to maximize the chances
of its survival. This issue is informed by predictions and observations of generally higher
extinction rates in smaller populations, and
lower probabilities of rescue by immigration in
more isolated patches (Hanski 1994). It was

originally phrased as the 'SLOSS' debate, i.e.
whether a single large or several small
(SLOSS) populations are better to protect the
species. Although simplistic, this formulation
captures the nub of the issue, and underlines
the relevance to conservation of spatial structure and metapopulation dynamics.

On the one hand, several small populations
may have a lower extinction risk than one large
one if the rate of dispersal is high enough and
the degree of spatial correlation of environments is low enough. This is because a single
large population will not benefit from uncorrelated environmental fluctuations; if it becomes
extinct, it cannot be recolonized. For example,
an important reason for establishing the wild
dog reserves discussed in Box 5.1 was to provide
a hedge against the possibility of a catastrophic
event hitting the single large Kruger population.

On the other hand, compared with a large
population, each of the small populations will
be more vulnerable to extinction due to
demographic stochasticity, higher mortality of

Box 5.1 Reintroduction of wild dogs in South Africa

Most metapopulations are the regional-scale expression of responses by individuals to patchiness in their habitat.
Persistence at the regional level is enhanced if individuals can retain some ability to move across the matrix to
prevent local extinctions or to recolonize empty patches. Here we describe a particularly extreme example of a
metapopulation, in which the habitat patchiness is caused by fences, and individuals have lost all intrinsic capacity to
mix freely between populations. The persistence of the metapopulation relies entirely on human-induced translocations, and corridors take the form of transportation vehicles.

A programme was initiated in 1997 to establish a second South African population of the endangered wild dog
Lycaon pictus apart from the only viable one in the Kruger National Park (Mills et al. 1998). As the Kruger population
fluctuates around 300 (Creel et al. 2004) it was thought prudent to bolster the small number of dogs in South Africa
and provide a hedge against the uncertainty of a catastrophic event hitting the Kruger population. At present South
Africa has no other protected area large enough to contain a self-sustaining population of wild dogs, so the strategy
has been to introduce them into a number of small widely scattered reserves separated by hundreds of kilometres
and to manage the various subpopulations as a single metapopulation.

Preliminary modelling of this wild dog metapopulation suggested that periodic, managed gene flow through
translocations should be implemented to reduce inbreeding and the resultant risks of meta- and subpopulation
extinction. The model indicated that by using a frequency of exchange based on the natural reproductive life span of
wild dogs (approximately 5 years) inbreeding could be reduced by two-thirds and population persistence could be
assured (Mills et al. 1998).

The guiding principle in reserve selection was to look for areas that reasonably can be expected to sustain at least
one pack of 10 to 20 animals. The average home range size for a pack is 537 km^2 in Kruger National Park (Mills &
Gorman 1997), which comprises a similar savannah woodland habitat to the habitat available in most of the
potential reserves for reintroduction. The range of sizes of the five reserves into which wild dogs have so far been

introduced for the metapopulation is 370–960 km². All reserves are enclosed with electrical fences, to protect the wild dogs and to minimize conflict with livestock farmers. Fences act as important barriers to the movements of the dogs, so that there is little emigration and even less immigration. The reserves are isolated from each other, with no possibility at present to establish corridors, and almost all movement of wild dogs between the reserves is conducted through artificial introductions and removals.

Apart from protecting the regional viability of the species, an important objective in the wild dog metapopulation management programme is to promote biodiversity conservation. Biodiversity is a broad concept incorporating compositional, structural and functional attributes at four levels of ecosystem organization: landscapes, communities, species and genes (Noss 1990). A biodiversity objective for wild dogs that may be especially difficult to achieve in a small reserve is to restore their ecological role as predator. Wild dog packs can produce large litters and more than double in size within a year, posing a particularly challenging situation for managers because of the rapidly escalating predation pressure, at least in the short term. This is exacerbated by the tendency for wild dogs to use fences as an aid to hunting (van Dyk & Slotow 2003), which may artificially increase kill rate. An important aspect of the programme is to research the viability of interactions between wild dogs and their prey in confined areas.

Following release of the first six to eight animals, the principle management strategy has been to continue to simulate the natural dynamics of wild dog packs by moving single sex groups between reserves as and when necessary, so as to maintain the genetic integrity of the metapopulation and, if necessary, to promote new pack formation as originally recommended (Mills et al. 1998). In the reserves, regular maintenance and daily patrolling of the fences is essential. In spite of this weaknesses do occur. Holes dug by other species such as warthogs (*Phacochoerus africanus*), flood damage along drainage lines and occasions when predators chase prey through a fence are among the ways in which breaches can occur. These are most likely to be exploited during dispersal events by the dispersing animals. Escapes are most likely to happen if there are no suitable dogs of opposite sex available with which dispersers can form a new pack, or if the reserve is too small to allow for the formation of another pack. The obvious solution to dispersers escaping from a reserve is to remove dogs before they break away, but it is difficult to know which dogs to remove and when. The preferred solution would be to remove dogs only after they have naturally split off from the pack. Managers decide on the removal of dogs when they are concerned about the impact of increasing numbers on the prey, or in order to decrease the risk of dogs escaping from a reserve. Behavioural observations may help to predict when a breakaway is about to occur and which dogs are involved, in which case management intervention can thus be applied pre-emptively based on this behavioural research.

Financial costs of the wild dog management programme have as much influence on strategy as do ecological imperatives. Costs include upgrading reserve fences, constructing a holding facility, radio-telemetric apparatus for monitoring the dogs, running vehicles, veterinary costs of capture, vaccination and transportation of the dogs, and liability insurance against escaped dogs causing damage to neighbours' domestic animals. Almost $380,000 was spent on wild dog conservation in South Africa between 1997 and 2001, of which c.75% was spent on establishing the metapopulation (Lindsey et al. 2005).

Despite the complexities outlined above, the extremely artificial nature of this metapopulation's spatial structure, and a general lack of knowledge about the dynamics of this species in small reserves, several aspects of this case are closely related to the metapopulation issues we will discuss in this essay.

dispersers and edge effects (smaller patches have a higher proportion of 'edge' to 'core' habitat). Thus, if they become extinct at the same time, or if the extinct ones cannot be recolonized from others, a metapopulation of several small populations may have a higher extinction risk than a single large population (see Akçakaya et al. (1999) for an example). In some cases, however, the choices are limited. In the wild dog case, for example, available habitat limited the size of the established populations to a maximum of three packs each, resulting in a mixture of several small populations and one large (Kruger) population.

There is no general answer to the SLOSS question. The answer depends not only on the degree of correlation and chances for recolonization, but also on other aspects of metapopulation dynamics, such as the configuration, size and number of populations, their rates of growth, density dependence, carrying capacities, etc.

Often the monetary or political cost of acquiring a patch for a reserve might not be related to its size; in other cases the size or carrying capacity of a patch might not be directly related to its value in terms of the protection it offers. A small patch that supports a stable population might contribute more to the persistence of the species than would a large patch that is subject to greater environmental variation or human disturbances. Each case requires individual evaluation, using all of the available empirical information to evaluate as many as possible of the potential impacts on the extinction time of the metapopulation. Predictions for individual cases, however, will always depend on a thorough understanding of the underlying dynamic processes of density dependence and interactions with the physical environment that drive the case-specific mechanisms (Doncaster & Gustafsson 1999). Although few of these processes can be observed directly in nature, the wider framework in which they operate is provided by generic models of the conceptual issues.

Wherever possible, design options should consider less extreme alternatives than SLOSS. A mixture of smaller and larger populations can hedge against uncertainty in the scale of future impacts, and it has potential genetic benefits. Unless the small populations act as sinks, they are likely to send out a greater proportion of emigrants as well as receiving more immigrants, than larger populations. For example, collared flycatchers (*Ficedula albicollis*) exhibit this higher turnover in smaller populations, which both reduces genetic drift and slows the evolution of adaptations to local conditions (Doncaster et al. 1997). The combination of small (habitat generalist) and large (habitat specialist) populations pre-adapts the metapopulation for future environmental changes.

A related question for reserve design concerns the optimum distribution of resources between patches. Is the species better protected by a more heterogeneous or more homoge-

neous distribution of resources? Temporal variability tends to stress populations near to extinction thresholds, so reducing their sizes (Hastings 2003). In contrast, spatial heterogeneity is likely to improve the predicament of such species across both population and metapopulation scales (Doncaster 2001). This effect arises because the abundance of rare consumers generally decreases disproportionately with degrading habitat quality, regardless of their particular functional response to limiting resources. For example, oystercatchers (*Haematopus ostralegus*) will abandon beds of mussels (*Mytilus edulis*) below a certain threshold of available shellfish set by their foraging efficiency (Caldow et al. 1999). The counter-intuitive implication for metapopulations is that the regional abundance of a target species can be raised by redistributing resources between patches even without any overall improvement to habitat quality, so that those of intrinsically higher quality are augmented to the detriment of others already below the giving-up density.

TRANSLOCATION AND REINTRODUCTION

Establishment of new populations through translocation and reintroduction actions requires many decisions: how often; how many individuals, of which age classes or sexes; from which population, to which existing population or formerly occupied habitat patch? Each decision is potentially a trade-off, because it may benefit one population while decreasing the size of another one. Metapopulation models can address these questions by finding strategies that maximize the overall viability of the metapopulation. This was especially important in the wild dog reintroduction case (Box 5.1), because almost all movement of wild dogs between the reserves is conducted through artificial introductions and removals. In this case, a metapopulation model with genetic structure would have helped to plan translocations in such a way as to reduce inbreeding and maintain population structure, but in the event a more needs-driven approach had to be taken in terms of supply and demand of suitable dogs, although always keeping in mind the genetic history of the individuals concerned.

Metapopulation models may be particularly important tools in decisions related to translocation and reintroduction of endangered species, because the status of these species discourages experimentation and makes a trial-and-error approach less desirable. Using a metapopulation model, McCarthy et al. (2004) assessed various options for establishing a new population of helmeted honeyeaters (*Lichenostomus melanops cassidix*) from a captive population. This bird is endemic to remnant riparian forests in southern Victoria, Australia. Extensive habitat destruction in the nineteenth century led to a dramatic decline, and by 1990 the only remaining population included 15–16 breeding pairs. As part of a recovery programme initiated in 1989, a captive colony was established to support the wild population and to establish populations in new areas (Smales et al. 2000). Because of uncertainties about the fates of individuals and the difficulty of integrating the available information from numerous different sources, the optimal release strategy is not immediately apparent. McCarthy et al. (2004) ran simulations to determine how the rate of release from the captive population affects the probability of success of the reintroduction over 20 years. The optimal strategy was to release individuals only when the captive population contained at least four adult males, and then to release 30% of the stock. The simulations suggested that the chance of success of the proposed reintroduction was moderately good, with little chance that the new population will have fewer than 10 males after 20 years (McCarthy et al. 2004). Although there were several factors that could not be modelled explicitly (e.g. whether the released birds would remain where they are released, would establish the same population behaviour, and would have the same vital rates as the current wild population), the modelling exercise provided valuable information that could not

have been obtained in any other way for this extremely rare species.

CONNECTIVITY AND HABITAT CORRIDORS

In addition to human-mediated dispersal through reintroduction and translocation, dispersal can be increased by conservation or restoration of the habitat lying between existing populations, sometimes called the 'landscape matrix'. Matrix restoration can reduce local extinctions by facilitating the 'rescue effect' of colonization, and it can increase the rate of recolonization following local extinction. One implementation of these efforts to increase the overall persistence of the species is the building or maintenance of habitat corridors. To answer the question 'Are corridors useful conservation tools?', we need to answer several subquestions that are intimately bound to metapopulation concepts.

1. Are the habitat corridors used by the target species? Use of a corridor depends not only on its habitat, but also its shape, particularly the width and length. For example, of the mammalian predators native to California, more species use creeks with wide margins of natural vegetation as corridors than use creeks with narrow or denuded margins (Hilty & Merenlender 2004). European hedgehogs (*Erinaceus europaeus*) dispersing across arable habitat use road verges as corridors, particularly on long-distance dispersals of as much as 10 km (Doncaster et al. 2001).

2. If used, do the corridors increase dispersal rate? Perhaps individuals using the corridor would have dispersed anyway; corridors are more likely to affect dispersal rate where dispersal is otherwise limited. For example, if it were possible to build corridors between the widely scattered wild dog reserves discussed in Box 5.1, the lack of natural connections and the pack-forming behaviour of the species suggest that such corridors would have increased dispersal between reserves. Corridors are likely to benefit fast-

reproducing species in the short term and slower reproducers in the long term, so their value depends on the time scale of conservation goals (Hudgens & Haddad 2003). Experimental fragmentation of moss banks has demonstrated rescue effects of artificial corridors for moss-living micro-arthropods. Figure 5.2 shows how corridors between moss fragments arrested declines in the abundance of most species (Gonzalez et al. 1998). It is worth noting that this experiment on an abundant fauna cost little to run, yet has provided invaluable quantification of the positive relation between abundance and distribution in connected landscapes, and of the breakdown of this relation in the ab-

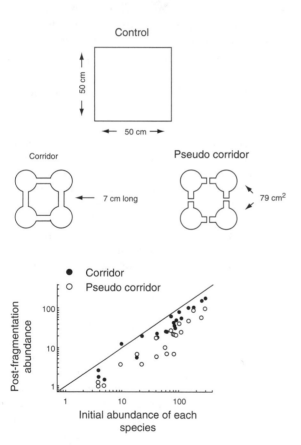

Fig. 5.2 Experimental fragmentation of moss banks into small patches reduces the abundance of micro-arthropods, but most species are saved from substantial decline by corridors connecting the fragments. (From Gonzalez et al. 1998. Reprinted with permission. Copyright 1998 AAAS.)

sence of corridors. Conservation needs more such field tests of metapopulation theory by ecologists.

3. Does an increased dispersal rate increase the overall viability of the metapopulation? Usually it does, by rescuing local populations from potential extinction. Colonists can also bring hybrid vigour to isolated populations that suffer from inbreeding depression (e.g. of *Daphnia*: Ebert et al. 2002). However, increased connectivity may also have 'anti-rescue effects' (Harding & McNamara 2002), with documented examples due to the spread of infectious diseases, or parasites or predators (Hess 1996; Grenfell & Harwood 1997), or gene flow reducing local adaptation (Hastings & Harrison 1994; Harrison & Hastings 1996). High dispersal can increase impacts of catastrophes (Akçakaya & Baur 1996), and losses to sink habitats. In other cases, the effectiveness of dispersal in reducing extinction risks depends on the correlation of environmental fluctuations experienced by different populations. If the correlation is high, all populations decline simultaneously, reducing recolonization rates of empty patches. For example, if a major climatic shift caused a region-wide decline in the prey base of the wild dogs discussed in Box 5.1, all populations would decline or become extinct and it would be difficult to recolonize them even with a well-planned translocation programme. If, on the other hand, the fluctuations are at least partially independent, some patches can act as sources of emigrants (Burgman et al. 1993). Extinction risks are often sensitive to spatial correlation in environmental fluctuations and the pattern of disturbance, as demonstrated by models for a variety of species, including the mountain gorilla (*Gorilla beringei beringei*; Akçakaya & Ginzburg 1991), spotted owl (*Strix occidentalis*; LaHaye et al. 1994) and Leadbeater's possum (*Gymnobelideus leadbeateri*; McCarthy & Lindenmayer 2000).

4. Do corridors have any other effects on metapopulation viability? Negative impacts may include increased mortality due to predation along the corridor. All Dutch highways constructed since 1990 have underpasses for

European badgers (*Meles meles*), and these are also used by other wildlife, including hedgehogs. This benefits the hedgehogs because they too are a frequent casualty of roads, but the benefit is undone because they are also a favoured food of badgers, into whose jaws they are channelled by the underpasses (Bekker & Kanters 1997). Costs such as these need to be weighed against the benefits of dispersal. The Dutch Government spends US$5 million each year on tunnels and fences for wildlife along its highways (Teodorascu 1997), so it makes economic sense to evaluate the combined effect of all changes on metapopulation viability.

5. Are there cheaper alternatives to corridors? These may involve seeding new habitable patches, or improving existing populations by augmenting net growth rates or carrying capacities. In simulations of metapopulations prone to local extinction events, the viability of the system is found to benefit more from reducing local extinction probabilities, particularly on patches with the lowest probabilities, than from increasing colonization probabilities (Etienne & Heesterbeek 2001). Birds using rainforest fragments show evidence of this response (Lens et al. 2002), but more empirical testing is needed of this model, as with most metapopulation models.

Although the importance of corridors has long been recognized, it is only with the use of metapopulation models that their advantages can be quantified, for example, in terms of increased persistence or viability of the species, and compared with advantages of alternative strategies.

SOURCES AND SINKS

When conservation is geared to local populations with dynamics of sources and sinks, management options must consider many interdependent factors. Two general issues arise:

1. How do source–sink dynamics affect metapopulations? The overall effect on metapopulation persistence of dispersal from sources to sinks depends on the cost to source population (increased risk of local extinction), the benefit to sink population (decreased risk of local extinction), and the changes with local population density in dispersal, survival or reproduction. In the presence of density dependence, the excess of deaths to births that is characteristic of a sink can be caused directly by the influx of immigrants rather than being an inherent property of the patch. The viability of such a 'pseudosink' consequently need not depend on the arrival of emigrants from sources. It may even benefit from a reduced influx, in contrast to a true sink which is rescued by immigration (Watkinson & Sutherland 1995). Management options will differ for true and pseudosinks because of this, yet the two types can be hard to distinguish in field surveys. For example, sources and pseudosinks in the highly fragmented Taita Hills forests of Kenya could be identified only from a combination of demographic, genetic and behavioural work (Githiru & Lens 2004). To sidestep these complications, sinks can be defined as populations whose removal would increase the overall viability of the metapopulation. This approach, however, requires modelling of the underlying dynamics of the metapopulation, and therefore more data.

2. Should sink populations be protected? This depends on various factors, the most important being what is meant by 'protected' and its alternatives. If 'protected' means that fecundity or survival may increase to the extent that the sink population can become self-maintaining (i.e. have a low risk of extinction even in the absence of dispersal from other populations), and the alternative is continuation as a sink population, then protection is probably justified (as Breininger & Carter (2003) demonstrated for the Florida scrub jay (*Aphelocoma coerulescens*)). If 'protected' means it is maintained as a sink population and the alternative is that individuals that would have dispersed to the sink end up in a habitat patch with higher survival or fecundity, then protection of the sink is probably not justified (as Gundersen

et al. (2001) demonstrated for root voles (*Microtus oeconomus*)). In the wild dog case (Box 5.1), for example, if mortality exceeds reproduction in one of the reserves as a result of a local decline in the prey base, then it would be justified to attempt to maintain this population by increasing the prey base in that reserve (e.g. by enlarging the reserve or by habitat improvement), but it might not be justified to attempt to maintain this population only by increasing translocations from healthier populations. Other considerations include whether the sink population can increase connectivity (as a 'stepping stone' between other populations), its contribution to total abundance and its function as a buffer against catastrophic events. Where conservation is aimed at culling an invasive alien, its regional decline can be hastened by allocating culls to sinks as well as sources (as for the European hedgehog introduced into the Scottish Western Isles; Travis & Park 2004). The point is that there are a lot of details, and generalizations are difficult if not impossible. The only way to address such questions is to develop case-specific models that incorporate all that is known about the dynamics of the metapopulation, including survival, fecundity and dispersal for all populations whether source or sink, as well as temporal variability and density-dependence in these parameters.

Do metapopulations need models?

The metapopulation concept lends itself to modelling because its core dynamic – of populations colonizing patches (and their potential local extinction) – bridges models of persistence at the levels of the individual and the community: of individuals consuming resources (and their eventual death), and of species colonizing niches (and their potential extinction: Doncaster 2000). Metapopulations encompass landscape-level processes of patches being formed, split and merged in habitat successions and disturbance events. At all of these scales, models

are used to pare away as much of the complexity inherent to nature as is necessary to reveal the underlying patterns and to explore the range of forces that shape these patterns. Models are particularly important to the conservation of metapopulations, because the regional focus and undesirability of experimental manipulations usually rules out any other methods of distinguishing causes of endangerment from secondary effects. Most of the issues and decisions regarding metapopulations concern interdependent factors, such as number of populations, spatial correlation, dispersal and density dependence. Because many of these factors involve interactions between populations, there is no simple way of combining models on dynamics of individual populations into regional-scale decisions. The only way to incorporate all these factors is to simultaneously include all populations and their interactions in one model, in other words, to use a metapopulation model. Models are particularly valuable tools in cases where the endangered status of species makes other (e.g. experimental) approaches difficult or impossible.

Models are also useful in evaluating management actions at large spatial scales, at which experiments may not be feasible. Frequently, management of a metapopulation means management of the species' habitat. Habitat management may take many forms, including controlling the rate and pattern of habitat alteration through the effects of grazers or harvest by humans. For example, Schtickzelle & Baguette (2004) used a structured metapopulation model to study the effect of grazing on the bog fritillary butterfly (*Proclossiana eunomia*) in south-eastern Belgium. This species has a very restricted habitat; both larval and adult stages feed on a single plant species that occurs mainly in wet hay meadows along rivers of some uplands scattered in western Europe. Grazing by large herbivores is sometimes used by conservation agencies to maintain early successional stages in wet hay meadows. The metapopulation model demonstrated that grazing substantially increases the extinction risk for the bog fritillary butterfly in south-eastern Belgium. Its predictions led to modifications in the management protocol of a nature reserve: several grazing regimes are being tested and half the area is now kept ungrazed.

Controlled timber harvest is another form of habitat management. Regan & Bonham (2004) developed a metapopulation model of the carnivorous land snail *Tasmaphena lamproides* inhabiting native forests in northwest Tasmania. This species is listed as threatened due to its small range, much of which is within timber production forest. The model was designed as a decision support tool for managers to explore the trade-offs between timber production requirements and conservation of the species under various management scenarios. Future use of the area involves converting native forest to eucalypt plantations, or harvesting native forest followed by burning to promote regeneration. Burning is thought to eliminate populations of this snail, but they reinvade native forest areas once the required habitat has formed with adequate level of litter and food sources. The metapopulation model combines geographical information system (GIS) data on the distribution of forests and demographic data on the dynamics of the species, and allows the investigation of alternative harvesting strategies which meet wood production needs in the long term but minimize population declines in the short term.

In aquatic systems, habitat management often involves water regimes and barriers such as dams. Changes in water regime were implicated in the severe decline of the European mudminnow (*Umbra krameri*) along the River Danube during the second half of the twentieth century (Wanzenböck 2004). Water regulation in the Danube has increased flow velocity and caused the river to cut a deeper channel, lowering the groundwater level in the surrounding floodplain. As a result, the original side channel used by the mudminnow has been transformed into a chain of disconnected, groundwater-fed ponds. A simple metapopulation model was used to demonstrate that reversing this declin-

ing trend in habitat capacity is critical to the mudminnow's persistence, and to recommend increasing the habitat availability for, and connectivity of, populations. To implement these recommendations, groundwater levels are being raised by opening some of the longitudinal dams bordering the main river and reconnecting some backwaters to the river. These conservation efforts began in the late 1990s, are continuing today, and their impact on the mudminnow is being monitored closely.

There are several different types of metapopulation models, each with their own set of assumptions and restrictions (detailed in Akçakaya & Sjogren-Gulve 2000; Breininger et al. 2002). Patch-occupancy models have the simplest demographic structure, describing each population as present or absent (e.g. within regional distributions of butterflies or other winged insects; Hanski 1994). Intermediate complexity is found in structured (or, frequency-based) models that describe each population in terms of the abundances of age classes or life-history stages (Akçakaya 2000a). These models incorporate spatial dynamics by modelling dispersal and temporal correlation among populations (e.g. of the land snail *Arianta arbustorum*; Akçakaya & Baur 1996). At the other extreme are individual-based models, which describe spatial structure within the location of territories, or of each individual in the population (e.g. of northern spotted owls (*Strix occidentalis caurina*); Lamberson et al. 1996; Lacy 2000). Some models use a regular grid where each cell can be modelled as a potential territory. For example, Pulliam et al. (1992) used this approach in a region managed for timber production to show that population sizes of Batchman's sparrow (*Aimophila aestivalis*) depended more strongly on mortality rates than on dispersal ability. Another approach uses a habitat suitability map to determine the spatial structure of the metapopulation (e.g. of the helmeted honeyeater (Akçakaya et al. 1995) and California gnatcatcher (Akçakaya & Atwood 1997)). All of these approaches have been applied to specific conservation management questions (Chapter 9). The appropriate choice depends on the complexity of the problem at hand, the assumptions of the model in relation to the ecology of the species (see below) and the data available.

Current limitations and dilemmas

Single-species focus

Most metapopulation applications focus on a single species, yet much of conservation management concerns communities. Even where a single species is targeted for conservation, its survival and fecundity will often depend on competition within the trophic level or predation from higher trophic levels. For example, in the case of wild dogs (Box 5.1), an important objective is to restore their ecological role as predator, which requires research into the viability of wild dog–prey interactions in confined areas. Metapopulation models tend to focus on single-species dynamics because these are better understood than foodweb and ecosystem processes. Adding an extra species to the system requires at least two extra dimensions in the analysis (to account for both exclusive and shared occupancy of suitable habitat), greatly increasing the number of parameters for estimation and thus model error. The general lack of understanding and data on multispecies interactions means that few empirical metapopulation studies have sufficient parameter estimates to model community dynamics. An astute use of simplifying assumptions, however, can bring theory within the grasp of empirical data.

Simple models have achieved some robust predictions for competitive coexistence by reducing the representation of competition to a binary distinction between competitively dominant and inferior (fugitive) species. For example, habitat destruction is predicted to disadvantage dominant species with slow dispersal to the benefit of fugitive species, and the early loss of dominants has most effect on

community structure because of their potential role as keystone species (Tilman et al. 1997). The dominant–fugitive dichotomy applies particularly to plant diversity in prairie grasslands. The generalized version of this patch-occupancy approach explores the full range of competitive asymmetries in regional coexistence, and without needing extra dimensions in the analysis if it can be assumed that both residents and colonists experience similar effects of density on survival (Doncaster et al. 2003). This model reveals that subdominant species with poor dispersal are the most sensitive to habitat degradation. Their loss from the community provides a useful early warning of regional disturbance and degradation, because it will have less impact on community structure than the subsequent disappearance of dominant and potentially keystone species. In general, faster reproducing communities (e.g. invertebrate assemblages) are both predicted and observed to have higher tolerance for differences in growth capacity, compared with slower reproducing communities (e.g. forest trees), which have higher tolerance for competitive interactions. Coexistence is even possible amongst tree species competing for identical resources in the same metapopulation, if they differ in their threshold conditions for switching from vegetative growth to seed production (e.g. Mexican rain forest trees; Kelly & Bowler 2002). These low productivity communities tend to be the most at risk from human induced disturbance, and therefore the most in need of predictive models.

Where conservation efforts are directed towards a community of species, a practical approach to dealing with the single-species limitation is to select a target species that is representative of the natural community, that is sensitive to potential human impact and whose conservation will protect other species (Noon et al. 1997). One danger here is that the target species and others may have different networks of habitat patches in the same region. For example, from a large-bodied predator's point of view, there may be a few large habitat patches, but for its small prey, there may be hundreds of distinct patches. Or, the degree of fragmentation may be different for each species depending on their habitat requirements. For example, roads fragment forest habitat for song-birds in direct proportion to their dependence on canopy-level vegetation for nesting and feeding (St Clair 2003). Endangered silver-studded blue butterflies (*Plebejus argus*) and sand lizards (*Lacerta agilis*) both disperse between heathland fragments, but the greater capacity for the butterflies to use areas between habitat patches (also called 'matrix') suggests they will benefit most from climate warming, at least in terms of increased patch connectivity and metapopulation stability (Thomas et al. 1999). The best strategy in such habitat and community studies is often to combine results from different target species (Root et al. 2003).

For addressing most conservation questions involving species in fragmented landscapes, metapopulation models often have less severe limitations than the available alternatives, such as rule-based methods, expert opinion, reserve-selection algorithms and habitat mapping. However, these alternatives have the potential of contributing to the realism of metapopulation models or of complementing them (Akçakaya & Sjogren-Gulve 2000; Breininger et al. 2002; Brook et al. 2002).

Definition and delineation of populations in a metapopulation

Most metapopulation approaches represent the landscape by discrete habitable patches within a surrounding matrix that may allow dispersal but does not support populations. To the extent that there exist areas where a species can reproduce and those where it cannot, this assumption is not unrealistic. However, it does require the definition of a population, and a method for identifying these areas (patches) in a given landscape.

A general definition of a population presents dilemmas, regardless of the metapopulation

context. Considering the difficulty of defining a species, a much more fundamental concept, this is perhaps not surprising. A biological population can be defined as a group of interbreeding (i.e. panmictic) individuals. Assuming that the distribution of a species is more-or-less continuous across parts of the landscape, the question of delineating a population can be rephrased as: how far apart must two individuals be in order to be considered to be in different populations? This depends on the movement distance, home range, or some other measure related to the possibility of interbreeding. This approach, combined with modelling and prediction of suitable habitat, is used in habitat-based metapopulation models to delineate populations (Akçakaya 2000b, 2005). In the wild dog metapopulation (Box 5.1), populations are easily defined by fenced reserves.

Assumptions of metapopulation models

All models assume certain constants, in order to interpret the dynamics of interest. The usefulness of any model therefore depends on the validity of its assumptions. Below we discuss recent approaches to improving the fit of metapopulation models to data.

COLONIZATION MATCHES EXTINCTION

Some metapopulation models assume that the metapopulation is in equilibrium with respect to the extinction and recolonization of patches (e.g. incidence function type of patch-occupancy models; Hanski 1999). There is little evidence to suggest that metapopulations of any species are in fact at equilibrium (Baguette 2004), and small, highly variable metapopulations are particularly unlikely to be so. However, metapopulations that persist over long time-scales must be under some form of density regulation at the regional scale, often assumed to be in colonization rate, which implies at least a deterministic attraction towards an equilibrium density of occupied habitat. Equilibrium models therefore can play a useful role as null hypotheses for analysing the processes that may threaten viability, such as habitat loss, exploitation and alien invasions. A poor fit of equilibrium models to the data can signal the need to account for other factors, such as competitive interference in addition to exploitation (Doncaster 1999), or multiple equilibria (Hanski et al. 1995), or it may result from random fluctuations. Null models test these alternatives parsimoniously by seeking to explain deviations from equilibrium predictions. Note that density regulation of local populations in a metapopulation does not guarantee the existence of an equilibrium at the metapopulation level. The metapopulation may still decline if the rate of local extinctions due to environmental fluctuations and demographic stochasticity exceeds colonization rates, because of factors such as limited dispersal, or Allee effects on small populations, or correlated environments.

The equilibrium assumption is sometimes mistakenly believed to apply to the metapopulation concept in general, yet several metapopulation models and approaches do not make this assumption (e.g. structured and individual-based models, described in Akçakaya & Sjogren-Gulve 2000). These are particularly useful for predicting the extinction probability of small metapopulations which may have unbalanced sex ratios or age structures, or low genetic variability, and which are most prone to environmental fluctuations (e.g. some coral reef fishes: Bascompte et al. 2002), or which may be declining (e.g. California gnatcatcher (*Polioptila californica*); Akçakaya & Atwood 1997).

INDEPENDENT DYNAMICS OF LOCAL POPULATIONS

Some metapopulation models assume that the dynamics of local populations are independent of each other. However, this assumption is violated in many metapopulations where local populations are affected by regional environ-

mental factors that impose a correlation. For example, fecundities of the California least tern (*Sterna antillarum browni*) are correlated across different subpopulations, presumably due to the effects of large-scale weather patterns such as the El Niño–Southern Oscillation that may simultaneously affect the food resources of many populations. The correlation coefficients average 0.32 (range 0–0.6), and decline with increasing distance between the populations (Akçakaya et al. 2003a). When correlations are based on population sizes rather than vital rates such as fecundity, it may be difficult to untangle the relative contributions of correlated environmental factors, dispersal, and trophic interactions to the observed spatial correlation in population dynamics (Ranta et al. 1999). However, it is clear that for many species, subpopulations experience spatially correlated dynamics (Leibhold et al. 2004). In these cases, results of simple models that assume independence may be misleadingly optimistic in their estimation of risks of extinction and decline. However, it is possible to make realistic and unbiased assessments by using models that incorporate dependencies or spatial correlations among populations (e.g. Harrison & Quinn 1989; Akçakaya & Ginzburg 1991; LaHaye et al. 1994).

STATIC HABITAT

Many metapopulation models assume a constant number and location of habitable patches, yet natural landscapes are inherently dynamic. Spatial structure changes according to seasons, climatic fluctuations and succession, as well as human impacts (urban sprawl, global climate change, agricultural expansion, etc.). The viability of a metapopulation will depend on its rate of patch turnover, as well as the static quantity and quality of suitable habitat (Keymer et al. 2000). Under habitat succession or age-dependent disturbance, for example, a metapopulation is predicted to persist for as long as the mean age of its constituent patches exceeds the average interval between colonization events (Hastings 2003). A metapopulation with a slow turnover of patches thus may persist even with a high extinction rate of local populations, and managers should be wary of underestimating its viability. Equally, management action aimed at increasing the lifespan of patches is likely to do more good than action focused directly on the survival of local populations on the patches.

Some metapopulation models incorporate community succession, which tends to be particularly patchy in time and space at its early stages and determines critical habitat for certain species (Johnson 2000; Hastings 2003). Other models incorporate changes in carrying capacity over time, either deterministically, for example to simulate forest growth, or stochastically to simulate the effects of random disturbances such as fires, or both, e.g. as a deterministic function of time since a stochastic disturbance event (Pulliam et al. 1992; Lindenmayer & Possingham 1996; Stelter et al. 1997; Akçakaya & Raphael 1998; Johst et al. 2002; Keith 2004). A recently developed approach addresses these issues by linking a landscape model and a metapopulation model (Akçakaya et al. 2003b, 2004, 2005).

An example of incorporating habitat change in metapopulation dynamics involves the woodland brown butterfly (*Lopinga achine*), which lays its eggs at the edges of glades of the partly open oak woodland pastures where its host plant *Carex montana* grows. The habitat quality for this species is related to the amount of bush and tree cover within the pastures and the occurrence of its host plant (Bergman 1999). As discussed above, grazing often helps maintain grassland habitats in successional stages that favour certain species. As grazing ceases, the essential habitat of this species (open glades with host plants) becomes overgrown and deteriorates. Using a metapopulation model, Kindvall & Bergman (2004) calculated long-term extinction risks under various landscape scenarios. An important aspect of this analysis was that the landscape

scenarios were dynamic; thus, the study integrated the changes in the habitat (as a result of succession and grazing) with changes in the metapopulation, and demonstrated the importance of landscape dynamics in affecting the viability of this species.

Conclusions

Metapopulation models have been essential to the management of many species. The listing of several species on the Endangered Species List in the USA, as well as the management and recovery plans for a number of species, were based in part on the analysis of their metapopulation dynamics. For example, the draft recovery plan for the Pacific coast population of the western snowy plover (*Charadrius alexandrinus nivosus*) included a metapopulation model (Nur et al. 1999), which highlighted the need for increased management of the species and its habitats. This population is listed as threatened in the USA, because habitat degradation caused by human disturbance, urban development, introduced beachgrass (*Ammophila* spp.), and expanding predator populations have resulted in a decline in active nesting areas and in the size of the breeding and wintering populations. Using a metapopulation structure that allowed estimates for demographic parameters to vary among subpopulations was considered an important aspect of this model. The metapopulation model predicted a high probability of decline under existing conditions, which included intensive management in some areas by area closures, predator exclosures and predator control. The model suggested that recovery at a moderate rate would be possible with a productivity of 1.2 or more fledglings per breeding male, but would require short-term intensive management and long-term commitments to maintaining gains. Other species for which metapopulation models have been used in recovery planning or listing include northern spotted owl (*Strix occidentalis caurina*), California spotted owl (*Gambelia silus*), south-western willow flycatcher (*Empidonax traillii extimus*), marbled murrelet (*Brachyramphus marmoratus*), Florida scrub jay (*Aphelocoma coerulescens*) and Florida panther (*Felis concolor coryii*). These cases, and several examples discussed throughout this essay, illustrate our answers to the two questions posed earlier in the essay: conservation needs metapopulation approaches, and metapopulations need models.

Many species live in naturally heterogeneous or artificially fragmented landscapes, and decisions on their conservation and management should consider metapopulation concepts and models. Models make assumptions, however, many of which await evaluation and should not be tested on our most treasured wildlife. The metapopulation literature is full of caveats to the effect that more empirical data are needed to distinguish between alternative processes and mechanisms. These data must come from field experiments, yet too often field ecologists are pulled towards the expediency of mission-oriented conservation with the result that we still lack a well-tried framework for

managing endangerment at the regional scale. The wild dog case (Box 5.1) illustrates how the principal function of metapopulation models in conservation – to evaluate alternative options and scenarios – depends on there being alternatives to choose from. Metapopulation models stimulated the original concept of linked reserves, and contributed to addressing potential problems of inbreeding at the planning stage (Mills et al. 1998). Options at the construction stage were severely limited by the small number of sites and animals available, favouring a pragmatic approach of adaptive management for this large social species with complex behavioural ecology. Population monitoring and autecological studies are now providing data for optimizing population sizes and translocation rates. Models will thus become increasingly important decision tools in the long-term management of the metapopulation. Despite these caveats and limitations, we believe current conservation efforts for many species would benefit from a more explicit and quantitative consideration of metapopulation dynamics.

There is nothing in this world constant, but inconstancy.
(**Jonathan Swift (1667–1745),** *A Critical Essay upon the Faculties of the Mind.*)

References

Akçakaya, H.R. (2000a) Population viability analyses with demographically and spatially structured models. *Ecological Bulletins* **48**: 23–38.

Akçakaya, H.R. (2000b) Viability analyses with habitat-based metapopulation models. *Population Ecology* **42**: 45–53.

Akçakaya, H.R. (2005) *RAMAS GIS: Linking Spatial Data with Population Viability Analysis*, Version 45.0. Applied Biomathematics, Setauket, New York.

Akçakaya, H.R. & Atwood, J.L. (1997) A habitat-based metapopulation model of the California Gnatcatcher. *Conservation Biology* **11**: 422–34.

Akçakaya, H.R. & Baur, B. (1996) Effects of population subdivision and catastrophes on the persistence of a land snail metapopulation. *Oecologia* **105**: 475–483.

Akçakaya, H.R. & Ginzburg, L.R. (1991) Ecological risk analysis for single and multiple populations. In *Species Conservation: A Population-Biological Approach* (Eds A. Seitz and V. Loeschcke), pp. 73–87. Birkhauser Verlag, Basel.

Akçakaya, H.R. & Raphael, M.G. (1998) Assessing human impact despite uncertainty: viability of the northern spotted owl metapopulation in the north-western USA. *Biodiversity and Conservation* **7**: 875–94.

Akçakaya H.R. & Sjögren-Gulve, P. (2000) Population viability analysis in conservation planning: an overview. *Ecological Bulletins* **48**: 9–21.

Akçakaya, H.R., McCarthy, M.A. & Pearce, J. (1995) Linking landscape data with population viability analysis: management options for the helmeted honeyeater. *Biological Conservation* **73**: 169–76.

Akçakaya, H.R., Burgman, M.A. & Ginzburg, L.R. (1999) *Applied Population Ecology: Principles and Computer Exercises using RAMAS EcoLab 2.0*, 2nd edn. Sinauer Associates, Sunderland, Massachusetts, 285 pp.

Akçakaya, H.R., Atwood, J.L., Breininger, D., Collins, C.T. & Duncan, B. (2003a) Metapopulation dynamics of the California least tern. *Journal of Wildlife Management* **67**(4): 829–42.

Akçakaya, H.R., Mladenoff, D.J. & He, H.S. (2003b) RAMAS Landscape: integrating metapopulation viability with LANDIS forest dynamics model. User Manual for version 1.0. Applied Biomathematics, Setauket, New York.

Akçakaya, H.R., Radeloff, V.C., Mladenoff, D.J. & He, H.S. (2004) Integrating landscape and metapopulation modeling approaches: viability of the sharp-tailed grouse in a dynamic landscape. *Conservation Biology* **18**: 526–537.

Akçakaya, H.R., Franklin, J., Syphard, A.D. & Stephenson, J.R. (2005)Viability of Bell's sage sparrow (*Amphispiza belli* ssp. *belli*): altered fire regimes. *Ecological Applications* **15**: 521–31.

Auld, T.D. & Scott, J. (1997) Conservation of endangered plants in urban fire-prone habitats. Proceedings: Fire Effects on Rare and Endangered Species and Habitats Conference, Coeur D'Alene, Idaho, pp. 163–71. International Association of Wildland Fire, Hot Springs, South Dakota.

Baguette, M. (2004) The classical metapopulation theory and the real, natural world: a critical appraisal. *Basic and Applied Ecology* **5**: 213–224.

Baguette, M. & Mennechez, G. (2004) Resource and habitat patches, landscape ecology and metapopulation biology: a consensual viewpoint. *Oikos* **106**: 399–403.

Bascompte, J., Possingham, H. & Roughgarden, J. (2002) Patchy populations in stochastic environments: critical number of patches for persistence. *American Naturalist* **159**: 128–37.

Bekker, G.J. & Canters, K.J. (1997) The continuing story of badgers and their tunnels: In *Proceedings Habitat Fragmentation and Infrastructure* (Eds K. Canters, A. Piepers & D. Hendriks-Heersma), pp. 344–53. DWW Publications, Delft.

Bergman, K-O. (1999) Habitat utilization by *Lopinga achine* (Nymphalidae: Satyrinae) larvae and ovipositing females: implications for conservation. *Biological Conservation* **88**: 69–74.

Bleich, V.C., Wehausen, J.D. & Holl, S.A. (1990) Desert-dwelling mountain sheep: conservation implications of a naturally fragmented distribution. *Conservation Biology* **4**: 383–90.

Breininger, D.R. & Carter, G.M. (2003) Territory quality transitions and source–sink dynamics in a Florida scrub-jay population. *Ecological Applications* **13**: 516–29.

Breininger, D.R., Burgman, M.A., Akçakaya, H.R. & O' Connell, M.A. (2002) Use of metapopulation models in conservation planning. In *Applying Landscape Ecology in Biological Conservation* (Ed. K. Gutzwiller), pp. 405–27, Springer-Verlag, New York.

Brook, B.W., Burgman, M.A., Akçakaya, H.R., O'Grady, J.J. & Frankham, R. (2002) Critiques of PVA ask the wrong questions: throwing the heuristic baby out with the numerical bathwater. *Conservation Biology* **16**: 262–3.

Burgman, M.A., Ferson, S.& Akçakaya, H.R. (1993) *Risk Assessment in Conservation Biology*. Chapman and Hall, London, 314 pp.

Caldow, R.W.G., Goss-Custard, J.D., Stillman, R.A., Durell, S.E.A.L.D., Swinfen, R. & Bregnballe, T. (1999) Individual variation in the competitive ability of interference-prone foragers: the relative importance of foraging efficiency and susceptibility to interference. *Journal of Animal Ecology* **68**: 869–78.

Creel, S., Mills, M.G.L. & McNutt, J.W. (2004) Demography and population dynamics of African wild dogs in three critical populations. In *Biology and Conservation of Wild Canids* (Eds D.W. Macdonald & C. Sillero-Zubiri), pp. 337–350. Oxford University Press, Oxford.

Dennis, R.L.H., Shreeve, T.G. & Van Dyck, H. (2003) Towards a functional resource-based concept for habitat: a butterfly biology viewpoint. *Oikos*, **102**: 417–26.

Doncaster, C.P. (1999) A useful phenomenological difference between exploitation and interference in the distribution of ideal free predators. *Journal of Animal Ecology* **68**: 836–8.

Doncaster, C.P. (2000) Extension of ideal free resource use to breeding populations and metapopulations. *Oikos* **89**: 24–36.

Doncaster, C.P. (2001) Healthy wrinkles for population dynamics: unevenly spread resources can support more users. *Journal of Animal Ecology* **70**: 91–100.

Doncaster, C.P. & Gustafsson, L. (1999) Density dependence in resource exploitation: empirical test of Levins' metapopulation model. *Ecology Letters* **2**: 44–51.

Doncaster, C.P., Clobert, J., Doligez, B., Gustafsson, L. & Danchin, E. (1997) Balanced dispersal between spatially varying local populations: an alternative to the source-sink model. *American Naturalist* **150**: 425–45.

Doncaster, C.P., Rondinini, C. & Johnson, P.C.D. (2001) Field test for environmental correlates of dispersal in hedgehogs *Erinaceus europaeus*. *Journal of Animal Ecology* **70**: 33–46.

Doncaster, C.P., Pound, G.E. & Cox, S.J. (2003) Dynamics of regional coexistence of more or less equal competitors. *Journal of Animal Ecology* **72**: 116–26.

Ebert, D., Haag, C., Kirkpatrick, M., Riek, M., Hottinger, J.W. & Pajunen, V.I. (2002) A selective advantage to immigrant genes in a *Daphnia* metapopulation. *Science* **295**: 485–8.

Etienne, R.S. & Heesterbeek, J.A.P. (2001) Rules of thumb for conservation of metapopulations based on a stochastic winking-patch model. *American Naturalist*, **158**: 389–407.

Figuerola, J. & Green, A.J. (2002) Dispersal of aquatic organisms by waterbirds: a review of past research and priorities for future studies. *Freshwater Biology* **47**: 483–94.

Githiru, M. & Lens, L. (2004) Using scientific evidence to guide the conservation of a highly fragmented and threatened Afrotropical forest. *Oryx* **38**: 404–9.

Gonzalez, A., Lawton, J.H., Gilbert, F.S., Blackburn, T.M. & Evans-Freke, I. (1998) Metapopulation dynamics, abundance, and distribution in a microecosystem. *Science* **281**: 2045–7.

Grenfell, B. & Harwood, J. (1997) (Meta)population dynamics of infections diseases. *Trends in Ecology and Evolution* **12**: 395–9.

Gundersen, G., Johannesen, E. Andreassen, H.P. & Ims, R.A. (2001) Source–sink dynamics: how sinks affect demography of sources. *Ecology Letters* **4**: 14–21.

Hanski, I. (1994) Patch occupancy dynamics in fragmented landscapes. *Trends in Ecology and Evolution* **9**: 131–5.

Hanski, I. (1999) Metapopulation dynamics. *Nature* **396**: 41–9.

Hanski, I. & Gilpin, M. (1991) Metapopulation dynamics : brief history and conceptual domain. *Biological Journal of the Linnean Society* **42**: 3–16.

Hanski, I. & Simberloff, D. (1997) The metapopulation approach. In *Metapopulation Biology: Ecology, Genetics and Evolution* (Eds I. Hanski & M. Gilpin), pp. 5–26. Academic Press, San Diego, CA.

Hanski, I., Pöyry, J., Pakkala, T. & Kuussaari, M. (1995) Multiple equilibria in metapopulation dynamics. *Nature* **377**: 618–21.

Harding, K.C. & McNamara, J.M. (2002) A unifying framework for metapopulation dynamics. *American Naturalist* **160**: 173–185.

Harrison, S. & Hastings, A. (1996) Genetic and evolutionary consequences of metapopulation structure. *Trends in Ecology and Evolution* **11**: 180–3.

Harrison, S. & Quinn, J.F. (1989) Correlated environments and the persistence of metapopulations. *Oikos* **56**: 293–8.

Hastings, A. (2003) Metapopulation persistence with age-dependent disturbance or succession. *Science* **301**: 1525–6.

Hastings, A. & Harrison, S. (1994) Metapopulation dynamics and genetics. *Annual Review of Ecology and Systematics* **25**: 167–88.

Hess, G. (1996) Disease in metapopulation models – implications for conservation. *Ecology* **77**: 1617–32.

Hilty, J.A. & Merenlender, A.M. (2004) Use of riperian corridors and vineyards by mammalian predators in Northern California. *Conservation Biology* **18**: 126–35.

Hudgens, B.R. & Haddad, N.M. (2003) Predicting which species will benefit from corridors in fragmented landscapes with population growth models. *American Naturalist* **161**: 808–20.

Johnson, M.P. (2000) The influence of patch demographics on metapopulations, with particular reference to successional landscapes. *Oikos* **88**: 67–74.

Johst, K., R. Brandl & S. Eber. (2002) Metapopulation persistence in dynamic landscapes: the role of dispersal distance. *Oikos* **98**: 263–70.

Keith, D. (2004) Population viability analysis of the endangered Australian heath shrub, *Epacris barbata*, subject to recurring fires and disease epidemics. In *Species Conservation and Management: Case Studies* (Eds H.R. Akçakaya, M.A. Burgman, O. Kindvall, et al.), pp. 90–103.. Oxford University Press, Oxford.

Kelly, C.K. & Bowler, M.G. (2002) Coexistence and relative abundance in forest trees. *Nature (London)* **417**: 437–40.

Keymer, J.E., Marquet, P.A., Velasco-Hernandez, J.X. & Levin, S.A. (2000) Extinction thresholds and metapopulation persistence in dynamic landscapes. *American Naturalist* **156**: 478–94.

Kindvall, O. & Bergman, K-O. (2004) Woodland brown butterfly *Lopinga achine* in Sweden. In *Species Conservation and Management: Case Studies* (Eds H.R. Akçakaya, M.A. Burgman, O. Kindvall, et al.), pp. 171–178. Oxford University Press, Oxford.

Lacy, R.C. (2000) Considering threats to the viability of small populations with individual-based models. *Ecological Bulletins* **48**: 39–51.

LaHaye, W.S., Gutierrez, R.J. & Akçakaya, H.R. (1994) Spotted owl meta-population dynamics in southern California. *Journal of Animal Ecology* **63**: 775–785.

Lamberson, R.H., McKelvey, K., Noon, B.R. & Voss, C. (1996) A dynamic analysis of Northern Spotted Owl viability in a fragmented forest landscape. *Conservation Biology* **6**: 505–12.

Lens, L., Van Dongen, S., Norris, K., Githiru, M. & Matthysen, E. (2002) Avian persistence in fragmented rainforest. *Science* **298**: 1236–8.

Liebhold, A., Koenig, W.D. & Bjornstad, O.N. (2004) Spatial synchrony in population dynamics. *Annual Review of Ecology Evolution and Systematics* **35**: 467–490.

Lindenmayer, D.B. & Possingham, H.P. (1996) Ranking conservation and timber management options for Leadbeater's Possum in southeastern Australia using population viability analysis. *Conservation Biology* **10**: 235–51.

Lindsey, P.A., Alexander, R., du Toit, J.T. & Mills, M.G.L. (2005) The cost efficiency of wild dog conservation in South Africa. *Conservation Biology* **19**: 1205–14.

Macdonald, D.W. & Sillero-Zubiri, C. (2004) *Biology and Conservation of Wild Canids*. Oxford University Press, Oxford.

McCarthy, M.A. & Lindenmayer, D.B. (2000) Spatially-correlated extinction in a metapopulation model of Leadbeater's possum. *Biodiversity and Conservation* **9**: 47–63.

McCarthy, M.A., Menkhorst, P.W., Quin, B.R., Smales, I.J. & Burgman, M.A. (2004) Helmeted Honeyeater (*Lichenostomus melanops cassidix*) in Southern Australia: assessing options for establishing a new wild population. In *Species Conservation and Management: Case Studies* (Eds H.R. Akçakaya, M.A. Burgman, O. Kindvall, et al.), pp. 410–20. Oxford University Press, Oxford.

Mills, M.G.L. & Gorman, M.L. (1997) Factors affecting the density and distribution of wild dogs in the Kruger National Park. *Conservation Biology* **11**: 1397–1406.

Mills, M.G.L., Ellis, S., Woodroffe, R., et al. (Eds) (1998) *African Wild Dog* (Lycaon pictus) *Population and Habitat Viability Assessment*. Conservation Breeding Specialist Group, Species Survival Commission, International Union for the Conservation of Nature and Natural Resources, Apple Valley, MN.

Noon, B.R., McKelvey, K.B. & Murphy, D.D. (1997) Developing an analytical context for multispecies conservation planning. In *The Ecological Basis for Conservation: Heterogeneity, Ecosystems, and Biodiversity* (Eds S.T.A. Pickett, R.S. Ostfeld, M. Shachak & G.E. Likens), pp. 43–59. Chapman and Hall, New York.

Noss, R.F. (1990) Indicators for monitoring biodiversity: a hierarchal approach. *Conservation Biology* **4**: 355–64.

Nur, N., Page, G.W. & Stenzel, L.E. (1999) Population viability analysis for Pacific coast Snowy plovers. Appendix D of 'Western Snowy Plover (*Charadrius alexandrinus nivosus*) Pacific Coast Population Draft Recovery Plan'. U.S. Fish and Wildlife Service, Region 1, 2001, Portland, Oregon, xix + 630 pp.

Pulliam, H. P., Dunning, J.B. & Liu, J. (1992) Population dynamics in complex landscapes: a case study. *Ecological Applications* **2**: 165–77.

Ranta, E., Kaitala, V. & Lindström, J. (1999) Spatially autocorrelated disturbances and patterns in population synchrony. *Proceedings of the Royal Society of London, Series B – Biological Sciences* **266**: 1851–6.

Regan, H.M. & Auld, T.D. (2004) Australian shrub *Grevillea caleyi*: recovery through management of fire and predation. In *Species Conservation and Management: Case Studies* (Eds H.R. Akçakaya, M.A. Burgman, O. Kindvall, et al.), pp. 23–35. Oxford University Press, Oxford.

Regan, T.J. & Bonham, K. (2004) Carnivorous land snail *Tasmaphena lamproides* in Tasmania: effects of forest harvesting. In *Species Conservation and Management: Case Studies* (Eds H.R. Akçakaya, M.A. Burgman, O. Kindvall, et al.), pp. 112–4. Oxford University Press, Oxford.

Regan, H.M., Auld, T.D., Keith D.A. & Burgman, M.A. (2003) The effects of fire and predators on the long-term persistence of an endangered shrub, *Grevillea caleyi*. *Biological Conservation* **109**: 73–83.

Rondinini, C. & Doncaster, C.P. (2002) Roads as barriers for hedgehogs. *Functional Ecology* **16**: 504–9.

Root, K.V., Akçakaya H.R. & Ginzburg, L.R. (2003) A multi-species approach to ecological valuation and conservation. *Conservation Biology* **17**: 196–206.

Schtickzelle, N. & Baguette, M. (2004) Metapopulation viability analysis of the bog fritillary butterfly using RAMAS/GIS. *Oikos* **104**: 277–90.

Smales, I., Quin, B., Krake, D., Dobrozczyk, D. & Menkhorst, P. (2000) Re-introduction of helmeted honeyeaters, Australia. *Re-introduction News* (Newsletter of the Re-introduction Specialist Group of the Species Survival Commission, International Union for the Conservation of Nature and Natural Resources) **19**: 34–6.

Smedbol, R.K. & Stephenson, R.L. (2004) Atlantic herring (*Clupea harengus*) in the northwest Atlantic Ocean: dynamics of nested population components under several harvest regimes. In *Species Conservation and Management: Case Studies* (Eds H.R. Akçakaya, M.A. Burgman, O. Kindvall, et al.), pp. 245–55. Oxford University Press, Oxford.

St Clair, C.C. (2003) Comparative permeability of roads, rivers, and meadows to songbirds in Banff National Park. *Conservation Biology* **17**: 1151–60.

Stelter, C., Reich., M., Grimm, V. & Wissel, C. (1997) Modelling persistence in dynamic landscapes: lessons from a metapopulation of the grasshopper *Bryodema tuberculata. Journal of Animal Ecology* **66**: 508–18.

Teodorascu, D. (1997) Infra Eco Network Europe – an open European platform for cooperation and exchange of information in the field of habitat fragmentation caused by transportation infrastructure: In *Proceedings Habitat Fragmentation and Infrastructure* (Eds K. Canters, A. Piepers and D. Hendriks-Heersma), pp. 442–50. DWW Publications, Delft.

Thomas, J.A., Rose, R.J., Clarke, R.T., Thomas, C.D. & Webb, N.R. (1999) Intraspecific variation in habitat availability among ectothermic animals near their climatic limits and their centres of range. *Functional Ecology* **13**(Supplement 1): 55–64.

Tilman, D., Lehman, C.L. & Yin, C. (1997) Habitat destruction, dispersal, and deterministic extinction in competitive communities. *American Naturalist* **149**: 407–35.

Travis, J.M.J. & Park, K.J. (2004) Spatial structure and the control of invasive alien species. *Animal Conservation* **7**: 321–330.

Van Dyk, G. & Slotow, R. (2003) The effects of fences and lions on the ecology of African wild dogs reintroduced to Pilansberg National Park, South Africa. *African Zoology* **38**: 79–94.

Wanzenböck, J. (2004) European Mudminnow (*Umbra krameri*) in the Austrian floodplain of the river Danube: conservation of an indicator species for endangered wetland ecosystems in Europe. In *Species Conservation and Management: Case Studies* (Eds H.R. Akçakaya, M.A. Burgman, O. Kindvall, et al.), pp. 200–7. Oxford University Press, Oxford.

Watkinson, A.R. & Sutherland, W.J. (1995) Sources, sinks and pseudo-sinks. *Journal of Animal Ecology* **64**: 126–30.

6

Managing biodiversity in the light of climate change: current biological effects and future impacts

Terry L. Root, Diana Liverman and Chris Newman

Now there is one outstandingly important fact regarding Spaceship Earth, and that is that no instruction book came with it.
(R. Buckminster Fuller (1895–1983), *Operating Manual for Spaceship Earth*, 1963.)

Introduction

Climate is one of the primary controls on species diversity and distribution globally, and past climate changes have surely modified biodiversity. Thus, predicted changes in global and regional climates as a result of increasing atmospheric carbon dioxide have tremendous implications for species and habitat conservation. As carbon dioxide increases are associated with human activities, principally the burning of fossil fuels and through deforestation, climate change poses a challenge to development, international environmental policy and resource consumption. This is particularly so in the developed world where, per capita, emissions of carbon dioxide are highest. Climate change must become an integral consideration in conservation, linking those concerned with non-human life on the planet with the polluting activities of its human inhabitants.

Many earlier predictions of global warming are becoming a reality as glaciers melt, hotter summers become more frequent, and in many places the distributions of species begin to shift. Britain's Chief Scientist, David King, sounded a dramatic warning when, in 2004, he identified climate change as a greater risk to society than terrorism (King 2004). At the Kyoto Climate Summit in 1997, dozens of eminent scientists issued a World Scientists' Call of Action. They stated, 'Climate change will accelerate the appalling pace at which species are now being extirpated, especially in vulnerable ecosystems. One-fourth of the known species of mammals are threatened, and half of these may be gone within a decade. Possibly one-third of all species may be lost before the end of the next century'. The recent Millennium Ecosystem Assessment Synthesis Report agreed that 'the balance of scientific evidence suggests that there will be a significant net harmful impact on ecosystem services worldwide if global mean surface temperature increases more than 2°C above pre-industrial levels or at rates greater than 0.2°C per decade'. It concludes that 'by the end of the century, climate change and its impacts may be the dominant direct drivers of biodiversity loss and the change in ecosystem services globally' (Millennium Ecosystem Assessment, 2005, p. 126). Consequently, plans for the next several centuries of biodiversity conservation must already take into account that the emissions of greenhouse gases due to the human activities will continue to increase the global temperature.

Lessons from patterns of palaeoclimatic change

This is far from the first time that global temperatures have changed – indeed, they have done so often, and radically, throughout geological history, thereby affecting the distribution of fauna and flora. For example, dramatic climatic events are implicated in mass extinctions at the conclusion of both the Palaeozoic and Mesozoic, 245 and 65 million years ago, respectively. During the Pleistocene Epoch, since the Olduvai–Matuyama boundary about 1.8 million years ago, there have been 32 cycles of cooling and warming, with annual mean global air temperature dipping to 5°C cooler than today's average of 14°C, bringing ice to much of the Northern Hemisphere. Intriguingly, early in the Pleistocene, mammal numbers and diversity stayed fairly stable. However, of more than 150 genera of megafauna (>44 kg) known to be alive 50,000 years ago, 97 were extinct by 11,000 years ago when the last glacial period concluded (Stuart, 1991). Theory has it that the knock-out punch to megafaunal biodiversity in the closing part of the Pleistocene Epoch was the combination of this climate change with the escalating pressures (in the forms of hunting and habitat change) brought about by growing populations of early humans. The contemporary parallels are obvious, with hunting, land-use change and other human activities making ecosystems more vulnerable to climatic change.

So if climate changes happen naturally, why all the fuss? First, this time it is changing much faster—an order of magnitude faster—than during most of the Pleistocene. Second, the resulting pressures are having an impact on ecosystems already clearly stressed by an ever-growing human footprint (as evidenced in diverse ways by every other essay in this book). Third, at least parts of society at large have committed to protecting biodiversity for reasons spanning economics to philosophy. Climate change is one of the most profound changes that humanity has brought to the planet and to its non-human inhabitants and some find this ethically uncomfortable.

The magnitude and nature of climate change

GREENHOUSE GAS EMISSIONS AND PREDICTED CLIMATE CHANGE

The role of 'greenhouse' gases is critically important for understanding the mechanisms underlying accelerated climate change. These gases, such as carbon dioxide (CO_2), methane (CH_4), nitrous oxide (NO) and water vapour, are normal atmospheric components essential for life. They trap solar energy, which warms the surface of the Earth from what would otherwise be around $-18°C$. Thus, the concern is not that the greenhouse effect exists (without it life would be in trouble), but that it is enhanced by human activities which result in the trapping of more solar energy causing the planet to warm further. Nobody seriously disputes that fossil fuel combustion has increased the concentration of atmospheric greenhouse gases, principally CO_2. According to the report on emission scenarios by the Intergovernmental Panel for Climate Change (IPCC 2001a), pre-industrial levels of carbon dioxide were in the region of 280 parts per million by volume (ppmv), whereas current levels are around 370 ppmv. This is the highest level of CO_2 in the past 400,000 years, which is as far back as accurate estimates can be made, and probably it is the highest level in the past 20 million years. By the

end of the twenty-first century the IPCC antici-
pates CO_2 concentrations to be anywhere from
490 to 1250 ppmv, depending on economic
development paths, population and technology.
Projections of future CO_2 concentrations under
various emission scenarios are used to drive complex atmospheric general circulation models
(GCMs), which are used to predict global climate.
The scientific consensus (IPCC, 2001a) suggests an
increase in globally averaged surface temperatures
of 1.4 to 5.8°C by 2100. An increase above 2°C,
which equates to levels greater than 450 ppmv,
will cause serious economic and possibly disastrous ecological impacts (Mastrandrea & Schnei-

der, 2004). Models are, of course, only models
(see Chapter 9), but the observational evidence
to suggest that these models are right is growing.

OBSERVED TEMPERATURE CHANGES

Since the late nineteenth century average
global surface temperatures have increased
about 0.6°C, with two-thirds of the increase in
the past 25 years (Fig. 6.1). A longer term record (Fig. 6.2) has been called the 'hockey stick'
plot because it shows a dramatic hook-like increase after 1975 following almost 1000 years

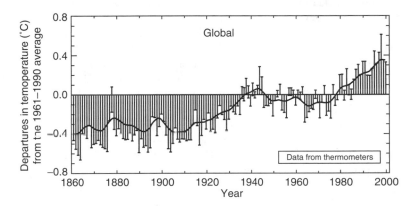

Fig. 6.1 Temperature deviations (°C) from the average temperatures between 1961 and 1990. These data are
collected from thermometers around the globe.

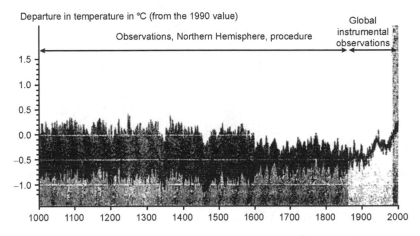

Fig. 6.2 Temperature deviations (°C) from the average global temperatures in 1990. These data are collected
from proxies, such as tree rings and ice cores, and from thermometers placed around the globe.

of fluctuations around a level or slightly decreasing trend (IPCC, 2001a; Mann et al. 1998).

All rigorous investigations of the average global air temperatures indicate significant increases in recent decades. As we go to press, the warmest year globally on record was 1998, closely followed in sequence by the ominously recent series: 2002 and 2003, then 2001, 1997,

---

Box 6.1 El Niño and climate change

During an El Niño event, the Equatorial undercurrent weakens, the surface water warms, macronutrients are reduced, primary production decreases (Chavez et al. 1999) and fish numbers diminish. In recent decades, however, the periodicity and magnitude of El Niño events have changed. El Niño events now occur two to seven times more frequently than they did 7000–15,000 years ago (Riedinger et al. 2002). Recent climate models show an increased El Niño pulse in the past three decades (Trenberth & Hoar 1996). The 1982–83 and 1997–98 El Niño events were the strongest recorded in the past 100 years and had severe biological impacts. Sea-surface temperatures and precipitation between 1965 and 1999 indicate that 1983 and 1998 were the hottest and wettest years on record for the Galapagos Islands.

Vargas et al. (2005, 2006) examined the impacts of El Niño activity on the population of Galapagos Penguins (*Spheniscus mendiculus*). Between 1965 and 2003, nine El Niño events were recorded of which two were strong (1982–83 and 1997–98); both were followed by crashes of 77% and 65% of the penguin population, respectively. Furthermore, increased frequency of weak El Niño events limits population recovery (Box Fig. 6.1).

In 2003 the penguin population was estimated to be at less than 50% of that prior to the strong 1982–83 El Niño event. Three causal mechanisms were identified: (i) shortage of food, (ii) unbalanced sex ratio and (iii) flooding of nests. For example, data from commercial fisheries indicated that the catch of mullets from the Galapagos during the 1997–1998 El Niño event was half that of the commercial catch in 1999 (Nicolaides & Murillo 2001) when there was no El Niño. Similarly, the catch of sardines along the coast of mainland Ecuador during the 1998 El Niño year was the lowest of the past two decades (Jácome & Ospina 1999). The Galapagos penguin has evolved in the presence of the environmental fluctuations caused by El Niño, and the associated negative effects probably have always affected their populations. However, the impacts of global warming will increase the frequency and intensity of these fluctuations, which will pose serious challenges for penguin conservation.

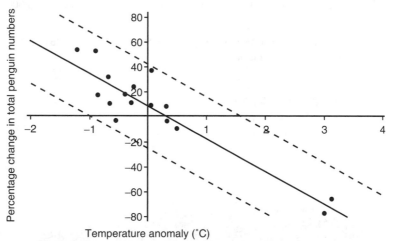

Box Fig. 6.1 Percentage change in penguin numbers in relation to the mean normalized sea-surface temperature (SST) anomalies for the period December–April that preceded each penguin count. We calculated changes in the penguin population for counts that were not more than 3 years apart ($n = 17$) ($F_{1,15} = 71.1$, $pL\ 0.001$, $b_{(adj)} = 0.81$). We also tested the relationship without the two strong El Niño events in 1983 and 1998 to determine that the relationship remained significant without these extreme values ($F_{1,13} = 10.2$, $p = 0.007$, $b_{(adj)} = 0.40$). Dotted lines are 95% confidence limits.

1995, 1990, 1999, 1991 and 2000. The warmth of 1997 and 1998 was exacerbated by a strong El Niño pattern of ocean heating in the South Pacific that had impacts around the world (see Box 6.1).

The temperature has not warmed uniformly around the globe. Some areas have been below the global average (e.g. south-eastern USA). The most pronounced warming has been in temperate and Arctic areas of Eurasia and North America between 40° and 70°N. Interestingly, urbanized and industrialized regions seem to have warmed less than expected owing to the countervailing role of air pollution. The particles can filter solar energy reaching the surface, producing a global 'dimming' effect, which may hide the true magnitude of temperature increases (Stanhill & Cohen, 2001).

OBSERVED ICE CHANGES

The 'fingerprints' of global warming – flowers blooming earlier in spring, sea-ice thinning and the like – are in substantial agreement with more direct measures from thermometers and satellites. One indirect measure of temperature is the melting of glaciers, which, for example, in Glacier National Park in Montana, USA are retreating so rapidly that they are projected to disappear by 2030 (Hall & Fagre, 2003). Those on Kilimanjaro and several Andean peaks are amongst the many rapidly following (Thompson et al. 2002). Glaciers have trapped information on the Earth's atmosphere for eons, so melting them is as irreparable (one might say sacrilegious) as the burning of the library at Alexandria. They are also critical water resources for the landscapes below them.

The poles are more sensitive to climate change than is the Equator. Arctic sea ice has decreased by 20% since 1988, and 87% of the Antarctic marine glaciers have retreated in the past 60 years (Stone et al. 2004; Cook et al. 2005). As melting continental glaciers flow into the sea, ocean levels rise, and much more

so because the warming water expands, reminiscent of mercury rising in a thermometer. Mean global sea level has been rising at a rate of 1–2 mm yr^{-1} over the past 100 years, significantly faster than the rate averaged over the past several thousand years. Indeed, the Greenland ice sheet has been melting at a rate equivalent to a 0.13 mm yr^{-1} increase in global sea level. Projected increases by 2100 range from 90 to 880 mm. At the higher end of this projection many densely populated areas, such as Bangladesh, would be submerged, and expanses of inland fresh water turned brackish, probably spurring mass migrations of people and terrestrial species. Warmer seas and sea-level rise will also affect the conservation of corals, mangroves and diverse marine and coastal ecosystems.

Although fraught with uncertainty, and depending on the model used, precipitation is also projected to increase with considerable regional variation, including increased rainfall in high and northern mid-latitudes in winter and decreases in winter rainfall in Australia, Central America and southern Africa. Although highly variable, land precipitation since 1900 has increased 2% on average. In most of the northern mid- to high latitudes precipitation has been rising at the rate of 0.5–1.0% per decade. Simultaneously, a decrease of 0.3% per decade has been observed in subtropical latitudes, although this appears to be a weakening trend. The extent of annual snow cover in the Northern Hemisphere has remained consistently below average since 1987, having decreased by 10% since 1966, mostly from a decline in spring and summer (IPCC, 2001a).

EVIDENCE OF ANTHROPOGENIC CLIMATE CHANGE.

Species are able to detect changes in temperatures and often adjust to them (Parmesan & Yohe, 2003; Root et al. 2003), but what is causing the temperature change to occur? Using wild animals and plants as temperature proxies,

Root et al. (2005) found that warming on a local scale, which is the scale that is important to species, can be attributed to human emissions. They compared the timings of life-histories (so-called phenological data) from species around the globe to temperatures modelled by a GCM. When natural forces alone, such as volcanoes or solar variations, are included as drivers in global GCMs over the past 100 years the predicted and observed temperatures do not match (Fig. 6.3a). When only anthropogenic sources, such as increased atmospheric dust particles, CO_2 and methane are considered, the modelled and actual values show a better match (Fig. 6.3b), but the best and most statistically significant match occurs when both natural and anthropogenic forces are included in the models (Fig. 6.3c). This strongly suggests that human activities are contributing significantly to the global warming of the atmosphere. This means that humans are indeed changing the temperatures at the local level. Plants and animals can detect this warming in our back gardens, and the warming can be attributed to humans using fossil fuels and burning tropical forests.

Uncertainty and the sceptics

Considering the weight of the foregoing evidence, it may seem surprising that so much airtime is given to the views of the small number of so-called contrarian 'scientists' disputing the interpretation of the evidence (McIntyre & McKitrick 2005). Some sceptics question the validity of the models and dismiss predictions of extinctions and other serious impacts as alarmist (e.g. www.marshall.org). Some maintain that there is no evidence of significant climate changes, whereas others acknowledge the changes but conclude they are not anthro-

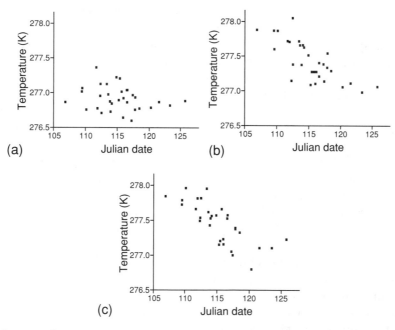

Fig. 6.3 For each year, the occurrence dates (Julian) of spring phenological traits are averaged over all Northern Hemisphere species exhibiting statistically significant changes in those traits ($n = 130$). These averages are plotted against: (a) the average modelled spring (March, April, May) temperatures including only natural forcings at each study location ($r = 0.22$, $p < 0.23$); (b) the same as (a) except including only anthropogenic forcings ($r = -0.71$, $p < 0.001$); (c) a combination of natural and anthropogenic forcings ($r = -0.72$, $p \leq 0.001$) (Root et al. 2005).

pogenic. Others, acknowledging the changes are unnatural, conclude that they are good for plant productivity and may beneficially 'green' the planet. Yet others look for an explanation in solar irradiance. The solar irradiance gambit rests on the observation that solar output varies cyclically, and may therefore theoretically contribute to temperature change. It is rebutted because irradiance on the Earth's surface is estimated to account for only 0.09 W m^{-2}, compared with 0.4 W m^{-2} warming from the insulation of greenhouse gases (IPCC, 2001a). The Earth's position and orientation relative to the Sun follow predictable cycles over long time periods (called Milankovitch cycles). Variations in the cycles are believed to be responsible for the Earth's ice-ages or glacial periods (Hays et al. 1976), when they bring a paucity of summer irradiance at northern latitudes during the summer months, allowing snow and ice to persist year round over an ever larger area. Milankovitch cycles, however, cannot explain the extent or rate of change in temperature over the past few decades. Additionally, if the Sun itself causes the warming, then the different vertical levels of the atmosphere would all show warming. They do not. The lower atmosphere is warming while the top atmosphere is cooling, which is exactly what you would expect if the warming is due to greenhouse gases.

One contrarian view is exemplified by the reconstruction of Northern Hemisphere temperatures for the past 1000 years using records of past climate captured in tree rings, corals and other proxies by Mann et al. (1999). This study produced the aforementioned 'hockey stick' curve that shows relatively stable temperatures until a significant rise in the twentieth century. Critics suggest that the increase is an artefact of the statistical techniques used and that the reconstructed curve is so unreliable that it fails to show past events that are well documented, such as the Little Ice Age – a period of cooling lasting from the mid-fourteenth to the mid-nineteenth centuries (McIntyre & McKitrick, 2003). Mann et al. (1999) rebuts this with the

argument that the hockey-stick curve is robust under alternative methodologies and, in any case, events such as the Little Ice Age and Medieval warming were regional and so not well reflected in global data sets (see www.real-climate.org).

Worst case scenarios and surprises: rapid climate change

In 1989 the National Science Foundation funded the Greenland Ice Sheet Project II (GISP2) to drill an ice core through the 2-mile depth of the Greenland ice sheet at a cost of $25 million. Simultaneously, a separate European project (GRIP) 20 miles away, drilled an independent, but corroborating core. By 1993, the two cores, detailing 110,000 years of climate history were ready to reveal their secrets, astonishing researchers with the apparent rapidity at which climate changes had occurred in the past. Warming and cooling of 8°C were evidenced frequently through the ice-core record, often flipping from extremes in as little as 10 years. This is what apparently occurred at the conclusion of the most recent ice-age, 11,600 years ago, with vast increases in snowfall leading to a doubling of accumulation within 3 years. Simultaneously, lake and ocean sediment data from Venezuela to Antarctica corroborate these rapid and extreme global temperature fluctuations. These data indicate at least 20 abrupt climate changes over the past 110,000 years.

The most dramatic scenarios for the near term are associated with the possibility of changes in the Atlantic thermohaline circulation associated with arctic melting and an influx of freshwater into the north Atlantic. At present currents in the Atlantic bring warm water north, warming the ocean and land areas such as north-west Europe by at least 5°C. The movement of the water is partially driven by a contrast in salinity between the less saline Southern Ocean, which receives freshwater from melting glaciers, and the more saline and denser waters of the northern Atlantic. A large change in the amount of freshwater from a melting Arctic can slow this circulation, causing much cooler temperatures in the north Atlantic. A sudden shutdown of the circulation pattern could cause a rapid cooling, similar to that shown in the ice-core records. Rahmstorf (1995) modelled the circulation pattern and concluded that this could happen with global average temperature changes from 2 to 5°C.

Changes in carbon and other biogeochemical cycles associated with warming could also trigger a reduction in the take up of carbon by oceans and terrestrial systems, or result in the large-scale release of methane, further increasing warming and creating positive feedbacks that could accelerate change (IPCC 2001a and www.stabilisation2005.com). Even without sudden and rapid changes, extreme events are likely to increase because a warmer atmosphere can hold more water and intensify the hydrological cycle. The IPCC reports an increase in the frequency and intensity of the extreme events associated with El Niño – the periodic warming of Pacific currents off the coast of Peru that produces droughts in the Andes and northeast Brazil, floods along the coast of western South America and declining marine productivity as warm waters replace colder nutrient-rich ones. Storms with heavy rain are becoming more frequent and intense in the Northern Hemisphere, yet as higher temperatures drive up evaporation the likelihood of drought and water shortages also increases (IPCC, 2001b).

The impacts of climate change on flora and fauna

Flora and fauna are responding fairly consistently with large-scale warming: flowers are blooming earlier, migrating birds are changing their arrival schedules, and some plant and animals are shifting their ranges northward (IPCC 2001b). For certain species, the consequences are quite dramatic. The yellow-bel-

lied marmot (*Marmota flaviventris*), a small mammal that lives in the alpine zone of North American mountains, is literally at risk of being 'squeezed' off the top of the mountain as temperatures increase and alpine habitat disappears (McDonald & Brown 1992), the same fate which Hersteinsson & Macdonald (1992) correctly predicted for arctic foxes (*Alopex lagopus*). In this case, however, the mechanism illustrates the sort of domino effect that should not surprise those familiar with the complexity of ecological communities: the northern limit of the red fox's (*Vulpes vulpes*) geographical range is determined by resource availability (and thus ultimately by climate), whereas the southern limit of the arctic fox's range is determined through interspecific competition with the red fox. If warming allows the red fox to thrive further north, or at higher altitude, and thus out-compete the arctic fox over more of its range, the arctic fox's distribution will become squeezed, paradoxically, in the face of ameliorating conditions. Pounds & Puschendorf (2004) suggest that 15–37% of a sample of 1103 land animals and plants could become extinct by 2050 as a result of climate change. For some there simply will be nowhere left with a suitable climatic regime, others will not be able to reach places where the climate remains suitable as warmer weather patterns shift polewards.

The different responses of species to climatic changes include:

1. shifts in the densities of species and their ranges, either poleward or upwards in altitude;
2. changes in the timing of events (phenology), such as when trees come into leaf or migrants arrive;
3. change (primarily loss) in genetic diversity;
4. morphological changes, such as longer wing length or larger egg sizes in birds;
5. behavioural changes such as relocation of bird nests;
6. extirpation or extinction (Parmesan & Yohe 2003; Root et al. 2003).

The most common or threatening of these changes are discussed below.

Density and range shifts

One of the most serious problems to face species around the globe is the combined or synergistic effect of climate change and habitat fragmentation (caused by urbanization, industrialization and agricultural development) (Root & Schneider 1993). Optimal conditions for the existence of a species can be defined as its 'fundamental range' (*sensu* Hutchinson, 1958), a major component of which is an appropriate bioclimatic envelope. As the climate warms many plants and animals will need to shift their ranges to remain within this envelope (e.g. by moving poleward, or ascending in altitude) – this is what happened during Pleistocene warm stages. Today such dispersals are much more difficult because, in most cases, individuals would face the generally impossible challenge of travelling across severely fragmented habitat. For instance, the quino checkerspot butterfly (*Euphydryas editha quino*), a resident of northern Baja California in Mexico, is being squeezed by temperature northwards from the southern boundary of its range, but urbanization in the area around San Diego, California is blocking its retreat. Such poleward range changes are widespread in temperate latitudes; in a sample of 35 non-migratory European butterflies, 63% have ranges that have shifted to the north by 35–240 km during the past 100 years, whereas only 3% have shifted to the south (Parmesan & Yohe 2003).

The responses of bird species to climate change are also likely to be highly variable (Harrison et al. 2003a,b). Some, such as the capercaillie (*Tetrao urogallus*) and red-throated diver (*Gavia stellata*), could decline with losses of suitable habitat, whereas others, such as turtle dove (*Streptopelia turtur*), yellow wagtail (*Motacilla flava*) and reed warbler (*Acrocephalus scirpaceus*), may expand their viable ranges. Several bird species, including willow tit (*Parus montanus*), nightingale (*Luscinia megarhynchos*) and nuthatch (*Sitta europaea*), respond well to moderate climate change but not to severe climate change, owing to their distributions in southern England either contracting significantly or becoming more fragmented.

The suggestion that birds might shuffle poleward and up in elevation will offer no solace to those striving to conserve species already at the poleward end of a continent, such as species in southern South Africa, or at the top of mountains. As climate change causes some species to redistribute polewards and upwards, the prospects are poor for those that already inhabit high latitudes or mountains. For example, denizens of what are called the 'Sky Islands' mountain ranges in the deserts of the southwestern USA survive only because they can thrive in the cooler and wetter climates at higher altitudes. The Sky Island complex contains 90 mammal species, 265 bird species, 75 reptile species and over 2000 plant species. Many species inhabiting the Sky Island range are also endemic, including six mammal subspecies and 60 snail species, nearly a third of all those found in the region. If these isolated mountain habitats disappear as a result of warming, the species that are unable to migrate, that is, the most unique and rare residents, will disappear.

The risk of 'falling off' the end of a continent is facing the numerous species in the highly speciose Fynbos in southern Africa, a region so rich in plant diversity that it qualifies as both a Biodiversity Hotspot (Myres et al. 2000) and a distinct floristic kingdom despite encompassing only 500,000 ha. Southward dispersal into the ocean is an unpromising option for its 7000 plus endemic species. Researchers at South Africa's National Botanical Institute predict a loss of Fynbos biome area of between 51% and 65% by 2050 (Midgley et al. 2002). At a chillier extreme, polar bears (*Ursus maritimus*) require sea ice on which to hunt seals all winter, thereby becoming sufficiently corpulent to fast through a relatively foodless Arctic summer. An adult female weighing 175 kg after weaning her cubs needs to gain at least 200 kg to have a successful pregnancy. If the sea ice forms later and melts earlier, the window of opportunity for hunting may be too brief for the bears to accumulate enough fat to breed, raise young, or even to survive themselves (Derocher et al. 2004).

No species exists in isolation. If species' distributions shuffle across the globe in response to climate change, there is a risk of tearing apart contemporary natural communities, and most importantly, uncoupling the predatory, competitive or beneficially coevolved relationships between species (Root & Schneider 1993). Faced with the same environmental change, species react differently, so the consequences of climate change may cascade. A disturbing example comes from the Monteverde Cloud Forest in Costa Rica, where Pounds (2002) found that submontane species are moving up to higher altitudes. This results in new encounters between species. The resplendent quetzal (*Pharomachrus mocinno*), for example, is a bird that nests in tree cavities. Until 1995, this species was not affected by the keel-billed toucan (*Ramphastos sulfuratus*). Coincident with increasing temperature, declining diurnal temperature range, and fewer days of montane mist, the toucans, formerly restricted to lowlands, have ascended the mountain to live alongside the quetzals in the cloud forest (Pounds et al. 1999). This situation proves problematic for the quetzal owing to the toucan's proclivity for predatorily poking its long bill into quetzal nests.

Climate change and phenology

Phenology – the study of the timing of such ecological events as when flowers bloom or when migrants arrive – has already revealed numerous shifts seemingly associated with climate change (see references cited in Appendix to Root et al. 2003). Changes have been observed in the timing of events such as maximum zooplankton biomass in the North Pacific (Mackas et al. 1998), peak insect abun-

dance in Europe (Sparks & Yates 1997) and New Zealand (White & Sedcole 1991), calling by frogs (which reflects timing of breeding) in North America (Gibbs & Breisch 2001), migration arrival and departure of birds in Europe (Bezzel & Jetz, 1995; Visser et al. 1998) and North America (Ball 1983; Bradley et al. 1999), breeding of birds in the UK (Thompson et al. 1986; Crick et al. 1997), Germany (Ludwichowski 1997) and North America (Brown et al. 1999; Dunn & Winkler 1999), and bud burst and blooming by trees in North America (Beaubien & Freeland 2000) and Asia (Kai et al. 1996).

The first calls of frogs and toads are a familiar harbinger of spring for many people in temperate parts of the world. Some amphibian species brave still-ice-crusted ponds at the first spring warming to begin courtship and the laying of eggs. Spring chorusing behaviour, which is associated with breeding activity, is closely linked to temperature (Busby & Brecheisen, 1997). Constituting one of the longest-running records of species' natural history, a study in England recorded the timing of first frog and toad croaks each year from 1736 to 1947 (Sparks & Carey, 1995). The date of spring calling for these amphibians occurred earlier over time, and was positively correlated with the annual mean spring temperature. For example, from 1980 and 1998, researchers found that the time of arrival of sexually mature common toads (*Bufo bufo*) at breeding ponds was highly correlated with the mean temperatures in the 40 days preceding their arrival (Reading, 1998). Similarly, two frog species, at their northern range limit in the UK, spawned 2 to 3 weeks earlier in 1994 than in 1978 (Beebee, 1995). Three species of newt similarly arrived 5 to 7 weeks earlier at breeding ponds.

Studies of migratory species such as birds are more complex. There is considerable documentation of changes in the spring arrival or breeding of birds in Europe (e.g. Berthold et al. 1995; Crick et al. 1997; Winkel & Hudde 1997; McCleery & Perrins 1998; Penuelas et al. 2002; Huppop & Huppop 2003) and more limited research in North America (e.g. Oglesby & Smith 1995; Bradley et al. 1999; Strode 2003). Although these studies used a combination of the biological observations, climate correlations and life-history information as described above, it is often difficult to determine which climate variables to associate with the observations of a species at one location. For instance, if a migrant species from Africa arrives earlier in the UK, has this been prompted by the conditions in Africa or along the migratory route?

Many aspects of breeding in some birds seem to be associated with temperatures. In southern Germany, the number of reed warblers (*Acrocephalus scirpaceus*) fledging early in the season increased significantly between 1976 and 1997, probably due to long-term increases in spring temperatures (Bergmann 1999). The spring arrival of this warbler was earlier in warm years. Also in Germany, Winkel & Hudde (1996) documented significant advances in hatching dates of nuthatches (*Sitta europea*) over the period 1970–1995. These advances correlated with a general warming trend. Migratory patterns of birds in Africa are also changing (Gatter 1992).

Differential shifts in the phenology of interacting species could easily disrupt the populations of all species involved. For instance, if each species in an obligatory mutualistic relationship responds differentially to climatic change, then the resulting asynchrony may be damaging, and perhaps catastrophic, to both. Even in non-obligatory relationships, such as between pollinators and plants, differential responses of species due to climate change, may lead to population declines.

If a scenario like that for the great tits and moth larva (see Box 6.2) occurs for species that control insect pests in an agriculture setting, there could be a boom in insect populations, resulting in a need for more pesticide control. In pasture and grassland ecosystems, for example, birds are important predators of grasshoppers. Models estimate that a single pair of savannah sparrows (*Passerculus sandwichensis*) raising their young consume approximately 149,000 grass-

hoppers over a breeding season. Considering typical bird densities, roughly 218,000 grasshoppers per hectare are consumed each season (Kirk et al. 1996). In many of these areas, the economic threshold for spraying insecticides occurs as densities reach approximately 50,000 grasshoppers per hectare (McEwen 1987). The birds are thus thought to keep current grasshopper populations at levels below which spraying would otherwise be required.

Management and policy implications

We have already hinted at some of the implications of climate change for conservation management – and climate change is a difficult challenge for policy making. If we wish to prevent our climate from changing we must find ways to reduce carbon emissions or to recapture carbon and other greenhouse gases from the atmosphere. If we unable or unwilling to reduce our dependence on greenhouse gases we will have to find ways to adapt to a warmer world and either accept the loss and change of ecosystems or manage them closely in the context of climatic change.

Carbon mitigation and the new carbon economy

The most direct way to prevent serious climate-change impacts on biodiversity is to slow or reverse the rate of global warming by either reducing greenhouse gas emissions or finding ways to recapture carbon from the atmosphere. Both options are being addressed internationally by the Framework Convention on Climate Change, most immediately through the Kyoto protocol that commits developed countries that adopt it to reducing their carbon dioxide emissions to 1990 levels by 2012. The Kyoto protocol sits at the centre of international debates about who should do what about climate change and when. For example, the current USA Government has been unwilling to ratify Kyoto because it does not require emissions

Box 6.2 Studies from Wytham Woods: from great tits to badgers

Oxford University's Wytham Estate embraces ancient woodland, plantation, grassland and mixed farmland. A microcosm of the countryside of lowland England, it has been used for ecological research for over 60 years, and two of these long-term data sets have revealed important effects of climate change (although, as is generally the way with the huge value of long-term data, this topic was not even in mind when the original data were gathered).

Long-term studies by the Edward Grey Institute of Ornithology have revealed that climatic change may be causing mismatching in the timing between the breeding of great tits (*Parus major*) in the UK and the hatching of caterpillars (Vannoordwijk et al. 1995; Visser et al. 1998). The tits do not seem to be shifting their clutch laying dates effectively. Because caterpillars are only abundant for a short period of time in the spring, females are under great pressure to lay early enough to take advantage of the richest flush of caterpillars, especially, the winter moth larva *Operophtera brumata*. The earliest breeders are generally much more successful, in terms of clutch size and survival. Late broods have to fledge and learn to feed while caterpillar abundance is rapidly diminishing (Haywood & Perrins, 1992). Simultaneously, laying early poses females with the problem of finding sufficient food to form their eggs a month before peak caterpillar abundance (Perrins, 1996); which seems to be so difficult that many birds breed later than they 'should'. Also, both sexes have to develop their reproductive systems from a regressed winter state in order to breed, which is another energetically expensive process. There is, however, a considerable advantage to the birds to breed as early as they can, because the earliest breeders tend to produce the most surviving offspring.

From 1970 onwards, tit laying-dates occurred earlier (Perrins & McCleery 1989), as caterpillar hatching dates, triggered by temperature, occurred earlier. With continued warming, the date of peak caterpillar abundance will probably continue to shift earlier, so much earlier that the tits may struggle to build nests, lay eggs and have the eggs hatch in time to take advantage of the caterpillars. Not least, earlier in the year nights are longer and colder, putting the tits under severe feeding constraints. The lack of the caterpillar availability could negatively affect the population size of this bird, and could greatly increase the population of caterpillars. In turn, greater caterpillar numbers could be detrimental to the trees.

The Wildlife Conservation Research Unit has also undertaken long-term ecological studies at Wytham, looking at the population dynamics of the Eurasian badger, *Meles meles*. The badger makes a good model species for testing the impacts of climate change owing to its wide geographical distribution, variable social system and, where available, favoured diet of earthworms (*Lumbricus terrestris*). Macdonald & Newman (2002) report that badger numbers at Wytham more than doubled between 1987 and 2001, with no change in population range, peaking in 1996 at 235 adults and 62 cubs (Box Fig. 6.2)

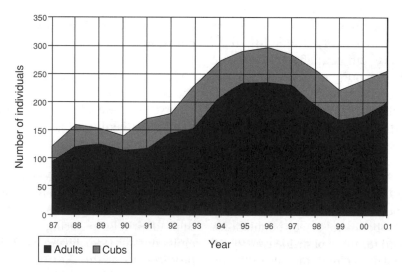

Box Fig. 6.2 Minimum number alive retrospective estimate for badger adult and cub numbers, 1987–2001.

Cub survival had a very significant affect on population size. In warmer, drier years cub survival was low (minimum 48.15%). In wetter years, and years with more wet days (both climatic factors influence earthworm availability), cub survival was much higher (maximum 94.74%), owing to a greater availability of earthworms under damp conditions.

Adult survival was not effected by annual temperature or absolute rainfall values, although the number of wet days showed a predictive trend, benefiting survival.

Developing this hydrological theme, volumetric soil water content (VSWC), although not a significant predictor of cub survival over the entire year, was strongly linked to cub survival in key spring months. Both lactating sows and their dependent offspring were very sensitive to food supply through the early part of the year. Badger cubs, weaned in May, start searching for food independently and thus effectively increase the number of foraging badgers in the population by up to one-third over a short time period.

Badgers preferentially eat earthworms. Of especial interest is the suggestion by the IPCC (2001a) that warming in annual mean temperature has occurred particularly as a result of night-time rather than day-time increases, thereby reducing the diurnal temperature range (DTR). This increases the availability of earthworms to nocturnal foraging badgers, as worms surface under mild, damp microclimatic conditions, and only in the absence of frost.

Warmer winters have been a particular feature of global warming (Brunetti et al. 2000). Northern Hemisphere annual snow-cover extent has consistently remained below average since 1987, and has decreased by about 10% since 1966 (IPCC, 2001a). Inspecting the relationship between winter temperatures, adult body-mass and subsequent cub productivity revealed a more insidious effect of climate change. In mild Januarys, both male and female badgers weighed up to 1 kg (c.10%+) more than in cold years. This weight gain cycle evolved precisely so that badgers can lay down a body-fat reserve should winter conditions turn harsh. In milder, wetter conditions, without ground frost or continuous snow cover (see Sagarin & Micheli 2001), badgers were able to continue to successfully forage for earthworms (and other food sources). A similar correlation between badgers' body-condition in January and the prediction of consequent offspring sex ratio was reported by Dugdale et al. (2003), as male cubs were favoured in milder years when adult females were heavier.

Adult female body mass in January was also a predictor of cub productivity and survival in the following spring. Badgers exhibit delayed implantation and while they mate post-partum in February–March, day-length (winter solstice) mediated by body condition (Woodroffe 1995) dictates implantation date. Gestation occurs through the winter; thus fat reserves are critical to embryonic and subsequent neonatal survival (Cheeseman et al. 1987). Warming trends in Fennoscandia (Carter 1998), affecting badger abundance and distribution (Bevanger &Lindstrom 1995), have allowed badgers to extend their distribution 100 km northwards in Finland since the mid-1940s, now as far north as the Arctic circle, with numbers in southern Finland doubling (Kauhala 1995a,b).

A climatic paradox appears to be developing. Mild, wet winters provide badgers with good earthworm foraging during a time of year when typically frozen ground forces them to live off their fat reserves. These conditions allow badgers to maintain better winter masses and lead to larger cub cohorts. Trends towards spring droughts, however, may not be so advantageous. The IPCC predicts that winters classified as 'cold' will become much rarer by 2020 and almost disappear by 2080. Simultaneously, hot dry summers will become much more frequent. These are scenarios in which the adult badger population likely could survive well, but could fail to produce enough surviving cubs to sustain their populations.

reductions in developing countries, including those that compete economically with the USA, such as China, and because many Americans think that the costs of carbon cuts will outweigh the benefits. Underlying such beliefs are cost-benefit analyses that place low or no value on biodiversity and that assume there are minimal economic opportunities in a lower carbon economy. Developing countries argue that they should not have to slow the growth of their economies by switching from coal and oil, given that the developed world has already based their economic success on the use of cheap fossil fuels. Again, this assumes that economic growth is only possible with fossil fuels rather than alternatives.

As a result of Russia's decision to ratify Kyoto, the treaty (which required 55 countries producing at least 55% of the emissions to sign) went into force in 2005, but without the participation of major emitters including the USA and Australia.

Unfortunately the Kyoto protocol as currently implemented only achieves a modest (2–5.2%) reduction in emissions, whereas to stabilize the climate at, for example, twice the pre-industrial levels of greenhouse gases, would require a 60% reduction in worldwide emissions. Even with such dramatic reductions, the planet would still experience some warming with associated impacts on ecosystems. The scale of mitigation requires not only aggressive reductions by the USA, but also the participation of major developing countries such as China, India and Brazil, whose current development paths will produce significant emissions over the next 50 years. The developing world is reluctant to reduce its energy use and development when they perceive the profligate per capita consumption of developed countries (see Chapter 18). One widely accepted proposal is to stabilize emissions at 450 ppmv through a process of 'contraction and convergence', permitting the developing world to grow economies and emissions while the developed world reduces emissions so that the two converge at a roughly equal per capita allocation by 2050 or 2100, perhaps as a result of trading in carbon permits. But given the wide range in current per capita emissions, from less than 1 ton per capita in most of Africa to more than 20 in the USA (2002 data from http:// cdiac.esd. ornl.gov/home.html), such convergence will prove hard to achieve in a world with such vast differences in consumption and lifestyles.

Kyoto provides an option for countries to meet their commitments by investing in energy efficiency or carbon sequestration (through reforestation) in the developing world through the UN Clean Development Mechanism (CDM) and in Eastern Europe through Joint Implementation (JI) options. Thus, countries and corporations can offset domestic emissions by a development project that plants forests or increases the efficiency of a power station (effectively allowing them to continue to burn fossil fuels by investing in carbon reduction more cheaply elsewhere). Carbon trading has provided a new investment opportunity as companies arrange for carbon reductions and sell the credit. One of the challenges to mitigation is to derive accurate estimates of the carbon savings and to ensure that the price of carbon reflects the costs of potential damages rather than speculation in a highly uncertain market. On its own reforestation or other land uses that sequester carbon to reduce emissions cannot balance the consumption of fossil fuels, and, often have serious implications for biodiversity. For example, sequestration through large-scale plantations is likely to reduce species diversity and require conversion of natural forests and grasslands to intensive carbon management.

So far, the international commitment to mitigation does not hold out much hope for preventing climate change because Kyoto will produce such a modest reduction by 2012. Some countries are already struggling to meet their Kyoto commitments and the USA and the developing countries currently outside the regime are likely to continue to increase emissions. Major investments are now being made in carbon capture options that might re-inject carbon into deep wells or oceans, but these are unlikely to move beyond pilot projects over the next decade. And policies to move energy use away from fossil fuels towards renewables or nuclear are already controversial, with wind power, for example, opposed by some ecologists and conservationists because of risks to species and landscape aesthetics and nuclear risks and waste management unacceptable to the public in many countries.

Researchers at Princeton University (Socolow et al. 2004) have offered a set of seven 'stabilization wedges' that they suggest produce a reduction of 200 billion tons of carbon between 2004 and 2054. This would be achieved through expansion and investments in: (i) energy conservation (especially transport fuel efficiency, and building construction); (ii) renewable energy, especially wind and solar; (iii) renewable fuels such as biofuels; (iv) enhanced natural sinks to capture carbon, such as well managed forests and soils; (v) nuclear energy; (vi) substitution of gas for other fossil fuels; and (vii) carbon capture within geological storage. Although these 'wedges' might reduce emissions and climate change, several of them, such as biofuels and nuclear, might have other implications for ecosystems and biodiversity (Chapter 18 concludes that such unhappy trades-off are facing the future of biodiversity conservation at every turn).

Adaptation

Adapting to climate change might seem the easier option, especially for natural systems that have coped with variations in climate over the millennia. But on a planet where humans are everywhere modifying and managing ecosystems, and given the rapidity of anthropogenic climate change, conservation for climate change adaptation is a complex technical, ethical and economic challenge.

The World Wildlife Fund has produced several reports proposing conservation strategies for climate change (e.g. WWF 2003), which is now accepted by most ecologists and conservation organizations – not least through the Millennium Ecosystem Assessment (2005) – as a major threat to biodiversity. The major proposals include:

1. Establishing protected areas that provide a margin for adaptation to climate change through, for example, north–south transects securing space for species to shift northwards or upwards as they adapt, conservation corridors that facilitate migration, or buffer zones that allow adjustment of range within the protected area. The World Wildlife Fund suggests that protected area creation and management could focus on potential refuges that might be more resilient to climate changes because they are in the core rather than the margins of climatic zones. Alternatively conservation might focus on the critical margins between climatic zones where there will be competition between species moving poleward at a faster rate than slower ones and where larger protected areas might thus be needed.

2. Reducing the non-climatic stresses on key species and ecosystems including land-use change, simplification, pollution, introduction of exotics, and hunting pressures in order to reduce vulnerability and maximize flexibility to cope with climate change. A reduction in ecosystem fragmentation is particularly important for adaptations in terms of the protected areas noted above, where continuous areas are needed for movement across the landscape.

3. Employing adaptive management strategies that can adjust to the onset of climatic changes and directly intervene to reduce its impacts and facilitate adaptation through, for example, assisted migration, species reintroduction, prescribed burning and control of invasive species.

Some of these strategies pose great challenges and unprecedented costs to conservation managers. For example, the management of species range and density is very complex. For species that need to move upslope on mountains to escape warmer temperatures lower down, one solution is to plan for reserves that include both lower and upper elevations, but this poses the conundrum faced by the competition of quetzals encountering toucans in the Costa Rica case, mentioned above. Wildlife managers are in a challenging situation: to increase the quetzal's chance of survival, do they begin to kill toucans? Such intertwined ethical and ecological dilemmas will continue to arise as historically unprecedented environmental changes such as these unfold.

The solution to northward movement of suitable habitats by setting up interconnected nature reserves that run north–south, or along altitudinal gradients, is also complicated. Unfortunately the human footprint of large swaths of agricultural and urban lands and private ownership of property in many regions means that contiguous reserves may not be possible without land purchase or appropriation. Routes for fauna to cross over or under highways may be needed to facilitate dispersal, similar to what was done for caribou (*Rangifer tarandus*) in the Arctic when the Alaskan pipeline was built (Smith & Cameron 1985). Costa Rica has established conservation corridors that join several protected areas including part of the Paseo Pantera (Panther Path) that would provide migration routes for the panther through Central America.

Such a programme would be costly and it is hard to protect large enough areas to make a difference. Many migratory birds use a series of staging areas along their north–south migration routes, meaning an appropriate reserve system would need to consider much of a hemisphere. Conversely, a small sedentary amphibian might spend its entire life in a small pond and have no means to migrate to the next wetland without assistance, even over short distances. And for some species, such as polar bears and many migratory geese, who rely on the presence of Arctic ice and large areas of tundra vegetation, it is hard to envisage a solution that would adequately protect their habitat from the serious changes projected for Arctic ecosystems.

Another outcome would be for the remnants of relatively immobile wildlife and plant communities to remain in existing isolated reserves and parks. Such 'habitat islands' probably would require management and manipulation, and examples of metapopulation management are already surfacing (see Chapter 5), as are those of small, isolated populations (Chapter 4). The resulting difficult distinctions along the spectrum from zoo to wilderness are raised in Chapter 18. Biodiverse and more complete communities are sometimes less vulnerable to the impacts of climate change than impoverished ones that lack keystone species (Power et al. 1996; Naem & Li 1997; Wilmers et al. 2002).

As in cases with range shifts, species experiencing discordant phenological shifts face peril. The opportunities for conservation and mitigation are more daunting – because they are so limited. The need for interventions seems likely to increase, perhaps 'helping' a species bypass an obstacle to reach new, more suitable habitat, or intervening to, for example, protect prey from predators – the morass of awkward judgments is unappealing. The reality is that spring plants and insects cannot be convinced to 'wait' for later migrants. The early bird gets the worm, so to speak! Migrants that do adapt their arrival times will be at a competitive advantage to later arrivals, claiming nesting sites and taking advantage of the optimal food sources. Unsurprisingly, most conservation legislation can cope poorly with change, and does so all the worse across national boundaries. Thus, even as dramatic and dangerous as climatic-induced range shifts are projected to be, phenological shifts, although difficult to anticipate and observe, have the potential to be an equal or greater conservation risk to many species.

Conclusion

Climatic change is an environmental challenge unprecedented in historical times because of its global scope and far-reaching implications for biodiversity and human society. Furthermore, the social responses to these challenges step out of the scientific realm and into decisions that must be made in the swirling waters of ethics, politics and theoretical uncertainty.

Projected future rapid climate change (Mastrandrea & Schneider 2004) could soon become a more looming concern, especially when occurring in concert with other already well-established stressors, particularly habitat fragmentation. Attention must be focused not only on each of these stressors by themselves, but the interactions between them. Change can best be managed, even ameliorated, if it is anticipated, and that necessitates understanding its causes and thereby predicting its scope and tenor. The study of climate change, and its interaction with numerous other complex factors that together impact biodiversity, is an immense, daunting, but urgent challenge for the twenty-first century.

Acknowledgements

We wish to acknowledge partial support for this work for TLR from the US Environmental Protection Agency and the Winslow Foundation.

CN gratefully acknowledges support from the Peoples Trust for Endangered Species and the Earthwatch Institute. Thanks to Stephen Schneider, Dena MacMynowski and Christina Buesching for many helpful comments on early drafts of this paper.

Facts do not cease to exist because they are ignored.
(Aldous Huxley, *Proper Studies*, 1927.)

References

Ball, T. (1983) The migration of geese as an indicator of climate change in the southern Hudson Bay region between 1715 and 1851. *Climatic Change* **5**: 85–93.

Beaubien, E.G. & Freeland, H.J. (2000) Spring phenology trends in Alberta, Canada: links to ocean temperature. *International Journal of Biometeorology* **44**: 53–9.

Beebee, T.J.C. (1995) Amphibian breeding and climate. Nature **374**: 219–20.

Bergmann, F. (1999) Long-term increase in numbers of early-fledged Reed Warblers (*Acrocephalus scirpaceus*) at Lake Constance (Southern Germany). *Journal Fuer Ornithologie* **140**: 81–6.

Berthold, P., Nowack, E. & Querner, U. (1995) Satellite tracking of migratory birds from Central Europe to South African winter quarters – a case-report of the white stork. *Journal für Ornithologie* **136**: 73–6.

Bevanger, K. & Lindstöm, E.K. (1995) Distributional history of the European badger *Meles meles* in Scandinavia during the 20th century. *Annales Zoologicae Fennici* **32**: 5–9.

Bezzel, E. & Jetz, W. (1995) Delay of the autumn migratory period in the blackcap (*Sylvia atricappila*) 1966–1993: a reaction to global warming? *Journal Fuer Ornithologie* **136**: 83–7.

Bradley, N.L., Leopold, A.C., Ross, J. & Huffaker, W. (1999) Phenological changes reflect climate change in Wisconsin. *Proceedings of the National Academy of Sciences of the United States of America* **96**: 9701–4.

Brown, J.L., Li, S.-H. & Bhagabati, N. (1999) Long-term trend toward earlier breeding in an American bird: a response to global warming? *Proceedings of the National Academy of Sciences of the United States of America* **96**: 5565–9.

Brunetti, M., Maugeri, M. & Nanni, T. (2000) Variations of temperature and precipitation in Italy from 1866 to 1995. *Theoretical and Applied Climatology* **65**: 165–74.

Busby, W.H. & Brecheisen, W.R. (1997) Chorusing phenology and habitat associations of the crawfish frog, *Rana areolata* (Anura: Ranidae), in Kansas. *Southwestern Naturalist* **42**: 210–17.

Carter, T.R. (1998) Changes in the thermal growing season in Nordic countries during the past century and prospects for the future. *Agricultural and Food Science in Finland* **7**: 161–79.

Chavez, F.P., Ryan, J., Luch-Cota S.E. & Niquen, M. (1999) Biological and chemical response of the equatorial Pacific Ocean to the 1997–98 El Niño. *Science* **286**: 2126–31.

Cheeseman, C.L., Wilesmith, J.W., Ryan, J. & Mallinson, P.J. (1987). Badger population dynamics in a high-density area. *Symposia of the Zoological Society of London* **58**: 279–94.

Cook, A.J. Fox, A.J., Vaughan, D.G. & J.G. Ferrigno, J.G. (2005) Retreating glacier fronts on the Antarctic Peninsula over the past half-century. *Science* **22**: 541–544.

Crick, H.Q., Dudley, C., Glue, D.E. & Thomson, D.L. (1997) UK birds are laying eggs earlier. *Nature* **388**: 526.

Derocher, A.E., Lunn, N.J. & Stirling, I. (2004) Polar bears in a warming climate. *Integrative and Comparative Biology* **44**: 163–76.

Dugdale, H.L., Macdonald, D.W. & Newman, C. (2003) Offspring sex ratio variation in the European badger, *Meles meles*. *Ecology* **84**(1): 40–5.

Dunn, P.O. & Winkler, D.W. (1999) Climate change has affected the breeding date of tree swallows throughout North America. *Proceedings of the Royal Society of London Series B – Biological Sciences* **266**: 2487–90.

Gatter, W. (1992) Timing and patterns of visible autumn migration: can effects of global warming be detected? *Journal Fuer Ornithologie* **133**: 427–36.

Gibbs, J.P. & Breisch, A.R. (2001) Climate warming and calling phenology of frogs near Ithaca, New York, 1900–1999. *Conservation Biology* **15**: 1175–8.

Hall, M.H. & Fagre, D.B. (2003) Modeled climate-induced glacier change in Glacier National Park, 1850–2100. *BioScience* **53**: 131–40.

Harrison, D., Vanhinsbergh, P., Fuller, R.J. & Berry, P.M. (2003a) Modelling climate change impacts on the distribution of breeding birds in Britain and Ireland. *Journal for Nature Conservation* **11**: 31–42.

Harrison, P.A., Berry, P.M. & Dawson, T.P. (2003b) Modelling natural resource responses to climate change (the MONARCH project): an introduction. *Journal for Nature Conservation* **11**: 1–4.

Hays, J.D., Imbric, J. & Shackleton, N.J. (1976) Variations in the Earth's orbit: pacemaker of the ice ages. *Science* **194**: 1121–32.

Haywood, S. & Perrins, C.M. (1992) Is clutch size in birds affected by environmental conditions during growth? *Proceedings of the Royal Society of London* **249**: 196–197.

Hersteinsson, P. & Macdonald, D.W. (1992) Interspecific competition and geographical distribution of red and arctic foxes *Vulpes vulpes* and *Alopex lagopus*. *Oikos* **64**: 505–515.

Huppop, O. & Huppop, K. (2003) North Atlantic Oscillation and timing of spring migration in birds. *Proceedings of the Royal Society of London* **270**: 233–40.

Hutchinson, G.E. (1958) Concluding remarks. *Cold Spring Harbor Symposium for. Quantitative. Biology* **22**: 415–27.

Intergovernmental Panel on Climate Change (IPCC) (2001a) *Climate Change 2001: the Science of Climate Change*. Cambridge University Press, New York.

Intergovernmental Panel on Climate Change (IPCC) (2001b) *Climate Change 2001: Impacts, Adaptations and Vulnerabilities*. Cambridge University Press, New York.

Jácome, R. & Ospina, P. (1999) La Reserva Marina de Galápagos. Un año difícil. In *Informe Galápagos 1998–1999* (Eds P. Ospina & E. Muñoz,), pp. 35–42. Fundación Natura-WWWF, Quito.

Kai, K., Kainuma, M. & Murakoshi, N. (1996) Effects of global warming on the phenological observation in Japan. In *Climate Change and Plants in East Asia* (Eds K. Omasa, K. Kai, H. Taoda & Z. Uchijima), pp. 164–73. Springer-Verlag, New York, NY.

Kauhala, K. (1995a) Mayran levinneisyys ja runsaus Suomessa. *Suomen Riista* **41**: 85–94.

Kauhala, K. (1995b) Distributional history of the European badger *Meles meles* in Scandinavia during the 20th century. *Annales Zoologicae Fennici* **32**: 183–91.

King, D. (2004) Climate change science: adapt, mitigate, or ignore? *Science* **303**: 176–7.

Kirk, D.A., Evenden, M.D. & Mineau, P. (1996) Past and current attempts to evaluate the role of birds as predators of insect pests in temperate agriculture. *Current Ornithology* **13**: 175–269.

Ludwichowski, I. (1997) Long-term changes of wing-length, body mass and breeding parameters in first-time breeding females of goldeneyes (*Bucephala clangula clangula*) in northern Germany. *Vogelwarte* **39**: 103–16.

Macdonald, D.W. & Newman, C. (2002) Badger (*Meles meles*) population dynamics in Oxfordshire, UK. Numbers, density and cohort life histories, and a possible role of climate change in population growth. *Journal of Zoology* **256**: 121–38.

Mackas, D.L., Goldblatt, R. & Lewis, A.G. (1998) Interdecadal variation in developmental timing of *Neocalanus plumchrus* populations at Ocean Station P in the subarctic North Pacfic. *Canadian Journal of Fisheries and Aquatic Science* **55**: 1858–93.

Mann, M.E., Bradley, R.S. & Hughes, M.K. (1998) Global-scale temperature patterns and climate forcing over the past six centuries. *Nature* **392**: 779–87.

Mann, M.E., Bradley, R.S. & Hughes, M.K. (1999) Northern Hemisphere temperatures during the past millennium: inferences, uncertainties, and limitations. *Geophysical Research Letters* **26**: 759–62.

Mastrandrea, M.D. & Schneider, S.H. (2004) Probabilistic integrated assessment of dangerous climate change. *Science* **304**: 571–5.

McCleery, R.H. & Perrins, C.M. (1998) Temperature and egg laying dates. *Nature* **391**: 30–1.

McDonald, K.A. & Brown, J.H. (1992)Using montane mammals to model extinctions due to global change. *Conservation Biology* **6**(3): 409–15.

McEwen, L.C. (1987) Function of insectivorous birds in a shortgrass IPM system. In *Integrated Pest Management on Rangeland: a Shortgrass Prairie Perspective* (Ed. J.L. Capinera), pp. 389–402. Westview Press, Jackson, TN.

McIntyre, S. & McKitrick, R. (2003) Corrections to the Mann et al. (1998) Proxy data base and Northern Hemispheric average temperature series. *Energy and Environment* **14**(6): 751–71.

McIntyre, S. & McKitrick, R. (2005) Hockey sticks, principal components and spurious significance. *Geophysical Research Letters* **32**(3): 12 February, LO 3710.

Midgley, G.F., Hannah, L., Millar, D., Rutherford, M.C. & Powrie, L.W. (2002) Assessing the vulnerability of species richness to anthropogenic climate change in a biodiversity hotspot. *Global Ecology and Biogeography* **11**(6): 445–51.

Millennium Ecosystem Assessment (2005) *Synthesis Report.* http: //www.millenniumassessment.org [Last accessed on 1 August 2005]

Myres, N.R., Mittermeier, A., Mittermeier, C.G., da Fonseca, G.A.B. & Kent, J. (2000) Biodiversity hotspots for conservation priorities. *Nature* **403**: 853–8.

Naem, S. & Li. S.B. (1997). Biodiversity enhances ecosystem reliability. *Nature* **390**: 507–9.

Nicolaides, F. & Murillo, J. C. (2001). Efectos de El Niño 1997–1998 y Post Niño 1990 en la pesca del bacalao (*Mycteroperca olfax*) y la lisa rabo negro (*Mugil cephalus*) de las Islas Galápagos. In *Informe Galápagos 2000 –2001* (Eds C. Falconí, R. E. Ruiz & C. Valle), pp. 104–9. Fundación Natura and WWF, Quito.

Oglesby, T.T. & Smith, C.R. (1995) Climate change in the northeast. In *Living Resources: a Report to the Nation on the Distribution, Abundance and Health of U.S. Plant, Animals and Ecosystems* (Eds E.T. LaRoe, G.S. Farris, C.E. Puckett, P.D. Doran & M.J. Mac), pp 38–56. Department of the Interior, National Biological Service, Washington, DC.

Parmesan, C. & Yohe, G. (2003) A globally coherent fingerprint of climate change impacts across natural systems. *Nature* **42**: 37–42.

Penuelas, J., Filella, I. & Comas, P. (2002) Changed plant and animal life cycles from 1952 to 2000 in the Mediterranean region. *Global Change Biology* **8**: 531–44.

Perrins, C.M. (1996) Eggs, egg formation and the timing of breeding. *Ibis* **138**: 2–15.

Perrins, C.M. & McCleery, R.H. (1989) Laying dates and clutch size in the great tit. *Wilson Bulletin* **101**: 234–53.

Pounds, J.A. (2002) Impacts of climate change on birds, amphibians and reptiles in a tropical montane cloud forest reserve. In *Impacts of Climate Change on Wildlife* (Eds R.E. Green, M. Harley, M. Spalding & C. Zöckler), pp. 30–32. Royal Society for the Protection of Birds, Sandy.

Pounds, J.A. & Puschendorf, R. (2004) Feeling the heat: climate change and biodiversity loss (editorial). *Nature* **427**: 107–9.

Pounds, J.A., Fogden, M.P.L. & Campbell, J.H. (1999). Biological response to climate change on a tropical mountain. *Nature* **398**: 611–5.

Power, M.E., Tilman, D., Estes, J.A., et al. (1996). Challenges in the quest for keystones. *BioScience* **46**: 609–20.

Press, F. & Siever, R. (1982) *Earth*, 3rd edn. W.H. Freeman and Co, New York.

Rahmstorf, S. (1995) Bifurcations of the Atlantic thermohaline circulation in response to changes in the hydrological cycle. *Nature* **378**: 145–9.

Reading, C.J. (1998) The effect of winter temperatures on the timing of breeding activity in the common toad *Bufo bufo. Oecologia* **117**: 469–75.

Riedinger M.A., Steinitz-Kannan M., Last W.M. & Brenner M. (2002.) A similar 6100 $C-^{14}C$ yr record of El Niño activity from the Galapagos Islands. *Journal of Paleolimnology* **27**: 1–7.

Root, T.L. & Schneider, S.H. (1993) Can large-scale climatic models be linked with multi-scaled ecological studies? *Conservation Biology* **7**: 256–70.

Root, T.L., Price, J.T., Hall, K.R., Schneider, S.H., Rosenzweig, C. & Pounds, J.A. (2003) Fingerprints of global warming on animals and plants. *Nature* **421**: 57–60.

Root, T.L, MacMynowski, D.P., Mastrandrea, M.D. & Schneider, S.H. (2005) Human-modified temperatures induce species changes: joint attribution. *Proceedings of the National Academy of Sciences* **102**: 7465–9.

Sagarin, R. & Micheli, F. (2001) Climate change in nontraditional data sets. *Nature* **294**: 811.

Smith, W.T. &. Cameron, R.D. (1985) Reactions of large groups of caribou to a pipeline corridor on the arctic coastal plain of Alaska. *Arctic* **38**: 53–7.

Socolow, R., Hotinski, R., Greenblatt, J.B. & Pacala, S. (2004) Solving the climate problem: technologies available to curb CO_2 emissions. *Environment* **46**: 8–19.

Sparks, T.H. & Carey, P.D. (1995) The responses of species to climate over two centuries: An analysis of the Marsham phenological record. *Journal of Ecology* **83**: 321–9.

Sparks, T.H. & Yates, T.J. (1997) The effect of spring temperature on the appearance dates of British butterflies 1883–1993. *Ecography* **20**: 368–74.

Stanhill, G. & Cohen, S. (2001) Global dimming: a review of the evidence for a widespread and significant reduction in global radiation with discussion of its probable causes and possible agricultural consequences. *Agricultural and Forest Meteorology* **107**: 255–78.

Stone, R.S., Belchansky, G., Drobot, S. & Douglas, D.C. (2004) Diminishing sea ice in the western Arctic Ocean. In *State of the Climate in 2003* (Eds D.H. Levinson & A.M. Waple). *Bulletin of the American Meteorological Society* **85**: S32–S33.

Strode, P.K. (2003) Implications of climate change for North American wood warblers (Parulidae). *Global Change Biology* **9**: 1137–44.

Stuart, A.J. (1991) Mammalian extinctions in the Late Pleistocene of northern Eurasia and North America. *Biological Review of the Cambridge Philosophical Society* **66**: 453–62.

Thompson, D.B.A., Thompson, P.S. & Nethersole-Thompson, D. (1986) Timing of breeding and breeding performance in a population of greenshank. *Journal of Animal Ecology* **55**: 181–99.

Thompson, L.G., Mosley-Thompson, E., Davis, M.E., Lin, P.-N., Henderson, K. & Mashiotta, T.A. (2002) Tropical glacier and ice core evidence of climate change on annual to millennial time scales. *Climatic Change* **59**: 137–155.

Trenberth, K.E. & Hoar, T.J. (1996) The 1990–1995 El Nino Southern Oscillation event: longest on record. *Geophysical Research Letters* **23**: 57–60.

Vannoordwijk, A.J., McCleery, R.H. & Perrins, C.M. (1995) Selection for the timing of great tit breeding in relation to caterpillar growth and temperature. *Journal of Animal Ecology* **64**: 451–458.

Vargas, H., Lougheed, C. & Snell, H. (2005) Population size and trends of the Galapagos Penguin *Spheniscus mendiculus*. *Ibis* **147**: 367–74.

Vargas, H., Harrison, S., Rea, S. & Macdonald, D.W. (2006) Biological effects of El Niño on the Galapagos Penguin. *Biological Conservation* **127**:107–14

Visser, M.E., Vannoordwijk, A.J., Tinbergen, J.M. & Lessells, C.M. (1998) Warmer springs lead to mistimed reproduction in Great Tits (*Parus major*). *Proceedings of the Royal Society of London Series B – Biological Sciences* **265**: 1867–70.

White, E.G. & Sedcole, J.R. (1991) A 20–year record of alpine grasshopper abundance, with interpretations for climate change. *New Zealand Journal of Ecology* **15**: 139–52.

Wilmers, C.C., Sinha, S. & Brede, M. (2002) Examining the effects of species richness on community stability: an assembly model approach. *Oikos* **99**: 363.

Winkel, W. & Hudde, H. (1996) Long-term changes of breeding parameters of Nuthatches *Sitta europaea* in two study areas of northern Germany. *Journal Fuer Ornithologie* **137**: 193–202.

Winkel, W. & Hudde, H. (1997) Long-term trends in reproductive traits of tits (*P.arus major*, *P. caeruleus*) and pied flycatchers (*Ficedula hypoleuca*). *Journal of Avian Biology* **28**: 187–90.

Woodroffe, R. (1995) Body condition effects implantation date in the European badger, *Meles meles*. *Journal of Zoology* **236**: 183–8.

WWF (2003) *Annual Report*. World Wide Fund for Nature, Gland, Switzerland.

7

Technology in conservation: a boon but with small print

Stephen A. Ellwood, Rory P. Wilson
and Alonzo C. Addison

The real question is not whether machines think, but whether men do?
(B. F. Skinner (1904–1990), *Contingencies of Reinforcement*, 1969, p. 288.)

Introduction

The microelectronics revolution of the late twentieth century left a world in which new technology might be expected to offer a solution to any problem. Conservation science is no exception, and as scientists strive to use the latest inventions to serve conservation goals as different as tracking the movement of whales across the globe to recording prey selection by sparrow hawks, we ask whether these devices really work. After all, it is one thing to use gleaming new systems built for commercial, consumer applications, but it is quite another to put them through the rigours, complexity and mud of the field, and yet expect to gather good quality, unbiased data. In this essay we illustrate some of the pitfalls that can bias data collection, or lead to unnecessary harm (to animal and environment), cost (to reputation and finances), and disappointment, while showing what can be possible when new technology really works. This essay is **issue**- rather than **technology**-driven; issues are timeless, whereas all technology has a limited shelf-life. Space considerations oblige us to present issues with few examples, which we have unashamedly selected from our own areas of interest (both within the text and Boxes 7.1 to 7.6). We do not provide a comprehensive list of all available technologies and we omit those used only in the laboratory. In fact, our remit is to consider electronic gadgets, used in the wild, and which enhance our understanding of animal biology while making life easier for the researcher, the study animal and the environment. However, within these limitations we do cover systems as diverse as animal tracking collars that telephone home to 'say' 'I'm migrating', to sensors that detect the latest meal of the wandering albatross as it circumnavigates the world. Finally, for simplicity's sake, we minimize jargon but provide worldwide web links and references for those who crave more (Table 7.1).

Crucial issues, problems and dilemmas

The evidence-based approach

Sutherland et al. (2004) describe how 'Evidence-based medicine' has revolutionized human healthcare over the past few decades, and that the evidence-based approach should

Box 7.1 Illustrations from animal tracking

Attaching position-tracking devices to wild animals has a five decade history. Amlaner & Macdonald (1980) and Kenward (2001) comprehensively explain the evolution of tracking devices, whereas Cooke et al. (2004) provide an excellent review of devices that record physiological variables. Modern tracking devices illustrate all the conundrums referred to in this essay. For example, the immobilization of the endangered black rhino, in order to fit tracking collars, has been shown to affect fecundity (Alibhai & Jewell 2001) and the subsequent high failure rate of the collars raised serious scientific, ethical and financial concerns (Alibhai et al. 2001). Clearly this was not conducive to conservation of the species.

Radio-tracking

The simplest trackers are radio-transmitters, with the process of determining an animal's movements being referred to as radio-tracking. Radio-transmitters can be attached by collar or harness, and then picked up at a distance (metres to kilometres) by an appropriate radio-receiver and antenna. Transmitters that last a matter of days weigh as little as 0.35 g, with larger batteries allowing deployments of years. Although the hardware costs are relatively low (tens of pounds) radio-tracking incurs high labour costs, because researchers are required to locate tagged animals (by a process of triangulation or visual observation). Furthermore, a combination of system inaccuracy, human error and physical disturbance caused by a human tracker can produce considerable device-induced error (DIE). However, radio-tracking has been used to solve important ecological and conservation problems. Pioneers in the field, such as David Macdonald, gathered data on habitat use by red foxes (*Vulpes vulpes*) (Macdonald et al. 1981) that were pivotal to Europe's successful eradication of rabies (Bacon & Macdonald 1981; Macdonald et al. 1981). Radio-tracking has recently jumped into the twenty-first century with the development of the Barro Colorado Island (BCI) project, in the Republic of Panama (http://www.princeton.edu/~wikelski/research). Here, a huge grid of static computer-controlled receiving towers is being built, rising above the jungle canopy, to track thousands of animals (including some insects) simultaneously.

The BCI project relies on a mixture of contact and non-contact systems to determine animal position. However, fully independent contact systems can also determine animal position. Devices such as 'Geolocators' and 'Dead-reckoners' plus the ARGOS (Advanced Research and Global Observation Satellites) and GPS (Global Positioning System) satellite systems fit this niche.

Geolocators

Small, lightweight (as little as 6 g) geolocators use time and day length to calculate position and then store it in an onboard memory. Geolocators can be deployed for months to years without a battery change. Their main drawback is inaccuracy, often estimating positions tens of kilometres away from true location, and of course the device must be recovered in order to access the data. However, for species that return predictably to known locations (even if they travel huge distances, such as grey-headed albatrosses, which may circumnavigate the globe in an astonishing 46 days (Croxall et al. 2005)), this is not a problem. Geolocators have demonstrated that, contrary to previous assumptions, wandering albatrosses do not spend their non-breeding years circumnavigating the globe. Instead they remain in preferred foraging areas, many of which are used by long-line fisheries, notorious for inadvertently snaring albatrosses; the conservation relevance is clear (Weimerskirch & Wilson 2000).

Dead-reckoners

Dead-reckoners contain an electronic compass and sensors that detect speed of travel. In combination these variables can be used to estimate position (in three dimensions) relative to a known starting point. Dead-reckoners are small and light, but can suffer from 'drift' as time from deployment augments inaccuracies in location calculation. Dead-reckoners are particularly useful where deployments are short, such as in the work by Davis et al. (2003), examining how Weddell seals capture their prey beneath the Antarctic ice.

Advanced Research and Global Observation Satellites

The ARGOS system is similar to radio-tracking, except that the receivers are located on satellites. Animals are fitted with a specialized radio-transmitter (jargon name = platform transmitter terminal (PTT)) that allows them to be tracked from space with data relayed to an Earth-based station. Transmitters are light (16–80 g) but costs are high (thousands of pounds per device plus charges for data download). Their location resolution is poor (hundreds to thousands of metres), but data can be accessed remotely, reducing the ethical, financial and scientific concerns associated with instrumenting an animal without obtaining any useful data. For long-distance movers, where weight is critical, PTTs may be the best option, e.g. PTTs were used in first documenting the incredible wide-ranging abilities of albatrosses, demonstrating that birds may cover thousands of kilometres in a single foraging trip (Jouventin & Weimerskirch 1990).

Global Positioning System

Where truly remote tracking, precision and accuracy are important, GPS is often the only solution. The GPS works in the opposite direction to radio-tracking and ARGOS, in that the animal wears the receiver. This contains a powerful miniature computer that uses data received from a constellation of orbiting satellites to calculate its position (by trilateration, similar to triangulation) to a resolution of 15–30 m (less than 1 m is possible with refinements). Unfortunately the hardware is power hungry and more bulky than other systems (30–1500 g), which has limited its use thus far to shorter deployments on medium to large animals.

Previously, 'poor signal conditions' (where the receiver's 'view' of the sky is blocked), could reduce accuracy considerably, or stop GPS receivers working. However, recent developments mean that GPS functions in most habitats except underground or water. Rodgers et al. (1995, 1996, 1997) provide a good explanation of the use of GPS on animals, and Hulbert & French (2001) and Frair etal. (2004) describe problems associated with GPS data. Hemson et al. (2005) used GPS on lions (*Panthera leo*) to elucidate their seasonal predation impact on domestic cattle (*Bos taurus*). This work provided the insight necessary to devise a way of keeping these two species apart, thus reducing lion–human conflict. Incidentally, Hemson et al. (2005) used GPS collars equipped with a radio-link that allowed them to access data remotely (perhaps reasoning that close proximity to lions could be detrimental to their health). The addition of other remote download systems, such as GSM (Global System for Mobile communication, i.e. the mobile phone network) and LEO (Low Earth Orbit) modems, may add weight, but does increase the chance of retrieving data and thus helps justify the procedure. Use of GPS with GSM download, coupled with 'geo-fencing' – of a virtual area predetermined by the researcher – opens up important areas of conservation research such as identifying dispersal or migration events. When an animal's position is calculated to be outside the virtual fence, then a text message or email is sent by way of notification and can be used to trigger more detailed observations. Other additions, such as automatic release systems, can be integrated into collar/harness design so that once batteries are dead the bearer does not have to suffer recapture, or risk wearing the device for the rest of its life.

be applied to conservation. They argue 'Much of current conservation practice is based upon anecdote and myth rather than upon the systematic appraisal of the evidence, including experience of others who have tackled the same problem.' They refer to conservation practices whereas we are considering tools>, but the arguments are the same. Evidence-based medicine is a simple concept in which professional judgment is combined with the scientific comparison of different treatments to determine the most effective measures for treating specific conditions (Sackett et al. 1996). If multiple treatments do not exist, then treatments are tested against a control or 'gold standard'. We believe that similar 'ground-truthing' should be applied to the use of gadgets in conservation. However, instead of thinking in terms of effectiveness, discussion of accuracy is more appropriate.

Device accuracy and sensor sensitivity

Accuracy and sensitivity are central to device choice. A device must be sensitive enough to detect specified changes, and to do so with sufficient accuracy. These issues revolve around matters such as a sensor's response to change and the ability of the recording device, to which it is connected, to document its output at an appropriate resolution. These matters are fundamental and are easily tested during suitable laboratory trials. However, they may have little to do with true accuracy.

Box 7.2 Digital camera traps: shedding light on predator–prey relationships

A study by Cresswell et al. (2003) exemplifies the correct choice of technology for gathering large quantities of high quality data, in an efficient low impact manner. Digital camera traps were installed at multiple locations, and automatically triggered to take a photograph when a lure (dead bird) was seized by a predator, e.g. sparrow hawk (*Accipiter nisus*), see Box Fig. 7.1. One researcher ran multiple set-ups simultaneously, effectively gathering 6423 hours of observations, over 45 days, recording 68 attacks.

Box Fig. 7.1 Sparrowhawk (*Accipiter nisus*) attacking model prey, captured by digital camera trap. (Picture courtesy of Will Cresswell.)

Box 7.3 Radio frequency identification (RFID)

Arguably one of the technologies set to make huge differences in conservation science is that of RFID. There are two categories of RFID: passive and active tags. Passive integrated transponders (PIT tags) are the grain-of-rice-sized identification 'chips' that can be injected, subcutaneously, into animals as small as mice, or attached to the leg rings of fledgling birds. The battery-free PITs each transmit a unique serial code that is detected by specialized reading units – each PIT costing £3.00 for the lifetime of the bearer. Although these tags can be read only from a few centimetres away, readers can be built into nests, feeders, tunnels, or mats on the ground. Beyond the initial injection, these gadgets have negligible impact on their owner, and can also be used to switch on other recording equipment automatically, such as video cameras, food dispensers and weighing scales. The opportunities for low impact, data-rich, field experiments are considerable (e.g. Boisvert & Sherry 2000). Active RFID tags (see http://www.wavetrend.co.uk), which do include a battery and can broadcast their identity plus sensor information at distances of up to 100 m, are already in use protecting valuable goods. These devices, if worn by an animal, have great potential as smaller, cheaper alternatives to conventional tracking systems, where information is required on proximity to resources (containing a receiver), as opposed to continuous positional tracking.

'True-accuracy?'

Imagine attaching a 1 kg tracking device to a 1 kg animal. The **device's accuracy**, i.e. its ability to pin-point its own position remains the same on or off the animal, but the overall **system accuracy** of the animal plus device will be extremely poor because the animal is unlikely to be able to move! Gadgets only really work when such **device induced error** (DIE)

Box 7.4 High-technology cameras: a window on a hidden world

Direct observation of the natural world has been a mainstay in the development of the sciences of ethology and ecology. Binoculars may get us 'closer' to subjects without disturbing them, but they do not allow us to see in the black of night, nor can they keep us alert long enough to see rare, transient events. For this reason various kinds of camera, beyond the simple film or digital stills/video variety, have been integrated into observational research.

The systems that have had the greatest impact are those that allow us to see, but not be seen, in the dark. In the early 1970s military development gave us infrared binoculars and the 'image intensifier', a telescope-like device that magnifies available light (from levels too dim to see) to generate a bright viewable image – see Macdonald (1987) for unique nocturnal observations of red foxes. Today, intensifiers are small and of sufficient quality to be incorporated into video cameras, allowing a permanent record to be made. Unfortunately their cost (£100s to £1000s) has prevented more widescale use. However, a cheaper system, 'infrared (IR) sensitive cameras,' has been developed. Infrared light, produced by the sun or special lamps, is invisible to the mammalian eye (Lythgoe 1979). In 1987, David Macdonald and the BBC Natural History Unit collaborated, using IR cameras, to record observations of wild foxes (Macdonald 1987). The resulting 'World About Us' documentary, 'Night of the fox', earned a place in that year's British Academy of Film and Television Awards (BAFTA) final. Today, the demand for closed-circuit television (CCTV) systems means that cheap cameras (< £200) are now available that produce full-colour images during the day, and then switch automatically to monochrome IR at night, requiring little unnatural illumination. The combination of similar cameras with time-lapse video recording, motion detectors and timer switches, has produced the autonomous system described by Stewart et al. (1997). It is capable of recording subsecond events, e.g. transitions in grooming behaviour (Stewart & Macdonald 2003), or detailed individual feeding behaviour (Baker et al. 2005a,b, in press), in complete darkness (Box Fig. 7.2). The continued miniaturization of devices, coupled with wireless technology, is making the long-term deployment of video cameras on animals possible. Such systems give the researcher an animal's-eye view of the world (e.g. Beringer et al. 2004).

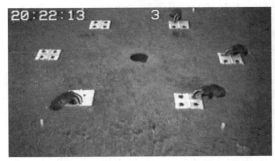

Box Fig. 7.2 Video stills of wild European badgers (*Meles meles*) taking part in a feeding experiment. (Pictures courtesy of Sandra Baker and Stephen Ellwood.)

Moving up a gear, thermal imaging cameras, sensitive to the IR radiation produced by animals, effectively decamouflage creatures normally too cryptic to be distinguished from their backgrounds. When used in conjunction with powerful surveying techniques such as distance sampling, thermal imagers may revolutionize the way we survey some animal populations (Laake et al. 1993; Gill et al. 1997; Ellwood 2003).

The main drawback of high-technology cameras is 'tunnel-vision', because observers are oblivious to everything outside the camera's field of view. Therefore considerable thought should be given to how representative the camera's view is of the system we are attempting to observe.

Box 7.5 Loggers used in marine research

On-board cameras

The attachment of cameras to the animals themselves, and avoiding the restrictions of static cameras, has become the major thrust behind the development of the 'Crittercam' system (Marshall 1998). Recent versions have been used to examine, for example, the work done during limb movement in cetacea and pinnipeds during travel (Williams et al. 2000). However, memory limitations are still problematic, making the unit unacceptably large for most animals. Use of less ambitious, but much smaller, digital cameras to store still images may be a useful compromise (Hooker et al. 2002).

Feeding

Given that feeding is such a fundamental part of life, it is little surprise that new devices are being used to study feeding behaviour, particularly of animals that feed at sea. Initially, the quantity of prey delivered by seabirds was assessed by automatically weighing chicks (and their nests) before and after feeding (Ricketts & Prince 1984; Gremillet et al. 1996). Further attempts to refine prey capture rates on site used ingested gadgets to log stomach or oesophagus temperature (ingestion of ectothermic prey by marine endotherms was detected via temperature reduction: Wilson et al. 1992; Ancel et al. 1997). The most recent development in food ingestion uses a magnetic field strength sensor fixed to one mandible of the study animal, and a small magnet glued to the other. When the jaw/beak is opened a reduction in magnetic field strength is experienced by the sensor and recorded by the logger at high frequency. Food ingestion is recorded by the change in jaw angle over time and can even be used to determine the mass of food ingested (Wilson et al. 2002).

Variations of this same sensor–magnet system include gadgets for measuring limb movement (Wilson & Liebsch 2003), digestion rate (Wilson et al. 2004), respiration rates and cardiac frequency (Wilson et al. 2004).

Where transmission telemetry meets logging

The development of loggers has been particularly prolific for use with marine animals because transmitters do not work under water. However, a major factor working against logging systems, particularly in the marine environment, is the problem of data recovery. Many marine species are too unpredictable in their movements for device recovery to be reasonably assured and this figures greatly in any cost-benefit analysis. The problem has been partially solved by linking ARGOS-based transmitters with loggers so that recorded data may be transmitted back to the researchers via satellite. This has provided some remarkable data on the ocean-basin movements and incredible diving capacities of critically endangered turtles (Hays et al. 2004), although the current size of such systems makes it difficult to justify deployment on species much smaller than seals.

An alternative solution, popular with fish biologists (whose study animals seldom break the water's surface unless *en route* to being eaten), is the 'pop-up tag', a logger that releases itself from its carrier after a prescribed time to float to the sea surface – from where it sends data to the ARGOS satellite. Although the satellite is not permanently overhead, the released tag remains permanently at the sea surface, thereby eliminating the problem of finding a time when the carrier and the satellite may communicate. Barbara Block and colleagues from Stanford have used such systems to study the ocean-basin movement and distribution of giant bluefin tuna (*Thunnus thynnus*), a species under intense commercial pressure (Block et al. 2005).

Loggers offer an incredible opportunity to study the detailed habits of animals that may be recovered in a predictable fashion. Researchers working with animals that must be recaptured, in order to access data, are automatically subject to stringent animal welfare regulations. Overly upset animals do not return to their capture sites and the data are lost! Were such a policing policy inherent in transmission telemetry applications we might see far fewer publications!

is negligible, although there is a wide spectrum of effects from total system collapse, as in the example above, to minimal impact; for example, penguins swim faster, as devices attached to them become smaller (Wilson et al. 1986). The challenge is to decide what level of DIE is acceptable, before true accuracy begins to suffer.

Box 7.6 Smart Dust: Sensor plus network-on-a-chip

In the late 1990s, anticipating an inevitable need to track and monitor environmental agents (from climatic to nuclear/biological/chemical) on the battlefield or cities under attack by foes, the US Defence Department funded the 'Smart Dust' project, their goal being to:

'... build a self-contained, millimetre-scale sensing and communication platform for a massively distributed sensor network. The device will be about the size of a grain of sand and will contain sensors, computational ability, bi-directional wireless communications, and a power supply, while being inexpensive enough to deploy by the hundreds.' (http://robotics.eecs.berkeley.edu/~pister/SmartDust/)

Each Smart Dust 'Mote' was to be composed of a solar cell to generate power, sensors custom-etched into silicon to capture specific information (e.g. acceleration, light, sound, temperature, presence of specific chemicals, etc), a tiny computer to store and send out information, and an optical communicator to respond when the sensor is interrogated by a base-station. Although no individual element of the project was new, it assembled an array of chip-level MEMS (microelectromechanical sensors) in one tiny integrated package (Box Fig. 7.3).

Box Fig. 7.3 (a) DeputyDust prototype with solar powered bi-directional communications and sensing (acceleration and ambient light). (b) GolemDust prototype showing device size (Courtesy Brett Warneke, University of California at Berkeley).

Smart Dust brings together all of the possibilities of the devices described earlier, from positional, physiological and environmental sensing, to data transmission back to a base-station. Just as GPS was initially deemed impossible, and too complex ever to work, Smart Dust has its detractors. The academic research finished in 2001, with real successes, but before reaching the 1 mm^3 'dust'-scale sought. Although power continues to be the nagging challenge, Smart Dust prototypes are already deployed for environmental monitoring of office temperature and lighting conditions (albeit with AA batteries!). Numerous companies are selling chip-level sensors for everything from temperature to light (the CCD at the heart of your digital camera), and even specific molecules. The researchers behind Smart Dust have gone on to launch Dust Inc (http://www.dust-inc.com), to sell communication modules to merge with the sensors (interestingly with financing from high-technology and venture capital firms, an agricultural conglomerate, and the CIA).

   Whatever its eventual name, and regardless of its origins, the Smart Dust concept shows immense promise for conservation biology. Given their size and flexibility this new breed of sensors can be animal-attached or left in a grid in an ecosystem to track environmental conditions. For now, microsensors are available to those with the expertise to select them and assemble the components into a deployable package, and Smart Dust exists in the form of larger motes (http://www.xbow.com).

Error in 'contact' and 'non-contact' devices

The level of potential DIE depends on whether or not a device is in contact with the system it is measuring (attached or unattached). As well as contact devices seriously biasing data collection (see above), there may also be considerable welfare issues. Ultimately, no matter howsmall a device is, it may affect animal performance, social ranking and/or fitness in some way. However, size is an important issue and any adaptations that add weight or bulk to the basic gadget will increase its impact. So, for example, contact systems can be subdivided into those that transmit data and those that record data, because the hardware, and power, required for transmission often increases the size of a device. In a nutshell, careful consideration must be given to the exact configuration, shape and size of any device attached to an animal.

Human error

Do we use gadgets to replace, or enhance, the ability of humans to record data? Human errors (e.g. observer bias, effects of fatigue, lack of systematic precision) are well known, and although new technologies may help to eliminate these, they can create other errors if the experimental design is poor. Failure to use a technology so that it collects an unbiased, representative data set is just as much human error as that generated by the entirely human-based system it was designed to replace. Box 7.4, on the use of remote video, deals with both sources of error. Here, automated cameras make observations that a human could not make accurately, but further human error is possible if the scientific questions asked are not focused purely on what can be answered within the limited field of view of a static camera.

Scientific, ethical, practical and financial considerations (cost-benefit analysis): effects on data quality and quantity

New technology is generally used in conservation to enhance the collection of data required to answer questions. It is important to have an a priori understanding of the quantity and quality (quality = statistical precision and true-accuracy) of data required so that they can be matched with an appropriate device. The evidence-based approach must be coupled with scientific, ethical, practical and financial considerations. The balance between quantity and quality is important because few, excellent quality data may not be sufficient to make reliable conclusions, and vice versa. With a specific application in mind, we need to ask the following:

1. What evidence is there that existing devices can provide the quality and quantity of data required to answer our questions?
2. If a technology has no competitors is it possible to carry out trials to determine its efficacy?
3. If the financial and scientific criteria can be met, what are the ethical and practical implications of deployment?

The process is analogous to a cost-benefit analysis, and should be based on real evidence rather than conjecture. For example, Fig. 7.1 shows the kind of questions that should be asked when choosing a device for attachment to a wild animal. Continuing this theme, the following crucial issues, problems and dilemmas can best be explained by reference to examples from the animal-tracking world. The types of technologies referred to are not important as they are fully explained in Box 7.1. Simply consider them as devices attached to an animal which allow its position to be recorded.

Table 7.1 Technical explanations on the Worldwide Web

		Technology	Source of example or explanation	
Attached	Transmitters	VHF	http://www.biotrack.co.uk/	Download interactive catalogue/radiotracking/
		ARGOS	http://www.npwrc.usgs.gov/perm/cranemov/location.htm/	Product-overview
		Passive RFID tags (PIT)	http://www.aimglobal.org/technologies/rfid/resources/papers/rfid_basics_primer.asp	
		Active RFID tags	http://www.wavetrend.co.uk/	Information/downloads
	Remote download	GPS-VHF	http://www.televilt.se/	System overview
		GPS-ARGOS	http://www.microwavetelemetry.com/index.php	Information/downloads
		GPS-GSM	http://www.oxoc.com/	
		GPS-Satellite	http://www.televilt.se/	
		Camera tags	http://www.smru.st-and.ac.uk/research/wild.htm	
		Sensor-ARGOS	http://smub.st-and.ac.uk/InstrumentationF.htm/srdls.htm	
		Zigbee	http://www.roke.co.uk/download/datasheets/ZigBee.pdf	
	Loggers	Physiological and environment	http://www.driesen-kern.de	
		GPS and DGPS	http://www.trimble.com/gps	
		Geolocators	http://www.an-arctica.ac.uk/About_BAS/Cambridge/Divisions/ALD/Engineering/projects/birdlogger.html	
			http://www.driesen-kern.de	
		Dead reckoning	http://www.driesen-kern.de	
Unattached		Infrared cameras/video	Stewart et al. (1997)	
		Film/digital stills/video camera trap	http://www.reconyx.com/	
			http://www.pixcontroller.com	
			http://www.cabelas.com	
		Image intensifier	http://science.howstuffworks.com/nightvision.htm	
		Thermal imaging	http://science.howstuffworks.com/nightvision.htm	

Understanding the limits of a chosen technology

Failure to understand the limitations of a technology can have serious consequences. For example, until recently, the Global Positioning System (GPS) tracking system could not generate positional data in heavily forested areas or deep gullies. Therefore, data were often curtailed or biased (Frair et al. 2004). This has led to disappointment (when a high proportion of fixes were missed) and misinterpretation (when errors were not accounted for) when steps could have been taken to mitigate against the problems with suitable validation (Girard et al. 2002; Johnson et al. 2002; Di Orio et al. 2003; Gau et al. 2004), or the use of an alternative technology.

'Drowning' in data!

A standard data logger (electronic memory) may record up to 64 million pieces of data during a single deployment over a few days – far more than conventional analytical software can cope with and, quite likely, more than required for analysis. If an animal has had to wear a device for longer than necessary, ethical questions arise. However, initial deployments that gather too much data are useful if they are used to determine optimum deployment times for future work.

Problems with reliability

Reliability (the ability of a device to remain functional for its predicted life) is a major issue, as it is unethical, not to mention expensive, to lumber an animal with a device that might not record useful data. New technology is often less reliable than old, tried and tested systems. For example, as a highly accurate

replacement to traditional radio-tracking, GPS is arguably the most complicated technology currently applied to animal tracking (Rodgers et al. 1995). It is, however, more susceptible to mechanical failure (an occupational hazard for anything attached to a wild animal) and soft/firmware problems (due to a rapidly changing electronics market requiring constant upgrades in computer code). Consequently, exhaustive field tests, ideally on captive animals, must be completed before a device may be deemed suitable for deployment in the wild.

Missing data and unknown influences: problems of overinterpretation.

Technology and experimental design must be carefully matched so that researchers understand the significance of missing data. In animal tracking, data are likely to be 'missed' in areas where the device cannot function properly, e.g. radio signals do not easily pass through dense vegetation (Frair et al. 2004; Gau et al. 2004). This could result in the complete absence of information from a particular habitat type and so, even if relatively few data are missing, they may be of great biological significance. Carefully designed validation experiments can help here, e.g. Frair et al (2004) placed GPS collars in different habitat types to test their ability to determine location in areas with different amounts and types of vegetation. The resulting habitat-dependent biases were considered predictable and subsequently could be controlled for.

Similarly, cause and effect relationships between device-recorded data and unseen occurrences cannot be inferred. There may be great temptation to overinterpret data gathered using high-technology systems. For example, Wilson & Wilson. (1990) assumed that long dives by penguins, as determined by pressure sensors, were indicative of prey capture. However, the later addition of prey-ingestion sen-

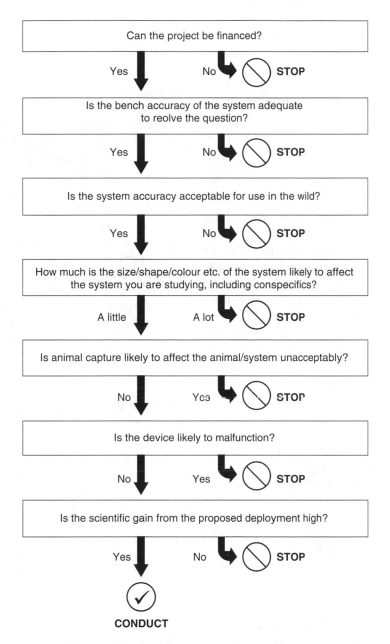

Fig. 7.1 Should the study go ahead? Does the conservation gain to the individual, species or science outweigh the costs?

sors showed that short dives were actually more characteristic of successful predation; the high speed chases involved rapidly using up oxygen, reducing the time an animal could remain submerged (R. P. Wilson et al., unpublished data).

Future requirements, directions and solutions

A lack of information and experts

The number of animal tracking devices manufactured (e.g. Hulbert & French 2001) far outnumbers those discussed in the literature (OxLoc Ltd, personal communication). This maybe due, in part, to the failure of devices going unreported, and this probably also happens with other technologies. Without knowledge of the extent of such 'missing data' the scientific community has very little on which to base technology choices, making cost-benefit analyses difficult. Equally worrying, the current peer review process generally favours referees chosen for their expertise in the species, or paradigm under investigation, rather than the technology used to record data. Consequently, a methodology that seems sound to the referee may be obviously flawed to the appropriate expert. The applications of technology, and methodologies used, in conservation biology are so diverse that standard peer review will struggle increasingly as no one can be a master-of-all-trades. A suggestion might be to have technical referees who are asked to comment on methodology only (minimizing their workload), and to create short sections, in existing journals, dedicated to reporting failures and the underlying reasons (see Gau et al. (2004) for a good example of problem reporting).

Wireless communications: device to device and device to person

The advent of worldwide wireless communications and access via the Internet means that data can be gathered locally and then accessed globally. Although we acknowledge that the addition of data transmission systems may raise ethical concerns, for some attached devices, adaptations that maximize the likelihood of data retrieval are generally good. Deploy-ment in dangerous, remote areas may be unacceptable where people are required for data retrieval. Similarly, loss of data before effective retrieval is at least annoying, if not unethical for attached devices. Fortunately, wireless data communications are advancing rapidly to the benefit of conservation. Hemson et al. (2005) illustrate the remote download of GPS data from lion. The ZebraNet system represents cutting edge technology with wild zebra (*Equus burchelli*) carrying gadgets that relay positional and physiological data to one another, automatically, and then via a longer range radio link to the researcher (http://www.princeton. edu/~mrm/zebranet.html). Smart Dust technologies such as Zigbee (see http://www. zigbee.org) promise to reduce the weight and cost of wireless communication, providing standardized protocols that allow a wide array of attached and unattached devices to 'talk' to each other.

Smart systems in the wild with desk-top control

Animal-attached devices are becoming smarter, packing more processing power into smaller packages. Already they communicate with each other, and with researchers, via the web (see above). Based on analysis of data received, instructions can be sent back to devices instructing how to proceed with further data gathering. When this second stage becomes automatic, possibly including the assessment of hundreds of signals from different interacting species, we will truly be able to observe and react at the ecosystem level. These days will come signalling a new and more potent era for conservation.

Software, hardware, miniaturization and power efficiency

We have discussed the risks of drowning in data, and the ethics of using bulky devices.

History dictates that computational power, i.e. hardware, software development and miniaturization will solve many of our current problems – even if this requires entirely new technologies. However, providing adequate power to remote devices will always be the greatest problem as we seek to do more and more with our gadgets. Unfortunately the development of electrical power sources has always lagged behind that of the equipment they supply.

Conclusions

In our brief consideration of the huge array of gadgetry available for studying animals in their environment, some major issues emerge which range from considerations of animal welfare to the quality of the data collected. The technology is complicated and the issues surrounding it no less so. One thing is certain though – we are on the verge of a fundamental breakthrough in understanding the biology of animals in the wild because we can take the laboratory, with all its technological capacity for quantification, into the field. The silicon chip industry, driven by a consumer market ever hungry for more, has given us eyes where we were blind, memories when we were fatigued and powers of information transmission that transcend the hindrances of space. Used judiciously, we can examine, quantitatively, and in minute detail, the world that our populations are slowly suffocating and use the data acquired to make powerful and useful decisions to protect and conserve. Used thoughtlessly, however, this technology may lead us astray just as political decisions may. The rest is up to us.

One machine can do the work of fifty ordinary men. No machine can do the work of one extraordinary man.
(Elbert Hubbard (1856 –1915) *A Thousand and One Epigrams***, 1911.)**

References

Alibhai, S.K. & Jewell, Z.C. (2001) Hot under the collar: the failure of radio-collars on black rhinoceros *Diceros bicornis*. *Oryx* **35**(4): 284–8.

Alibhai, S.K., Jewell, Z.C. & Towindo, S.S. (2001) Effects of immobilization on fertility in female black rhino (Diceros bicornis). *Journal of Zoology* **253**(3): 333–45.

Amlaner, C.J. & Macdonald, D.W. (Eds) (1980) *A Handbook on Biotelemetry and Radio Tracking: Proceedings of an International Conference on Telemetry and Radio Tracking in Biology and Medicine, Oxford, 22–27 March 1979.* Pergamon Press, Oxford.

Ancel, A., Horning, M. & Kooyman, G.L. (1997) Prey ingestion revealed by oesophagus and stomach temperature recordings in cormorants. *Journal of Experimental Biology* **200**(1): 149–54.

Bacon, P.J. & Macdonald, D.W. (1981) Habitat and the spread of rabies. *Nature* **289**(5799): 634–5.

Baker, S.E., Ellwood, S.A., Watkins, R. & Macdonald, D.W. (2005a) Non-lethal control of wildlife: using chemical repellents as feeding deterrents for the European badger *Meles meles*. *Journal of Applied Ecology* **42**(5): 921–31.

Baker, S.E., Ellwood, S.A., Watkins, R. & Macdonald, D.W. (2005b) A dose–response trial with ziram-treated maize and free-ranging European Badgers *Meles meles*. *Applied Animal Behavioural Science* **93**(3–4): 309–21.

Baker, S.E., Johnson, P.J., Slater, D., Watkins, R.W. & Macdonald, D.W. (In press) Learned food aversion with and without an odour cue for protecting untreated baits from wild mammal foraging. *Applied Animal Behaviour Science*.

Beringer, J., Millspaugh, J.J., Sartwell, J. & Woeck, R. (2004) Real-time video recording of food selection by captive white- tailed deer. *Wildlife Society Bulletin* **32**(3): 648–54.

Block, B.A., Teo, S.L.H., Walli, A., et al. (2005) Electronic tagging and population structure of Atlantic bluefin tuna. *Nature* **434**(7037): 1121–7.

Boisvert, M.J. & Sherry, D.F. (2000) A system for the automated recording of feeding behavior and body weight. *Physiology & Behavior*, **71**(1–2): 147–51.

Cooke, S.J., Hinch, S.G., Wikelski, M., et al. (2004) Biotelemetry: a mechanistic approach to ecology. *Trends in Ecology and Evolution* **19**(6): 334–43.

Cresswell, W., Lind, J., Kaby, U., Quinn, J.L. & Jakobsson, S. (2003) Does an opportunistic predator preferentially attack non-vigilant prey? *Animal Behaviour* **66**: 643–8.

Croxall, J.P., Silk, J.R.D., Phillips, R.A., Afanasyev, V. & Briggs, D.R. (2005) Global circumnavigations: tracking year-round ranges of nonbreeding albatrosses. *Science* **307**(5707): 249–50.

Davis, R.W., Fuiman, L.A., Williams, T.M., Horning, M. & Hagey, W. (2003) Classification of Weddell seal dives based on 3–dimensional movements and video-recorded observations. *Marine Ecology – Progress Series* **264**: 109–22.

Di Orio, A.P., Callas, R. & Schaefer, R.J. (2003) Performance of two GPS telemetry collars under different habitat conditions. *Wildlife Society Bulletin* **31**(2): 372–9.

Ellwood, S.A. (2003) The deer of Wytham Woods: a managed decline in numbers. *Deer* **12**(8): 461–63.

Frair, J.L., Nielsen, S.E., Merrill, E.H.,et al. (2004) Removing GPS collar bias in habitat selection studies. *Journal of Applied Ecology* **41**(2): 201–12.

Gau, R.J., Mulders, R., Ciarniello, L.J., et al. (2004) Uncontrolled field performance of Televilt GPS-Simplex (TM) collars on grizzly bears in western and northern Canada. *Wildlife Society Bulletin* **32**(3): 693–701.

Gill, R.M.A., Thomas, M.L. & Stocker, D. (1997) The use of portable thermal imaging for estimating population density in forest habitats. *Journal of Applied Ecology* **34**(5): 1273–86.

Girard, I., Ouellet, J.P., Courtois, R., Dussault, C. & Breton, L. (2002) Effects of sampling effort based on GPS telemetry on home-range size estimations. *Journal of Wildlife Management* **66**(4): 1290–300.

Gremillet, D., Dey, R., Wanless, S., Harris, M.P. & Regel, J. (1996) Determining food intake by great Cormorants and European shags with electronic balances. *Journal of Field Ornithology* **67**(4): 637–48.

Hays, G.C., Houghton, J.D.R. & Myers, A.E (2004) Pan-Atlantic leatherback turtle movements. *Nature* **429**: 522.

Hemson, G., Johnson, P., South, A., Kenward, R., Ripley, R. & Macdonald, D. (2005) Are kernels the mustard? Data from global positioning system (GPS) collars suggest problems for kernel home-range analyses with least-squares cross-validation. *Journal of Animal Ecology* **74**(3): 455–63.

Hooker, S.K., Boyd, I.L., Jessopp, M., Cox, O., Blackwell, J., Boveng, P.L. & Bengtson, J.L. (2002) Monitoring the prey-field of marine predators: Combining digital imaging with datalogging tags. *Marine Mammal Science* **18**(3): 680–97.

Hulbert, I.A.R. & French, J. (2001) The accuracy of GPS for wildlife telemetry and habitat mapping. *Journal of Applied Ecology* **38**(4): 869–78.

Johnson, C.J., Heard, D.C. & Parker, K.L. (2002) Expectations and realities of GPS animal location collars: results of three years in the field. *Wildlife Biology* **8**(2): 153–9.

Jouventin, P. & Weimerskirch, H. (1990) Satellite tracking of wandering albatrosses. *Nature* **343**(6260): 746–48.

Kenward, R.E. (2001) *A Manual for Wildlife Radiotagging*. Academic Press, London.

Laake, J.L., Buckland, S.T., Anderson, D.R. & Burnham, K.P. (1993) *DISTANCE Users Guide*. Colorado Co-operative Fish and Wildlife Research Unit, Colorado State University, Fort Collins, CO 80523, USA.

Lythgoe, J.N. (1979) *The Ecology of Vision*. Clarendon Press, Oxford.

Macdonald, D.W. (1987) *Running with the Fox*. Unwin Hyman, London.

Macdonald, D.W., Bunce, R.G.H. & Bacon, P.J. (1981) Fox populations, habitat characterization and rabies control. *Journal of Biogeography* **8**(2): 145–51.

Marshall, G.J. (1998) CRITTERCAM: an animal-borne imaging and data logging system. *Marine Technology Society Journal* **32**(1): 11–17.

Ricketts, C. & Prince, P.A. (1984) Estimation by use of field weighings of metabolic-rate and food conversion efficiency in albatross chicks. *Auk* **101**(4): 790–5.

Rodgers, A.R., Rempel, R.S. & Abraham, K.F. (1995) Field trials of a new GPS based telemetry system. In *Biotelemetry XIII. Thirteenth International Symposium on Biotelemetry* (Eds C. Cristalli, J. Amlanar & M.R. Neuman), pp. 173–78, Williamsburg, Virginia, 26–31 March.

Rodgers, A.R., Rempel, R.S. & Abraham, K.F. (1996) A GPS-based telemetry system. *Wildlife Society Bulletin* **24**(3): 559–66.

Rodgers, A.R., Rempel, R.S., Moen, R., et al. (1997) GPS for Moose telemetry studies: a workshop. *Alces* **33**: 203–9.

Sackett, D.L., Rosenberg, W.M.C., Gray, J.A.M., Haynes, R.B. & Richardson, W.S. (1996) Evidence based medicine: what it is and what it isn't. *British Medical Journal* **312**: 71–2.

Stewart, P.D. & Macdonald, D.W. (2003) Badgers and badger fleas: strategies and counter-strategies. *Ethology* **109**(9): 751–64.

Stewart, P.D., Ellwood, S.A. & Macdonald, D.W. (1997) Remote video surveillance of wildlife: an introduction from experience with the European badger *Meles meles*. *Mammal Review* **27**(4): 185–204.

Sutherland, W.J., Pullin, A.S., Dolman, P.M. & Knight, T.M. (2004) The need for evidence based conservation. *Trends in Ecology and Evolution* **19**(6): 305–8.

Weimerskirch, H. & Wilson, R.P. (2000) Oceanic respite for wandering albatrosses. *Nature* **406**(6799): 955–6.

Williams, T.M., Davis, R.W., Fuiman, L.A., et al. (2000) Sink or swim: strategies for cost-efficient diving by marine mammals. *Science* **288**(5463): 133–6.

Wilson, R.P. & Liebsch, N. (2003) Up-beat motion in swinging limbs: new insights into assessing movement in free-living aquatic vertebrates. *Marine Biology* **142**(3): 537–47.

Wilson, R.P. & Wilson, M.P. (1990) Foraging ecology of breeding *Spheniscus* penguins. In *Penguin Biology* (Eds L.S. Davis & J.T. Darby), pp. 181–206. Academic Press, San Diego, California.

Wilson, R.P., Grant, W.S. & Duffy, D.C. (1986) Recording devices on free-ranging marine animals – does measurement affect foraging performance. *Ecology* **67**(4): 1091–3.

Wilson, R.P., Cooper, J. & Plotz, J. (1992) Can we determine when marine endotherms feed – a case-study with seabirds. *Journal of Experimental Biology* **167**: 267–5.

Wilson, R.P., Steinfurth, A., Ropert-Coudert, Y., Kato, A. & Kurita, M. (2002) Lip-reading in remote subjects: an attempt to quantify and separate ingestion, breathing and vocalisation in free-living animals. *Marine Biology* **140**: 17–27.

Wilson, R.P., Scolaro, A., Quintana, F., et al. (2004) To the bottom of the heart: cloacal movement as an index of cardiac frequency, respiration and digestive evacuation in penguins. *Marine Biology* **144**(4): 813–27.

8

Animal welfare and conservation: measuring stress in the wild

Graeme McLaren, Christian Bonacic and Andrew Rowan

And the poor beetle, that we tread upon
In corporal sufferance finds a pang as great
As when a giant dies.
(William Shakespeare (1564–1616), *Measure for Measure*, **1623.)**

Introduction

Why should there be a essay on animal welfare in a book about conservation biology? This essay aims to answer this question. An inspection of papers published in the journal *Animal Welfare* reveals that welfare scientists vary widely in their backgrounds and aims, for example, some want to examine the effects of transportation on cattle, others study animal consciousness, or aim to enrich the lives of zoo animals. Underlying most of this work is a conviction that the welfare of individual animals is important and worthy of consideration. We will have answered our original question if we can convince you, a reader of a book on conservation, to share this conviction. Our aim is to demonstrate that welfare science can, sometimes does and certainly should make an important contribution to practical conservation biology. We also briefly examine animal rights and consider why this sister discipline to animal welfare has as yet had little impact in conservation biology.

Welfare and conservation: incompatible competitors?

When Charles Elton wrote what was arguably the first scientifically based work on conservation, *The Ecology of Invasions by Animals and Plants*, he suggested that a case for conservation could be made on three grounds: to promote ecological stability; to provide a richer life experience; and because it is the right relation between humans and other living things. The first two of these effectively constitute the bulk of the argument made by the modern conservation movement. Of the final point, Elton also wrote 'there are some millions of people in the world who think that animals have a right to exist and be left alone, or at any rate they should not be persecuted and made extinct as a species.' If the first two of Elton's arguments are practical or quality of life arguments, the third is an ethical argument in that it relates to moral judgement. Although this may not have been his intention, Elton provides us with a starting point for the discussion of the ethics of wild animal conservation: according to this father of ecology, wild animals, at least in their natural environment, have a **right** to exist and be left alone. It is interesting that the leading philosophers of the animal rights movement take just this view, essentially believing in a policy of non-intervention towards wild animals whenever possible (e.g. Singer 1976). Conservation biologists on the other hand, have largely ignored this rights-based view. But what exactly are animal rights and, how do these relate to animal welfare?

The animal rights movement has its own large and developing literature, with a range of philosophies, and it is important to understand that this field is distinct from the scientific study of animal welfare. Indeed, many welfare scientists disagree with the conclusions reached by some proponents of the animal rights movement (and **vice versa**). However, most welfare scientists and animal rights philosophers agree that at least vertebrates can suffer (e.g. by feeling pain or fear), and can be referred to as sentient. This means that both the animal rights movement and welfare scientists accept that we should give proper **consideration** to a sentient animal before carrying out any activity that could negatively affect its welfare, and that we should do this for the animal's **own sake**. Only living animals that can suffer can have welfare considerations, and in this essay we do not consider death itself as a welfare issue, although the manner of death certainly is a welfare issue. Death as a part of conservation is considered in Chapter 15). For some animal rights philosophers, the logical extension of these views is that sentient animals and humans should be given equal consideration (Singer 1976). For others this means (at least in the utilitarian sense) that animals should not suffer **without good reason**. What 'good reason' means depends on your viewpoint and for some this loose definition (which in effect would allow any level of suffering given sufficient justification) is unacceptable. To give an animal rights (in this sense a moral right rather than a legal right) is therefore to accept that the

animal should be given consideration in decisions that may negatively affect it. How far this consideration can be taken, and the question of whether or not animals can have rights in the same sense as humans, are matters well beyond the scope of this essay.

If conservation biologists accept that it is desirable for an animal's welfare to be given consideration (by accepting that sentient animals have even a minimal right not to suffer) then it will be helpful to have relevant scientific insight into the animal welfare implications of conservation practice. However, even if this argument is not accepted, welfare science will still be a relevant field in conservation. From a completely practical perspective, welfare science could have use even if animals are considered not to have any rights. For example, improved welfare may increase the success of captive breeding and release programmes – irrespective of the animals' right not to suffer.

Thus the case for animal rights is not an essential component in the argument for welfare as a part of conservation. However, we believe that by accepting that it is desirable for animals not to suffer, welfare science can have an even wider sphere of usefulness in conservation biology. Conservationists have typically avoided animal rights or welfare-based arguments, sometimes seeing them as less powerful than, or even irrelevant to, other (ecological or quality of life) arguments, or even as tactically harmful. For example, by arguing that the rights of animals are important conservationists might fear they could alienate those people directly affected by conservation decisions, if the people believe their rights have been neglected (e.g. see Hambler 2003). Typically, welfare is emphasized in conservation judgements only if a species is so rare that the well-being of individuals is a prerequisite to the survival of their population or species, as was the case with the Californian condor. Some animal rights philosophers have attacked conservationists for their concentration on species rather than individuals (Regan 2004), but interventions carried out for the welfare of individual animals may be seen as inappropriate by conservationists if they affect natural processes. This view certainly does not imply callousness – the judgment that it is inappropriate to intervene is not incompatible with despairing for the victim of these natural processes.

Perhaps the potential confusions in these philosophically very difficult issues has contributed to the fact that welfare science has yet to enter mainstream conservation (another contributory factor has been the inadequacy of practical tools whereby to apply welfare science to conservation – but we will show below how that is changing). However, a consideration of the welfare of individual animals may be significant in achieving conservation aims, for example, in the reintroduction and relocation of endangered species. Furthermore, the very fact that conservationists have tended to eschew deep thinking about welfare, is why we should grasp this issue here – the concept of Key Topics essays being to expose, and tackle, the awkward dilemmas that permeate conservation.

In this essay, we consider the role of animal welfare science in conservation biology, rather than the role of animal rights. This is because we believe, given the current extinction crisis, that welfare is of immediate relevance to conservation biology. We do not have space to elaborate the philosophical analysis of the role of animal rights in conservation, a key topic for the future.

Is welfare relevant to conservation practice?

By the time Elton was writing in 1958, human impacts on the environment were already so great, and species extinction so widespread, that interventions were necessary to protect many species. Given the continued and extensive human impact upon the environment since that time, and the current extinction crisis, conservation is now largely a science of intervention and management. However, it is clear that conservation interventions such as trapping, translocation and radio-collaring have an impact on their subjects: even observation may have an effect (Macdonald & Dawkins 1981). We believe that as conservation biology battles with the increasingly complex challenges of the twenty-first century, new issues will arise that will ensure that welfare science plays a much more important and elevant role in conservation biology.

An example of the dilemmas that are likely to face conservation biologists in the future comes from the creation of fenced reserves for wild dogs (*Lycaon pictus*) in South Africa (van Dyk & Slotow 2003). Wild dogs appear to do well in these reserves, even using the boundary fence as an aid to capture prey (an advantage to the dogs, but an ethical worry to the managers). However, these isolated reserves offer no opportunity for natural dispersal, and therefore this must take place through translocation of individuals between reserves (see Chapter 5). Similarly, future management of these reserves may require interventions involving other species within the reserve, including the control of potential predators such as lions. These reserves can be highly successful and it is likely that similar strategies will be used to conserve other endangered species. However, at diverse levels the management of wild dog reserves raises welfare issues. At what point, for example, should conservationists intervene to save wild dogs injured by lions? Should lion numbers be controlled to promote wild dogs? Should prey species be introduced? What are the welfare implications of translocating individuals to new reserves?

Attitudes of conservation biologists to animal welfare

What are the attitudes of conservation biologists to animal welfare? One insight, at least by the standards of 1988, was revealed when President Reagan commissioned an ice-breaker to create a passage to open water for three ice-bound whales trapped in Barrow, Alaska. Of marine biologists surveyed, only one-third viewed this action as positive, and then largely for the reason that it focused public attention on an endangered species (Anonymous 1990). Thus conservation biologists appear to place, in an ethical sense, a strong emphasis on wild-life being left alone (even if this means death): in the Kruger National Park in South Africa, animals injured and harmed through natural causes are left unattended by Park veterinarians, while those harmed by human actions are, when they were found, treated and cared for (D. Grobler, personal communication 2001).

Contrasting with these views is the more general concern of the wider public (although perhaps a very recent, largely western, view) whose sympathy for, and interest in, wildlife in general, has led to the creation of many organizations that deal directly with the welfare of individual wild animals. The people who create wildlife hospitals and similar centres have a 'right to live' bias that can dismay conservation biologists, particularly when introduced species are found in wildlife hospitals (although such programmes may encourage local populations to have a greater awareness of conservation issues; e.g. Drews 2003). Conversely, the 'right to be left alone' bias of conservation biologists (more precisely expressed as the conviction that natural processes should run their course without human intervention) dismays many who care deeply about animals. This conflict of ethics appears to be at the heart of the conflict between welfare and conservation – the individual versus the population, and the right to live against the right to be left alone. Of course this dichotomy is simplistic (Rowan 1996), but serves to highlight a legitimate tension between the priorities of welfare and conservation. Establishing a working relationship (ideally a synthesis) between welfare and conservation will require the development of a shared ethical viewpoint that can be used as a starting point for intervention decision-making.

An ethical framework for conservation intervention

Accepting that intervention, with potential animal welfare implications, is necessary for much of conservation biology, a combined welfare and conservation ethic could be that the rights to exist and be left alone should only be waived if there is sufficient justification. A conservation perspective might be that the intervention should take place only if there is sufficient justification in terms of conservation benefit to demonstrably outweigh any welfare implications. A welfare perspective would place the emphasis on the individual: the intervention should take place only if there is sufficient justification in terms of individual benefit to demonstrably outweigh any welfare implications. How can these positions be reconciled? Remembering that species and populations are made up of individuals helps us to reach agreement. Many conservation interventions ultimately benefit or protect individual animals. Even where this may not appear to be the case, such as with the control of introduced or pest species, ultimately, the aim of such conservation interventions is to benefit one or more species, comprised of individual animals. The spectrum of opinion regarding the level of equivalence between the rights of other animals and humans that it is inevitable and obvious that the killing of one species to protect another (whatever the conservation justification) will be unjustifiable to some shades of opinion, and sometimes justifiable to others. However, this should not prevent welfare science from becoming a part of conservation biology.

As wildlife comes under increasing human pressure and is further squeezed into managed refuges, as natural processes within our environment are increasingly managed and affected by humans, and as the line between nature reserve and zoo blurs, judging the balance between the rights to exist and be left alone and the need to intervene will become harder – and the need for validated welfare measures greater. Accepting the reality that conservation biology is a science of intervention and management, as much as, if not more than, protection and preservation, it is therefore pertinent to ask how great are these impacts, can they be reduced and are they justifiable? Ethically and also scientifically (because stress may distort interpretation), animal welfare is part of conservation. The fact that welfare is measured on an individual basis, whereas some wider goals of conservation are measured on populations does not necessarily (as some believe) make them incompatible, although people may differ in the emphasis they attribute to welfare when evaluating conservation strategies. If welfare is to be weighed in this balance, then it must be measured, and so the nub of this essay is to ask how such measures can be made. We also explore the relationship between the sciences of conservation and animal welfare and ask: how they might combine for mutual benefit?

Measuring stress: an evidence-base for welfare in conservation biology

So how can welfare be defined and measured? Animal welfare is most often defined as the ability of an animal to cope with a given situation or environment (e.g. Fraser & Broom 1980). The ability to cope is normally measured by assessing how stressed an animal is: stress occurs when an animal perceives a threatening situation. The brain is the starting point for the stress response, a sweeping physiological event that brings about metabolic, neuroendocrine, immune and behavioural changes (e.g. Dhabhar et al. 1995; Beerda et al. 1996; Bateson & Bradshaw 1997; Rushen 2000; Möstl & Palme 2002; McLaren et al. 2003; Goymann & Wingfield 2004; Romero 2004; Sands & Creel 2004). The stress response is not inherently damaging:

ultimately its function is to keep the animal alive in times of danger. However, it can be costly, both in terms of energy and because of the potential for some components of the stress response (particularly parts of the immune response) to cause physical damage. The biological cost of mounting a stress response, in terms of both energy and damage, depends upon the frequency and intensity of the stressor (Laugero & Moberg 2000; Moberg 2000; Montes et al. 2003, 2004), and here we argue that this biological cost should be the key currency used in determining the impact of human interventions on wild animals.

Types of stress and their consequences for the animal

Stress may be, broadly, chronic or acute. Acute stress is a response to a short-term stressor, after which it wanes rapidly and is followed by homeostasis. Chronic stress is prolonged and results in a long-term or possibly permanent shift in an animal's physiology in response to either repeated or constant stress (for a review see Moberg 2000). Acute stress brings about a 'flight or fight' style response, typified by the production of hormones such as cortisol and adrenaline, but also accompanied by other physiological and immunological changes: these are rapidly reversible, and in some cases (e.g. cortisol) are controlled by their own negative feedback system. Acute stress responses are adaptations to combat immediate danger. Chronic stress has greater potential to affect survivorship, growth and reproduction. In conservation, interventions that might be acute stressors include trapping and short-term transport, whereas chronic stressors could include captivity and translocation.

A stressor has a welfare implication when an animal's coping mechanisms impose such a large biological cost that other non-stress functions are impaired, or when the coping mechanisms themselves become harmful. Mounting a costly stress response could therefore affect an animal's future fitness and survival. For example, female rhinos subjected to repeated immobilization suffer from reduced fertility that is most likely to be stress-related (Alibhai et al. 2001).

To cover the costs associated with stress an individual must either use up stored biological reserves or, if these are insufficient, it must divert resources that would otherwise be used by other bodily functions – a situation defined as distress (Moberg 2000). This model predicts that the impact of a given stressor will differ between individuals, depending not only on the level of their stored reserves, but also upon their current pattern of energetic expenditure. The costs of mounting a stress response may be greater (and so have greater welfare implications) during energetically costly periods such as growth and reproduction, or in animals suffering from parasites or other diseases. Moberg (1996, 2000) argues that the key to determining the welfare implications of a stressor is to determine its biological cost, in terms of the utilization of biological reserves and resource diversion, and that this may be a more relevant measure of stress than specific physiological or behavioural changes.

Measures of stress using physiological approaches therefore can be broadly divided in two:

1. cost measures – those based on the biological cost of stress;
2. defence measures – those based on the magnitude of the defence system raised against the stressor.

Each is valid, but they differ in applicability. For conservation science, the crucial question is whether an animal has suffered a cost that will impair its fitness or survival. In principle, the greater the demonstrable cost of the conservation intervention, the greater is the need for it to be justified in terms of conservation benefit.

Current issues and approaches in stress measurement

Non-invasive measures of stress

Because, by definition, non-invasive measures of stress obviate capture and handling, they avoid the associated risks to the animal, the distortions of basal levels and, potentially, save time and money.

FAECAL CORTISOL ANALYSIS

The steroid hormone cortisol (and the related corticosterone) is released into the bloodstream in response to stress, and provides a measure of the defence system raised in response. The sample can be obtained from blood (normally invasively) or non-invasively from faeces, which provide a daily summary of metabolized, and thus actively used, hormone (Möstl & Palme 2002). However, hormones in faeces have experienced breakdown processes that complicate assay procedures (Möstl & Palme 2002), and necessitate validation, which involves both assay validation (concerning accuracy), followed by physiological validation (concerning interpretation).

Buchanan & Goldsmith (2004) should be consulted for the technical aspects of assay validation techniques. In addition, extracting the hormone from the sample, and storage, both risk bias and also require validation. Physiological validation is concerned with understanding the relationship between the production and excretion of the hormone, as illustrated by Touma et al.'s (2003) study of laboratory mice injected with [3]H – labelled corticosterone. Their faeces and urine were collected over the next five days and faecal ([3]H – labelled) corticosterone metabolites identified using high performance liquid chromatography (HPLC). Over 20 different metabolites were detected, and these differed between, but not within, sexes. Of the total excretory corticosterone, males excreted relatively more (73%) than did females (53%) via faeces than urine. The HPLC fractions were then screened with four different enzyme-immunoassays: one using a commercial corticosterone antibody, one using an 'in house' corticosterone antibody, and two antibodies designed specifically to bind to known (but different) faecal metabolites of corticosterone. Only one assay (a specific faecal metabolite assay) detected radioactive metabolites with high intensity. The other assays showed only minor reactivity.

These results have major implications. First, validation is crucial (in this case it revealed the sex differences). Second, the types of metabolite excreted must be understood, as these may not be constant either between individuals or even within an individual's lifetime. In practice, and despite the obvious advantages of non-invasive faecal hormone analysis as a measure of stress, the careful validation described by Touma et al. (2003) may be impossible in the field. One practical approach would be to use HPLC as described above on fresh faeces to identify likely corticosterone metabolites, and then screen fresh faeces using enzyme or radio-immunoassay for those metabolites. For samples from known individuals it would thus be possible to look for differences in metabolite production. We explored this approach for badgers, *Meles meles*, and found that a known cortisol metabolite, 11-oxoetiocholanolon (11-oxo-T), appeared in the faeces of both sexes (Fig. 8.1). To learn whether this metabolite increases in the faeces in response to stress we could have used adrenocorticotrophic hormone (ACTH) challenge (e.g. Bonacic et al. 2003): ACTH is the starting point for a stress response and causes the release of cortisol/corticosterone. Of course, there is a judgement in the costs and benefits of using this approach because it deliberately induces stress. Instead, we chose to use faeces from badgers that were

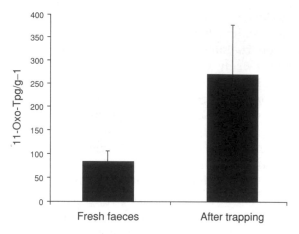

Fig. 8.1 The cortisol metabolite 11-Oxo-T was identified in badger faeces using high performance liquid chromatography (HPLC), and was found in increased levels in faeces from animals that were subjected to capture and transport (stress), in comparison with fresh faeces taken from the animals' latrines in the period before trapping ($n = 10$ for each group). This represents two steps in the process of validating this metabolite as a measure of the stress response in badgers (other steps are detailed in the text).

already being trapped and handled as part of a wider study (McLaren et al. 2003). We were able to test whether the trapping and handling stress was associated with an increase in 11-oxo-T in badger faeces. It was (Fig. 8.1). Therefore, 11-oxo-T is a promising metabolite for use in studies of adrenal activity and stress in wild badgers.

In summary, before using faecal hormone metabolites to measure stress in wild animals one must:

1. obtain as much information as possible about the actual metabolites that are found in faeces;
2. determine if there are differences between individuals;
3. be aware of the time delay that occurs between the stress response and the appearance of metabolites in faeces;
4. ensure all assays are fully validated;
5. take care with the interpretation of results.

BODY WEIGHT CHANGE AS A MEASURE OF STRESS

Body weight is a minimally invasive potential measure of the cost of stress. Energetically costly stress responses cause body weight loss when fat and other energy stores are utilized (e.g. McLaren et al. 2004). For example, adult rats subjected to a moderate stressor of 3 hours restraint for three consecutive days suffer a reduction in body weight and also reduce their food intake (Harris et al. 1998; Zhou et al. 1999). During the three days of restraint, mean body weight loss was approximately 5% (c.20 g) of initial weight, and much of this loss occurred one day after the first period of restraint, although restrained animals continued to have lower mean body weights than controls for as long as 40 days after the stressors were applied. There were three stages to weight loss (Harris et al. 2002): a period of weight loss during restraint; a period of reduced food intake following the end of restraint; and a following period of normal food intake but reduced body weight. The mechanisms responsible for these observed patterns are complex and probably involve interactions between stress-related hormones and other hormones that affect food intake, including growth hormone and prolactin (Harris et al. 2002).

Body weight change, therefore, has the potential to be used as a measure of the cost of capture and handling stress in wild mammals, especially where animals are routinely captured (e.g. see Tuyttens et al. 2002; McLaren et al. 2004). McLaren et al. (2004) subjected wild wood mice (*Apodemus sylvaticus*) and bank voles (*Clethrionomys glareolus*) to one of two handling regimes, one of which was putatively more stressful (trapping, handling and anaesthesia) than the other (trapping and handling). This experiment revealed that wood mice subjected to the minimal stressor gained weight overnight (probably because of the high quality food used to bait the trap), whereas wood mice subjected to the more intensive stressor

lost weight overnight, albeit a very small amount: mean of 0.46 g (SE = 1.17 g). Furthermore, there were differences between species – bank voles gained weight in response to both treatments – perhaps because bank voles have a relatively higher daily energy budget and lower production efficiency than do wood mice and may have less scope to reduce food intake in response to stress.

This study indicates not only that body weight change can be an indicator of stress, but also that the **absence** of weight loss does not necessarily indicate that there has been no biological cost: animals might maintain weight but divert resources from other bodily functions such as reproduction.

REMOTE MONITORING DEVICES AND ANIMAL BEHAVIOUR

Blood samples can be taken from free-ranging animals using remote blood sampling devices such as a 'DracPac' (Cook et al. 2000). In such systems a blood sampling system is attached to the animal using a harness and programmed to take blood samples as required. Such systems can provide detailed blood biochemistry data from free-ranging animals, although this technology lends itself to some species more than others (particularly large species held in captivity). Limitations include: the large size of the sampling unit; the requirement (in some cases) for the animal to be captured in order to retrieve the sample; and the risk to the animal from the device itself. Studies using remote blood sampling devices on farmed red deer (*Cervus elaphus*) illustrate both the value and limits of the technique. A DracPac device allowed the collection of highly detailed data on the daily circadian rhythm of cortisol production in red deer, as well as revealing cortisol responses to stressful activities such as transport (Cook et al. 2000). However, this device weighs approximately 2 kg, and is very intrusive for use in wild animals. There is no doubt that these devices can reveal significant detail in the responses of animals to stress, but at present it isunlikely that such devices would be deemed suitable for conservation welfare research. Other, less intrusive devices measure heart rate and body temperature, and are less invasive, but still require an animal to carry a potentially stressful burden (see Tuyttens et al. 2002).

Ultimately, one of the goals of the stress response is to bring about a change in behaviour, and thus behaviour should provide a measure of welfare. However, inferring welfare from behavioural observations is problematic because of the complexity of a behavioural response and ignorance of its physiological causes (Rushen 2000). Nevertheless, advances in using behavioural measures of welfare in farm animals (Dawkins 2004) have demonstrated the potential for these methods to be used in other situations. For example, using a combination of ecological and behavioural techniques, it was demonstrated that free-range hens are attracted to trees and are more likely to utilize outdoor space when trees are present (Dawkins et al. 2003). Because hens will use outdoor space if they feel safe to do so (with tree cover) this indicates that trees are important to their welfare, and that their welfare suffers if they are absent. Potentially, even relatively simple data on animal habitat selection and activity could be used to monitor welfare, if, as in the case of the hens, it can be interpreted appropriately.

Measuring stress in captured animals

HORMONAL AND OTHER BLOOD PARAMETER MEASURES

Capture and handling are often unavoidable in practical conservation (and most wildlife research), and almost inevitably cause stress. However, if the stress can be measured protocols can be selected to minimize it. Bonacic & Macdonald (2003) studied the traditional practice of capturing, transporting and shearing wild vicuna (*Vicugna vicugna*), an Andean camelid with fine, valuable wool. This study provided a revealing example of the direct interaction between conservation and welfare. The vicunas were facing extinction and were protected with the promise that locals would one day make money out of their fleeces. However, would shearing involve excessive welfare implications for the vicuna, and could the stress of alternative protocols be measured to reveal the least damaging? Bonacic & Macdonald (2003) found that capture and transport significantly affected haematological parameters including neutrophil:lymphocyte ratio, glucose, cortisol, creatine kinase (an enzyme important in energy regulation) and aspartate aminotransferase (an enzyme that regulates glutamate, an important neurotransmittor). However, they detected no haematological differences between vicunas that were then sheared and those that were not sheared. This suggests that shearing did not add to the burden of the other two stressors. An additional group of vicuna was shorn 12 days after capture, and these showed no significant differences in their haematological parameters in comparison with a control group (handled for sampling but not shorn), with the exception of cortisol and aspartate aminotransferase, which were higher in sheared animals. Shearing itself therefore does not appear to be an additional stress burden, but clearly capture and transport are stressors for vicuna.

Teasing apart the stressful components of a trapping and handling procedures is important. The use of vicuna for their fleece has a demonstrable impact on their welfare, but may become the commercial incentive that allows them to exist at all. In the balance of judgements, the welfare measures allow us to understand fully the impacts of capture, transport and shearing and balance this against the fact that vicuna survival may depend upon this intervention.

On another level this study also demonstrated the species-specific nature of the stress response. Comparisons of basal cortisol, and cortisol responses to ACTH challenge in different ungulate species revealed wide differences in both parameters (Bonacic & Macdonald 2003), indicating the difficulties of comparing stress responses between species, and the impossibility of inferring basal cortisol values of one species from those of another.

IMMUNE RESPONSES AS MEASURES OF STRESS

Both acute and chronic stress affects the immune system, causing changes in circulating immune cell number and composition, as well as affecting immune cell activity and responsiveness (e.g. Dhabar et al. 1995; Goebel & Mills 2000; Ellard et al. 2001; Bonacic & Macdonald 2003; McLaren et al. 2003; Mian et al. 2003). Because stress-induced immune changes are mostly associated with the down-regulation or suppression of the immune system (immunosuppression), stress can affect the ability of the immune system to deal with infection, leading to disease (e.g. Råberg et al. 1998; Moberg 2000). Stress-induced immune changes are therefore highly important in determining the welfare consequences of a stressor, and therefore immunosuppression can be used as a measure of the cost of stress (McLaren et al. 2003). One such approach has been based on a component of the stress response of some leukocytes, particularly neutrophils, which involves the release of reactive oxygen species (ROS) – potent oxidizing chemicals that act as bactericidal agents (McLaren et al. 2003; Mian et al. 2003; Montes et al. 2004). However, ROS can also cause tissue damage (e.g. Boxer & Smolen 1988) and the activation of neutrophils has been reported to be potentially detrimental to health (e.g. Kruidenhier & Verspaget 2002).

Given the potential for ROS to cause tissue damage, it is not surprising that the release of ROS in response to stress is strictly controlled; only a subpopulation of neutrophils is activated to produce ROS, and the size of the sub-population is related to the intensity of the stressor (the activated subpopulation can be quantified using a simple blood smear staining technique, called the Nitroblue Tetrazolium Test (NBT); Montes et al. 2004). Recent work has demonstrated that the non-activated neutrophils and other leukocytes in the circulation suppress their production of reactive oxygen species, and this form of immunosuppression can be used as the basis for a new measure of stress. The approach is technically simple: a small sample of whole blood (10 μl) is challenged with phorbol myristate acetate (PMA), a chemical that normally causes leukocytes to produce ROS. The resulting ROS production is measured over a period of about 30 minutes as luminol (a chemical that produces light in the presence of ROS) enhanced chemiluminescence (Fig. 8.2; McLaren et al. 2003). This technique has been called leukocyte coping

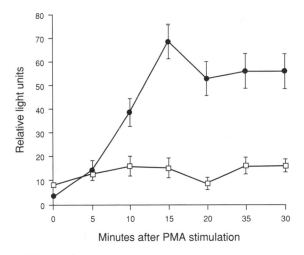

Fig. 8.2 The leukocyte coping capacity (LCC) response of transported (lower line) and non-transported badgers ($n = 8$ in both groups, means presented with standard error bars). Transport brings about a marked reduction in LCC, and in the method described here, reaction reactive oxygen species (the basis of the measure of LCC) react with a chemical to produce light which is then measured in relative light units (see McLaren et al. 2003). PMA, phorbol myristate acetate.

Table 8.1 A summary of the available methods for assessing stress and welfare. This is not intended to be an all-inclusive list. The decision of which type of measure to use will depend on (as a minimum) the study species, equipment availability and the nature of the stressor imposed

Measure of stress	Acute	Chronic	Issues that these techniques could be used to solve
Faecal cortisol metabolites	Increased levels in faeces followed by return to pre-stress levels. There is delay between the stressor and the appearance of metabolites in faeces: timing depends largely on gut passage time	Elevated levels in faeces over long periods	Long-term monitoring of translocated or captive individuals; acute stress responses
Behavioural observations	Flight or fight style responses, but these may be of limited value if the animal is in captivity or restrained	Over long periods behaviour may change considerably, possibly affecting reproduction, foraging and survival	Long-term monitoring
Body weight	Short-term, small weight losses may occur, depending on nature of stressor and condition of animal	Long-term changes variable; reduction in food intake and slow growth of juveniles. Body weight may not be lost if energy is diverted away form other functions (e.g. reproduction)	For acute stress response, of most use when individuals are likely to be recaptured as part of the study design. Longer term monitoring may be possible, either through regular capture or through non-invasive methods
Remote monitoring	Increase in heart rate, blood pressure, body temperature	Long-term remote monitoring of blood samples is probably a welfare issue in itself; longer-term monitoring of changes in heart rate, blood pressure and body temperature are potentially useful, although this may also be welfare issue	Issues with remote monitoring units restrict use over the long-term; can be best used for acute stress responses. It is advantageous to have basal and adrenocorticotrophic hormone (ACTH) response values for some parameters
Blood parameters	Widespread changes: plasma cortisol and adrenalin, changes in cell number and distribution.	Changes over long period are complex, but may include elevated cortisol and changes in energy metabolism	Long-term and acute stress monitoring, but requirement to capture individual at regular intervals may restrict use
Immune parameters	Changes in leukocyte activation, changes in cell number and distribution.	Immunosuppression over the long term characterised by increased incidence of parasites and disease	

capacity (LCC). Put simply, this technique directly measures the capacity of circulating leukocytes to produce ROS.

Leukocyte coping capacity can reveal significant information about an animal's physiological status during and after stressful events (McLaren et al. 2003). Short-term psychological stressors can produce measurable changes in neutrophil activation (Fig. 8.2; Ellard et al. 2001; McLaren et al. 2003; Mian et al. 2003; Montes et al. 2003, 2004). Even students 'experimentally exposed' to a horror film, *The Texas Chainsaw Massacre,* exhibited increased levels of neutrophil activation, as well as a whole suite of other changes that indicated that watching this film was stressful (Mian et al. 2003).

In an example of weighing a conservation option, McLaren et al. (2003) and Montes et al. (2004) examined the effect of transporting wild badgers (*Meles meles*) from their site of capture to a fully equipped field laboratory (about 10 minutes on a trailer pulled by an all-terrain quad bike). The field laboratory has equipment and support that make processing badgers easier and safer for both human operators and the animals, but does transport have a significant welfare impact on badgers? The NBT was used to measure the effect of transport, and revealed that transport did bring about neutrophil activation. Transported badgers also showed a lower LCC response than a control group of badgers that had been trapped, but not transported (Fig. 8.2; McLaren et al. 2003). It was also demonstrated that resting the animals for 30 minutes or more in a quiet, dark place after transport reduced their levels of neutrophil activation. This work indicates that (perhaps unexpectedly) transport stress is additional to capture stress and that transport affects their immune system. In evaluating the balance of options, it was considered that transport should be continued, but with the addition of a resting period between transport and anaesthesia.

Solutions: a validated approach to measuring stress in the wild

The stress response affects nearly all of an animal's physiology: and as such many different measures of stress could be used. The decision to use a particular measure of stress will depend on several factors (see Table 8.1). Table 8.1 summarizes some of the possible measures that could be used, and in most situations a combined approach using two or more of these measures would be recommended. For example, where a non-invasive approach is required, it may be possible to use body weight monitoring alongside non-invasive hormone monitoring. Extended periods of elevated cortisol production combined with weight loss or slow growth, or with increased disease incidence, would indicate that stress is causing welfare problems.

This sort of general monitoring approach has one major problem – the intervention must be carried out before the welfare impact can be assessed. This situation will be improved through the publication of research results, allowing conservation biologists to better predict the likely impact of their work. Nevertheless, by monitoring the welfare of their study animals, biologists will be in a position to amend or even cease intervention work. Ultimately, however, welfare science must be able to provide biologists with the tools and predictive power to be able to plan successful conservation interventions. We judge that conservationists have a duty to evaluate whether the animal welfare implications of their work are outweighed by conservation benefits. The potential for welfare and conservation to combine to produce effective science and decision making is clear: the success of this combination will be judged upon the technical developments produced by welfare science and ultimately their successful transfer into conservation action.

The welfare of each is bound up in the welfare of all.
(Helen Keller, 1880-1968)

References

Alibhai, S.K., Jewell, Z.C. & Towindo, S.S. (2001) Effects of immobilization on fertility in female black rhino (Diceros bicornis). *Journal of Zoology* **253**: 333–45.

Anonymous. (1990) The conference in Monaco. *Anthrozoos* **3**: 197–201.

Bateson, P. & Bradshaw, E.L. (1997) Physiological effects of hunting red deer (*Cervus elaphus*). *Proceedings of the Royal Society of London Series B – Biological Sciences* **264**: 1707–14.

Beerda, B., Schilder, M.B.H., Janssen, N. & Mol, J.A. (1996) The use of saliva cortisol, urinary cortisol, and catecholamine measurements for a noninvasive assessment of stress responses in dogs. *Hormones and Behavior* **30**: 272–9.

Bonacic, C. & Macdonald, D.W. (2003) The physiological impact of wool-harvesting procedures in vicunas (*Vicugna vicugna*). *Animal Welfare* **12**: 387–402.

Bonacic, C., Macdonald, D.W. & Villouta, G. (2003) Adrenocorticotrophin-induced stress response in captive vicunas (Vicugna vicugna) in the Andes of Chile. *Animal Welfare* **12**: 369–85.

Boxer, L.A. & Smolen, J.E. (1988) Neutrophil Granule Constituents and Their Release in Health and Disease. *Hematology-Oncology Clinics of North America* **2**: 101–34.

Buchanan, K.L. & Goldsmith, A.R. (2004) Noninvasive endocrine data for behavioural studies: the importance of validation. *Animal Behaviour* **67**: 183–5.

Cook, C.J., Mellor, D.J. Harris, P.J., Ingram, J.R. & Matthews, L.R. (2000) Hands-on and hands-off measurement of stress. In *The Biology of Animal Stress Basic Principles and Implications for Animal Welfare* (Eds G.P. Moberg & J.A. Mench), pp. 123–46. CABI Publishing, Wallingford.

Dawkins, M.S. (2004) Using behaviour to assess animal welfare. *Animal Welfare* **13**: S3–S7.

Dawkins, M.S., Cook, P.A., Whittingam M.C., Mansell, K.A. & Harper, A.E. (2003) What makes free-range broilers range? *In situ* measurement of habitat preference. *Animal Behaviour* **66**: 151–60.

Dhabhar, F.S., Miller, A.H., McEwen, B.S. & Spencer, R.L. (1995) Effects of stress on immune cell distribution – dynamics and hormonal mechanisms. *Journal of Immunology* **154**: 5511–27.

Drews, C. (2003) The state of wild animals in the minds and households of a neotropical society: the Costa Rican case study. In *The State of the Animals II: 2003* (Eds D.J. Salem & A.N. Rowan), pp. 193–205. Humane Society Press, Washington, DC.

Ellard, D.R., Castle, P.C. & Mian, R. (2001) The effect of a short-term mental stressor on neutrophil activation. *International Journal of Psychophysiology* **41**: 93–100.

Fraser, A.F. & Broom, D.M. (1980) *Farm Animal Behaviour and Welfare*, 3rd edn. Bailliere Tindall, London, p. 256.

Goebel, M.U. & Mills, P.J. (2000) Acute psychological stress and exercise and changes in peripheral leukocyte adhesion molecule expression and density. *Psychosomatic Medicine* **62**: 664–70.

Goymann, W. & Wingfield, J.C. (2004) Allostatic load, social status and stress hormones: the costs of social status matter. *Animal Behaviour* **67**: 591–602.

Hambler, C. (2003) *Conservation*. Cambridge University Press, Cambridge.

Harris, R.B.S., Mitchell, T.D., Simpson, J., Redmann, S.M., Youngblood, B.D. & Ryan, D.H. (2002) Weight loss in rats exposed to repeated acute restraint stress is independent of energy or leptin status. *American Journal of Physiology-Regulatory Integrative and Comparative Physiology* **282**: R77–R88.

Harris, R.B.S., Zhou, J., Youngblood, B.D., Rybkin, I.I., Smagin, G.N. & Ryan, D.H. (1998) Effect of repeated stress on body weight and body composition of rats fed low- and high-fat diets. *American Journal of Physiology-Regulatory Integrative and Comparative Physiology* **44**: R1928–38.

Kruidenier, L. & Verspaget, H.W. (2002) Review article: oxidative stress as a pathogenic factor in inflammatory bowel disease – radicals or ridiculous? *Alimentary Pharmacology and Therapeutics* **16**: 1997–2015.

Laugero, K.D. & Moberg, G.P. (2000) Energetic response to repeated restraint stress in rapidly growing mice. *American Journal of Physiology – Endocrinology and Metabolism* **279**: E33–E43.

Macdonald, D.W. & Dawkins, M. (1981) Ethology – the science and the tool. In *Animals in Research: New Perspectives in Animal Experimentation* (Ed. D. Sperlinger), pp. 203–224. Wiley, Chichester.

McLaren, G.W., Macdonald, D.W., Georgiou, C., Mathews, F., Newman, C. & Mian, R. (2003) Leukocyte coping capacity: a novel technique for measuring the stress response in vertebrates. *Experimental Physiology* **88**: 541–6.

McLaren, G.W., Mathews, F., Fell, R., Gelling, M. & Macdonald, D.W. (2004) Body weight change as a measure of stress: a practical test. *Animal Welfare* **13**: 337–41.

Mian, R., Shelton-Rayner, G., Harkin, B. & Williams, P. (2003) Observing a fictitious stressful event: Haematological changes, including circulating leukocyte activation. *Stress – the International Journal on the Biology of Stress* **6**: 41–47.

Moberg, G.P. (1996) Suffering from stress: an approach for evaluating the welfare of an animal. *Acta Agriculturae Scandinavica Section A – Animal Science* **27**(Supplement): 46–9.

Moberg, G.P. (2000) Biological response to stress: implications for animal welfare. In *The Biology of Animal Stress Basic Principles and Implications for Animal Welfare* (Eds G.P. Moberg & J.A. Mench), pp. 1–21. CABI Publishing, Wallingford.

Montes, I., McLaren, G.W., Macdonald, D.W. & Mian, R. (2003) The effects of acute stress on leukocyte activation. *Journal of Physiology* **54P**: 88P.

Montes, I., McLaren, G.W., Macdonald, D.W. & Mian, R. (2004) The effect of transport stress on leukocyte activation in wild badgers. *Animal Welfare* **13**: 355–9.

Möstl, E. & Palme, R. (2002) Hormones as indicators of stress. *Domestic Animal Endocrinology* **23**: 67–74.

Råberg, L., Grahn, M., Hasselquist, D. & Svensson, E. (1998) On the adaptive significance of stress-induced immunosuppression. *Proceedings of the Royal Society of London Series B – Biological Sciences* **265**: 1637–41.

Regan, T. (2004) *The Case for Animal Rights*, 2nd edn. University of California Press, Berkeley, CA.

Romero, L.M. (2004) Physiological stress in ecology: lessons from biomedical research. *Trends in Ecology and Evolution* **19**: 249–55.

Rowan, A.N. (Ed.) (1996) *Wildlife Conservation, Zoos and Animal Protection: Examining the Issues*. Tufts Center for Animals and Public Policy, School of Veterinary Medicine, North Grafton, MA.

Rushen, J. (2000) Some issues in the interpretation of the behavioural responses to stress. In *The Biology of Animal Stress Basic Principles and Implications for Animal Welfare* (Eds G.P. Moberg & J.A. Mench), pp. 23–42. CABI Publishing, Wallingford.

Sands, J. & Creel, S. (2004) Social dominance, aggression and faecal glucocorticoid levels in a wild population of wolves, *Canis lupus*. *Animal Behaviour* **67**: 387–96.

Singer, P. (1976) *Animal Liberation: Towards an End to Man's Inhumanity to Animals*. Jonathan Cape, London.

Touma, C., Sachser, N., Mostl, E. & Palme, R. (2003) Effects of sex and time of day on metabolism and excretion of corticosterone in urine and feces of mice. *General and Comparative Endocrinology* **130**: 267–78.

Tuyttens, F.A.M., Macdonald, D.W. & Roddam, A.W. (2002) Effects of radio-collars on European badgers (*Meles meles*). *Journal of Zoology* **257**: 37–42.

Van Dyk, G. & Slotow, R. (2003) The effects of fences and lions on the ecology of African wild dogs reintroduced to Pilanesberg National Park, South Africa. *African Zoology* **38**: 79–94.

Zhou, J., Yan, X.L., Ryan, D.H. & Harris, R.B.S. (1999) Sustained effects of repeated restraint stress on muscle and adipocyte metabolism in high-fat-fed rats. *American Journal of Physiology – Regulatory, Integrative and Comparative Physiology* **277**: R757–66.

9

Does modelling have a role in conservation?

Mark Boyce, Steven Rushton and Tim Lynam

Historians of science often observe that asking the right question is more important than producing the right answer.

(E.O. Wilson, *Cosilience*, 2003.)

Introduction

Modelling is an imprecise term that means different things to different people. In its widest sense it involves describing the functioning of a system numerically. But modelling should describe not only the functioning of a system, it should do so with the smallest number of system components or variables necessary. Amongst conservation applications, models have been used to map species distributions (Guisan & Zimmermann 2000), evaluate the consequences of alternative harvesting policies (Hunter & Runge 2004), predict the course of disease (Dobson & Meagher 1996) and to anticipate the risk of extinction (Boyce 1992a; Ludwig 1999). At its most incisive, modelling can be used to predict the future behaviour – in this context – of an ecological system.

Can modelling help practical conservation? Starfield (1997) suggested a list of misconceptions that have led to reluctance to adopt modelling as a conservation tool. For example, models often are considered to be difficult and to require a complete understanding of the system and sufficient data to describe it. If this parody of modelling were correct there might be an unbridgeable gap between the modeller's abstract mathematics and the conservationist's need for action in the face of complex nature. In reality, modelling has diverse uses. At a heuristic level, modelling can generate hypotheses on how to manage target species – and can do so cheaply, on the basis of general ecological principles, and without copious amounts of data. Conservation applications can involve simulation 'experiments' that would be impossible to undertake in the field. This combination of computer experimentation and hypothesis generation can be used to enhance the efficiency of field research, while reducing its cost. Such models may reveal general principles that can underpin management, but nonetheless are sometimes greeted with hostility by conservation managers who find the technique alien and who, anyway, are less interested in generalities than specific guidance. This parallels May's (1973) dichotomy between strategic models, which seek to 'grasp general principles', and tactical models, which strive for 'pragmatic understanding of real systems'. In this essay we focus primarily on the uses of tactical modelling in conservation while keeping in mind the inspiration that the most powerful tactical models require strategic components at the core.

Perhaps the most common objective in conservation is to manage the environment to ensure persistence of biodiversity. By developing models that isolate environmental components that identify where species are found, we can gain insight into management actions that most likely will ensure persistence. Indeed, we suggest that identifying what, where and how to conserve has been the greatest impact of modelling on conservation. A large number of modelling studies have sought to quantify relationships between species incidence or abundance and the environmental characteristics of the localities where they are found – an approach broadly classified as 'associative' (Rushton et al. 1997). Associative models attempt to characterize relationships between the distribution of species and environmental features, without explicitly modelling the demographic processes, such as reproduction, mortality and dispersal, responsible for the underlying spatial disposition of the population. Usually associative models relate distributional data for a species to measures of its environment (Manly et al. 2002). They contrast with 'process-based' approaches, which we consider more fully below, where the modeller uses the underlying demographic and movement processes to predict distribution, abundance and dynamics.

Associative approaches

A number of modelling approaches exist for linking the distribution of animals or plants to explanatory variables, typically habitat attributes, land use or environmental features. Methods differ in the complexity of the mathematical formulation of the linkage. Most simply, linkages might assume nothing about the species distributions or the underlying explanatory variables. The observed distribution of the organism is simply overlaid on, and compared with, maps of environmental data, and coincidences between the observed distribution and the variables are used to create 'rules' defining where the organism occurs in relation to the known distribution of the key environmental variables. The landscape is thus **classified** into areas with and without the species of interest. This approach was revolutionized in the early 1990s when satellite imagery became widely available, and when coupled with geographical information systems (GIS) to store and manipulate spatial data, it became relatively easy to identify areas with suitable habitats without even visiting the landscape. Aspinall & Veitch (1993) used a land-cover map of Scotland derived from aerial photographs to predict the distribution of habitats used by red deer (*Cervus elaphus*).

Although widely used, these habitat-classification methods are simplistic and often are presented with no estimate of the accuracy of model predictions. We can do much better using generalized linear modelling (GLM) techniques (Crawley 1993) that partition variation in a response variable (species data) into that due to variation in a set of predictor variables (such as habitat covariates) and residual variation (error). We could imagine the response variable being the abundance of a species that is predicted by key determinants of abundance, e.g. food, cover, predator abundance. In addition to the response variable and the explanatory variables, the modeller must decide upon an appropriate link function that relates the predictor variables to the expected value of the response variable and an 'error structure'. Selection of an appropriate link function depends on the 'error structure', which allows us to evaluate how well the model fits the data. The most familiar GLM approach is linear regression, where variation in a continuous response variable (e.g. density) is related to predictor variables (e.g. vegetation height, distance to water). The error structure is assumed to be normal, i.e. errors in estimating population density are normally distributed relative to the vegetation. If the errors derived from fitting a linear regression do not follow the normal distribution (e.g. if errors at low population density are greater than at high density) then the overall regression model is not valid. Errors may approximate one of several different families of distributions, and tests can be adapted to suit each. For example, tests to assess the adequacy of the error model are straightforward for the normal distribution, but incompetently these tests are often not done or not reported. An alternative might be a Poisson distribution which often captures the error distribution of counts of organisms, in which cases models can be adapted to deal with this form of randomness (e.g. Rushton et al. 1994). Using a model with the wrong error structure is akin to putting petrol in a diesel engine.

Although counts of animals and population density estimates are easy to model, the data can be costly and difficult to collect. Often conservationists know only that an organism is present (= 1) or absent (= 0), and such incidence data are beyond the scope of linear regression because the error model cannot possibly meet the criterion of normality. The solution lies in logistic regression, which has been used to investigate species habitat relationships across a broad range of plant and animal species (see review in Guisan & Zimmermann 2000). As with linear regression where we assume normally distributed errors, with logistic regression we assume a binomial error structure, without which the model is not valid. In the cases of Poisson and binomial error models the data may be 'overdispersed', i.e. where there is more variation than expected by random error, and there are three reasons that such poor fits to data might occur. First, key predictors for the model might have been omitted: there is often a tacit, but perhaps unfounded, assumption that conservationists know what is important (an assumption likely to be diminishingly reliable as fewer and fewer biology students are good naturalists). Second, habitats in the models are defined by the biologist, whose perception might not match the experience of the species under study, not least because they experience the environment at different scales. Humans may more intuitively identify and measure the key predictor covariates for a woodland squirrel than for a larval ground beetle. Third, the error model may not be appropriate for the data. In Rushton et al.'s (1994) analysis of species–habitat relationships the data were overdispersed partly because they were derived from nest counts along river corridors. Furthermore, some of the birds studied were territorial, so of course the errors left after allowing for habitat preferences would not be randomly distributed (Yasukawa et al. 1992). Overdispersion in counts of organisms also can occur when individuals aggregate, whereupon a solution might be to use a negative binomial model (see Hilborn & Mangel 1997); but beware, the biological processes that lead to aggregation (e.g. mating) might vary through time and with the density of the population so different models might be required at different seasons and in different areas.

A key fact of associative models is that the underlying processes are fundamentally spatial. Further, the presence of a species at a point in the landscape is dependent not only on suitable habitat, but also on processes such as territoriality and dispersal, which also are explicitly spatial. The need to take account of spatial 'processes' is one of the main reasons why process-based models have been developed (see below), but another approach has been to adapt associative approaches to accommodate spatial realities. One such approach, by Augustine et al. (1993), includes the presence of organisms nearby as a predictor variable in an 'autologistic' model; an approach that can accommodate spatial aggregation or autocorrelation. Although this is not the place to dissect their shortcomings, the harsh reality is that despite their elegance as a means of investigating spatial dependence, autologistic models have limited capacity for prediction and thus are not much use for the practice of conservation.

With increased availability of software packages and the burgeoning of large spatial data sets, particularly derived from remotely sensed imagery, GLMs have become widely applied to conservation issues. It has become seductively easy to undertake analyses without due consideration of the underlying biology. This has provoked a backlash which, amongst statisticians, questioned the fundamental premise behind hypothesis testing in this form of associative modelling (Eberhardt 2003). Scrutiny focuses on identification of the most parsimonious model or, more specifically, on the criteria used to assess whether variables should be retained in a model. A fundamental aspect of scientific procedure is that, when analysing the results of a randomized experiment, the null hypothesis that there is no difference between two treatments is assessed on the basis of an F statistic (or some other statistic) with an associated probability level (usually $P < 0.05$). In the analogous situation with a GLM, potential predictors can be added or removed sequentially from the model on the basis of whether or not their inclusion is significant at a similar critical threshold of $P < 0.05$. Burnham & Anderson (2002) argue that critical thresholds are arbitrary and that this approach to excluding variables from the model could lead to 'throwing the baby out with the bathwater'. What matters is the magnitude of the effect that potential predictors have, and the view is growing that the stakes are too high for this to be subsumed in an arbitrary cut-off point which ignores everything that does not reach a particular level of statistical certainty.

Burnham & Anderson (2002) argue that we should scrap our traditional hypothesis-testing approach when building models; instead we should be comparing multiple alternative models postulated by the ecologist to capture key elements of the system under study. Alternative models must be biologically plausible and the process entails challenging these models with data to find the models that best explain variation in the data. Specifically, model selection is based on finding the model(s) with the smallest values of Akaike's Information Criterion (AIC), a metric borrowed from information theory. The information–theoretic AIC is –2 multiplied by the log-likelihood for each model plus twice the number of parameters in the model – in effect it creates a

penalty for including parameters in models. Akaike's Information Criterion can be used to determine Akaike weights for each model – these are the weights of evidence in favour of each model, given the other models being considered. This approach aspires to find the best of a suite of models (which might be no better than the best of a bad lot). William of Ockham (fourteenth century philosopher, educated at Oxford) claimed that the simplest explanation is often the best explanation, and we aspire to finding the model that explains with the fewest parameters absolutely necessary. However, the quest for one, **best**, model may not always be prudent: where several plausible alternatives exist then a suite of good models can be the basis for inference and prediction (and there are methods for averaging their outputs). For example, we might know that vegetation is an important habitat component but we might not be able to justify one set of vegetation measurements over another. So we might accept a limited set of alternative vegetation measurements that perform well.

The case of the dormouse (*Muscardinus avellanarius*), a rodent associated with ancient woodland and declining in Britain, can illustrate this information–theoretic approach. The southerly distribution of dormice in the UK is attributed in part by climate and the availability of suitable woodland habitats. We collated 100 samples of records of dormice from 100 randomly selected 10-km grid squares from Arnold (1993), altitude from Ordnance Survey 1:50000 maps and woodland cover from the Countryside Information System Geographical Database Version 5.23. The altitude, geographical co-ordinate and woodland cover data for each 10-km square were used as predictors in logistic regression with presence of dormouse as the dependent variable (presence = 1, random squares = 0). We calculated the corrected Akaike information criterion (AIC_c) following the methods of Burnham & Anderson (2002). Log-likelihood, AIC_c, Akaike weights and evidence ratios for the 16 models are shown in Table 9.1. Clearly, no one model is best; three

Table 9.1 Evaluation of a series of models for investigating the factors determining the presence of dormice in 10 km squares of the National Grid for the UK. The sample data comprise 100 randomly selected squares from the National Grid. Predictor variables were geographical position (easting and northing coordinates in the National Grid), altitude and the availability of woodland habitat. Note that no one model adequately explains the variation in the dormouse record data and that the first three include four variables

Model	Log likelihood	AICc	Delta AICc	Akaike weights	Evidence ratio
Northing easting woodland	59.862	70.500	0.000	0.384	1.000
Northing easting	63.058	71.479	0.979	0.235	1.631
Northing easting altitude woodland	59.832	72.735	1.256	0.205	1.874
Northing	66.984	73.234	2.734	0.098	3.923
Northing easting altitude	62.943	73.581	3.081	0.082	4.667
Northing woodland	65.194	73.615	3.115	0.081	4.746
Northing altitude woodland	64.468	75.106	14.606	0.000	910.161
Northing altitude	66.842	75.263	14.763	0.000	1605.798
Easting woodland	75.008	83.429	22.929	0.000	95261.052
Easting altitude woodland	74.357	84.995	24.495	0.000	208459.266
Altitude woodland	82.951	91.372	30.872	0.000	5054938.862
Woodland	85.767	92.017	31.517	0.000	6978538.019
Easting	86.512	92.762	32.262	0.000	6208765.367
Easting altitude	86.373	94.794	34.294	0.000	27976524.781
Altitude	95.774	102.024	41.524	0.000	1039338198.554

have similar Akaike weights and use varying combinations of the four predictor variables, suggesting that all four are important. Had we, instead, adopted a stepwise analysis of these data, using a threshold criterion of $P < 0.05$ (and also $P < 0.15$), it would have led to the exclusion of all variables except 'northing', resulting in a loss of information and a less useful model for prediction.

Paradoxically, although models are created to fill in gaps in knowledge, it is simultaneously true that they can be as good only as the information on which they are based. A certain amount must be known about the ecological system before a realistic associative model can be formulated, because these approaches can identify only the best amongst a set that has been identified a priori. Identifying biologically plausible alternative models requires a naturalist who understands natural history sufficiently to know the variables that can predict distribution and abundance. The approach argues against 'data dredging' for putative relationships between species incidence and potential

predictor variables, and for a deep understanding of the system being studied. One would hope that the information–theoretic approach is less likely to yield spurious models because only those based on an understanding of the fundamental ecology and conservation objectives will be considered. But we must be diligent to explore alternative models for fear that by limiting the alternatives we inadvertently entrench preconceived notions about how the system works.

Process-based models

Although useful for identifying patterns in the distribution and abundance of organisms across landscapes, associative approaches sometimes fall short of providing what conservationists require in terms of identifying the conditions necessary to ensure that species persist, largely because they do not account for time. The need to evaluate the impacts of the environment on species persistence through time can be met by dynamic process-based models.

The idea behind process-based models is that the likelihood of a species surviving in a particular landscape hinges on the behaviour of individuals (foraging movements, territoriality, dispersal, etc.) and life-history processes (births and deaths). In process-based models habitats provide the template on which population processes occur, and the distribution of organisms amongst habitats emerges as an output of the model. These approaches are more complex than associative methods because they attempt to simulate underlying ecological processes. The wide range of processes and the complexity of their interactions offers scope for developing many approaches; two broad categories are population-based or individual-based models (depending on the level of detail at which life-history processes are modelled). In individual-based models the processes of mortality and reproduction are estimated at the level of the individual and the overall effects on the population are derived from summation of the life histories of each individual. In population-based models mortality and reproduction are instead applied to groups of organisms, such as cohorts or age classes, that constitute components of the population across a landscape.

Process-based modelling requires assumptions about the effects of various biotic and abiotic factors on the dynamics of fecundity, mortality and movement. Both individual and population-level models ultimately play out in space, and this spatial structure might be obtained from an associative modelling exercise (Boyce & McDonald 1999). Process-based models are of most use in combination with the enormous capacity of a GIS to store and retrieve so-called spatially referenced data, that is, the detail of what occurs where. The outputs of such combined approaches – a simulation of how a population of individuals might be predicted to behave in a virtual, but nonetheless realistic, landscape, then can become the input of yet a further class of models, for population viability analysis (PVA).

Population viability analysis

The nub of PVA is the estimation of the probability of persistence (1 – probability of extinction) for a population over some arbitrary time interval, say 100 years, into the future. Factors such as environmental variability, catastrophies, chance genetic events and many others contribute to the risk of extinction, and amongst them is 'demographic stochasticity', which might amount to a run of bad luck in the availability of reproductive males and females (Ellner & Fieberg 2003). A major consideration in developing a PVA is defining the details of chance – known as stochasticity – in the system.

Early PVAs focused on demographic projections based on estimates of age-specific rates of survival and reproduction, and the variances of these rates (Boyce 1992a). However, reliably estimating the necessary vital rates, their variances and covariances for stochastic demography requires dauntingly huge samples, a requirement almost inevitably impossible to meet for the rare species to which PVAs are most urgently applied. Insofar as the outcome is unreliable, estimates of persistence probabilities and the value of the whole exercise is questionable (Ludwig 1999). Yet, a PVA still can be useful, despite uncertainty, insofar as it allows the merit of alternative management options to be ranked (Ellner & Fieberg 2003). And sometimes the process of building the model can be as instructive as the model itself, by clarifying our understanding of how the system works.

An alternative to making demographic projections is to combine habitat models with population models (Boyce & McDonald 1999) to construct habitat-based PVAs, e.g. for spotted owls (*Strix occidentalis*) in the northwestern USA (Boyce et al. 1994). Here resource selection functions (RSF) were used to distribute the population in space and forest-management scenarios were explored to assess the effects on future population persistence. Metapopulations

(Chapter 5) also can be accommodated, using a habitat-based PVA in which subpopulations are governed by local extinction and recolonization (Akçakaya et al. 2004). An example is a habitat-based PVA for the bog fritillary butterfly (*Proclossiana eunomia*), in which the consequences of grazing and global warming were simulated, with results embraced by management (Schtickzelle & Baguette 2004). Although demographic projections tend to be unreliable, robust predictions of distribution and abundance sometimes can be made using habitat models (Fielding & Bell 1997; Boyce et al. 2002). Consequently, habitat-based PVAs may be less burdened by statistical uncertainty than are demographic models, making such models a better bet for conservation planning (Sutherland et al. 2004).

Multiple-species models

Whereas PVA has been attempted for many single-species applications, conservation problems involving two or more interacting species are rare, not least because readily available software has hitherto been unable to cope with such interactions. However, Rushton et al. (1997) developed a spatially explicit model for simulating the dynamics of red (*Sciurus vulgaris*) and grey squirrels (*S. carolinensis*) in the UK, where the red squirrel's decline correlated with the spread of the introduced grey squirrel. This model successfully predicted changes in the distribution of the two species in both the UK (Rushton et al. 1997, 2000) and Italy (Lurz et al. 2003), and was extended by Lurz et al. (2003) to assess the suitability of conservation areas for red squirrels. In the case of Kidland Forest (2050 ha in Northumberland, UK), a commercial woodland devoid of grey squirrels, the model provided a dynamic assessment of carrying capacity in each compartment of the forest based on simulating the production of seeds by different age classes of each of the constituent species in the forest. This facilitated anticipating the impacts of felling and restock-

ing plans, which revealed the risk of a drastic reduction in food (Fig. 9.1). The model also suggested that the proposal to include 15 ha of oak within the forest would lead to substantial colonization by grey squirrels. Managers therefore revised plans to increase the retention of seed-producing species, and abandoned the oak plantation. The success of this case study has led to a comparable modelling exercise being extended to over 80,000 ha of forest in the north of England. Further refinement of this model has investigated the impacts of controlling grey squirrels on the likely persistence of red squirrels (Rushton et al. 2002). In the early 1990s immunocontraception was proposed as the solution for controlling grey squirrels (Moore et al. 1997). The model compared lethal trapping of adult grey squirrels with using immunocontraception in two contrasting landscapes: one where a small isolated population of red squirrels was surrounded by a large population of grey squirrels, and another where red squirrels were still abundant and grey squirrels were dispersing into the area.

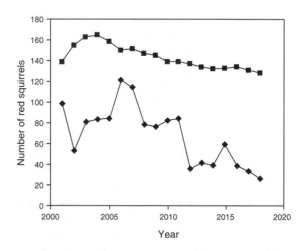

Fig. 9.1 Predicted red squirrel population size in Kidland Forest under the forest management plan projected from 2000 (♦) and the predicted population size following modification of the felling and restocking plan (■) to enhance red squirrel persistence in the forest. The design plan has now been implemented.

The model predicted that red squirrels would become extinct in both landscapes in the absence of population control of grey squirrels. Immunocontraception was successful only where invasion by grey squirrels still was low. In short, in the face of the grey squirrel's capacity to disperse, immunocontraception offered little promise unless applied, at punitive expense, to a large proportion of the target and surrounding population (Fig. 9.2). The costs of undertaking the modelling that led to these insights were at least one order of magnitude less than a field trial to test the same options. Although models cannot supplant field trials, they can assist managers in deciding whether the costs of field trials appear warranted.

Another multispecies example involved a predator–prey model anticipating the consequences of wolf (*Canis lupus*) recovery in Yellowstone National Park, USA to the dynamics of four species of ungulates (Boyce 1992b). This model was instrumental in securing wolf reintroduction in the park in 1995 and helped to shape the recovery programme. Predictions of the model were largely consistent with the observed abundance of elk (*Cervus canadensis*) following wolf reintroduction.

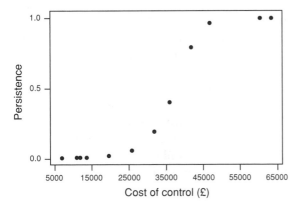

Fig. 9.2 Predicted likelihood of persistence of red squirrels in relation to the annual cost of controlling grey squirrels in Redesdale Forest, UK, using immunocontraception delivered in a bait.

When to model and with what

Both associative and process-based modelling often depend critically on the availability of data. To do a good job of modelling with either approach requires substantial data, but the data needs are of quite different sorts. Further, associative models assume a strong, but static, causal link between species occurrence and the predictor variables. Models based on survey data (i.e. 'snapshots' of species distributions) are inevitably limited as conditions change. Records of an animal at a site do not necessarily mean that it is part of an established population (it may be passing through), thus distorting apparent habitat associations. Dynamic landscapes and changing patterns of habitat selection can be modelled but there are few examples where conservation biologists have achieved this level of complexity. In choosing predictor variables, the most valuable are those that link the incidence of the species to the presence of an essential habitat attribute rather than a close surrogate. For generalist species, suitable surrogate variables can be difficult to identify. Both the choice of response variables and predictor variables have implications for any sampling or monitoring scheme.

A more important criticism of associative models is that they usually take no account of ecological processes such as the impact of predators and spatial and temporal changes in the type of habitats used, time lags in organisms' responses, or the degree of habitat saturation (Wiens 1989). Such limitations can be overcome using conditional logistic regression, as illustrated by modelling the effect of temporal variation in snow depth on use of habitats by elk in Yellowstone National Park (Boyce et al. 2003), or polar bears (*Ursus maritimus*) selecting habitats on ice flows (Arthur et al. 1996). Random landscape locations were drawn from within buffers surrounding each observed location, and environmental attributes specific to the time of the observed location can be included amongst habitat covariates.

Process-based approaches are based on a different philosophy than associative models, and consequently face different constraints in their use in conservation. Firstly, they are inherently more complex because they attempt to emulate system behaviour through the cumulative effects of many population processes. Each of the life-history processes modelled has to be described mathematically and parameterized to encompass the range of variation observed in the field. Fieldwork is sufficiently difficult that for most species describing even the fundamental processes of mortality and fecundity can be impractical. Thus, process-based models might be inappropriate for poorly known species. Even where data are available, the life-history processes may be subject to variation, which is difficult to characterize systematically, thwarting modelling for species with highly stochastic population dynamics – in which case Monte Carlo (i.e. random) simulations almost always are essential to investigate system functioning. Yet, much of conservation practice is about anticipating uncertainty, so such stochastic modelling would seem fundamental.

Because we seldom understand ecological systems completely, applications of models need to be adaptive and management must be able to change as new information becomes available (Lynam et al. 2002). For example, a population model for territorial birds based on basic principles (Lande 1987) focused attention on prereproductive dispersal of the northern spotted owl (*Strix occidentalis caurina*); and because of the owl's threatened status this model was instrumental in the development of the current system of forest management in the Pacific Northwest of the USA (FEMAT 1994). However, data were insufficient to know how well the model truly captured the dynamics of spotted owl populations and subsequent data call to question the utility of the dispersal model (Meyer et al. 1998). Although Lande's model was strategic by identifying a structure that potentially could have important management implications, we later learned that it was largely inappropriate. In another example where uncertainty in model structure has frustrated managers, demographic models for the viability of wild dog (*Lycaon pictus*) populations in Africa (Vucetich & Creel 1999) highlighted the pivotal role played by adult mortality in determining persistence. However, a more recent analysis, using a different form of sensitivity analysis, gave a completely different picture and suggested that pup mortality might be as important as, if not more important than, adult mortality (Cross & Beissinger 2001). The challenge is to honestly admit uncertainties without destroying credibility.

In this article we show that there is considerable scope for using modelling in conservation. Importantly, the process of modelling ensures that the biological features underpinning the model are identified because, whether associative or process-based, modelling entails a statement of hypotheses that then can be evaluated and challenged. Models therefore reveal underlying biology and thereby help to manage conservation efforts more effectively. Models have been used to anticipate the distribution and abundance of a proposed grizzly bear (*Ursus arctos*) restoration programme in Idaho and Montana (Boyce & Waller 2003), and to synthesize large amounts of complex data as in Grundy's (2003) study of weeds. Although we have illustrated examples of the utility of modelling, the gap between practitioner and modeller has to be bridged by improving cooperation between those who practice conservation and those who model it. Often this is not an issue of conflicting methods or ideologies but simply one of improving communication between allies working on the same cause.

'What is the meaning of Life, the Universe and Everything?'. 'Forty-two', said Deep Thought, with infinite majesty and calm.
(Douglas Adams (1952–2004), *The Hitch Hiker's Guide to the Galaxy*, 1979.)

References

Akçakaya, H.R., Burgman, M.A., Kindvall, O., et al. (2004) *Species Conservation and Management: Case Studies*. Oxford University Press, Oxford.

Arnold, H. (1993) *An Atlas of British Mammals*. Institute of Terrestrial Ecology, Banchory, Scotland.

Arthur, S.M., Manly, B.F.J., McDonald, L.L. & Garner, G.W. (1996) Assessing habitat selection when availability changes. *Ecology* **77**: 215–27.

Aspinall, R. & Veitch, N. (1993) Habitat mapping from satellite imagery and wildlife survey data using a Bayesian modelling procedures in a GIS. *Photogrammetric Engineering and Remote Sensing* **59**: 537–43.

Augustine, N.H., Mugglestone, M.A. & Buckland, S.T. (1993) An autologistic model for the spatial distribution of wildlife. *Journal of Applied Ecology* **33**: 339–48.

Boyce, M.S. (1992a) Population viability analysis. *Annual Review of Ecology and Systematics* **23**: 481–506.

Boyce, M.S. (1992b) Wolf recovery for Yellowstone National Park: a simulation model. In *Wildlife 2001: Populations* (Eds D.R. McCullough & R.H. Barrett), pp. 123–138. Elsevier Applied Science, London.

Boyce, M.S. & McDonald, L.L. (1999) Relating populations to habitats using resource selection functions. *Trends in Ecology and Evolution* **14**: 268–72.

Boyce, M.S. & Waller, J.S. (2003) Grizzly bears for the Bitterroots: predicting potential abundance and distribution. *Wildlife Society Bulletin* **31**: 670–83.

Boyce, M.S., Meyer, J.S. & Irwin, L.L. (1994) Habitat-based PVA for the northern spotted owl. In *Statistics in Ecology and Environmental Monitoring* (Eds D.J. Fletcher & B.F.J. Manly), pp. 63–85. Otago Conference Series No. 2, University of Otago Press, Dunedin, New Zealand.

Boyce, M.S., Vernier, P.R., Nielsen, S.E. & Schmiegelow, F.K.A. (2002) Evaluating resource selection functions. *Ecological Modelling* **157**: 281–300.

Boyce, M.S., Mao, J.S., Merrill, E.H., et al. (2003) Scale and heterogeneity in habitat selection by elk in Yellowstone National Park. *Écoscience* **10**: 421–31.

Burnham, K.P. & Anderson, D.R. (2002) *Model Selection and Multimodel Inferences. A Practical Information–Theoretic Approach*, 2nd edn. Springer-Verlag, New York.

Crawley, M. (1993) *GLIM for Ecologists*. Blackwell Scientific Publications, Oxford.

Cross, P.C. & Beissinger, S.R. (2001) Using logistic regression to analyze the sensitivity of PVA models: a comparison of methods based on African wild dog models. *Conservation Biology* **15**: 1335–46.

Dobson, A. & Meagher, M. (1996) The population dynamics of brucellosis in the Yellowstone National Park. *Ecology* **77**: 1026–36.

Eberhardt, L.L. (2003) What should we do about hypothesis testing? *Journal of Wildlife Management* **67**: 241–7.

Ellner, S.P. & Fieberg, J. (2003) Using PVA for management despite uncertainty: Effects of habitat, hatcheries, and harvest on salmon. *Ecology* **84**: 1359–69.

FEMAT (1993) *Forest Ecosystem Management: an Ecological, Economic and Social Assessment*. Report of the Forest Ecosystem Management and Assessment Team, US Department of Agriculture Forestry Service, US Department of Commerce National Oceanic and Atmospheric Administration, US Department of Commerce National Marine Fisheries Service, US Department of the Interior Bureau of Land Management, US Department of the Interior Fish and Wildlife Service, Washington, DC.

Fielding, A.H. & Bell, J.F. (1997) A review of methods for the assessment of prediction errors in conservation presence/absence models. *Environmental Conservation* **24**: 38–49.

Grundy, A.C. (2003) Predicting weed emergence: a review of approaches and future challenges. *Weed Research* **43**: 1–11.

Guisan, A. & Zimmermann, N.E. (2000) Predictive habitat distribution models in ecology. *Ecological Modelling* **135**: 147–86.

Hilborn, R. & Mangel, M. (1997) *The Ecological Detective. Confronting Models with Data*. Mongraphs in Population Biology, 28. Princeton University Press, Princeton, NJ.

Hunter, C.M. & Runge, M.C. (2004) The importance of environmental variability and management control error to optimal harvest policies. *Journal of Wildlife Management* **68**: 585–94.

Lande, R. (1987) Extinction thresholds in demographic-models of territorial populatons. *American Naturalist* **130**: 624–635.

Ludwig, D. (1999) Is it meaningful to estimate a probability of extinction? *Ecology* **80**: 298–310.

Lurz, P.W.W., Geddes, N., Lloyd, A.J., Shirley, M.D.F., Rushton, S.P. & Burlton, B. (2003) Planning a red squirrel conservation area: using a spatially explicit population dynamics model to predict the impact of felling and forest design plans. *Forestry* **76**: 95–108.

Lynam, T., Bousquet, F., Le Page, C., et al. (2002) Adapting science to adaptive managers: Spidergrams, belief models, and multi-agent systems modeling. *Conservation Ecology* **5**(2): Article No. 24.

Manly, B.F.J., McDonald, L.L., Thomas, D.L., McDonald, T.L. & Erickson, W.P. (2002) *Resource Selection by Animals: Statistical Design and Analysis for Field Studies*, 2nd edn. Kluwer Academic, Dordrecht, The Netherlands.

May, R. (1973) *Stability and Complexity in Model Ecosystems*. Monographs in Population Biology 6. Princeton University Press, Princeton, NJ

Meyer, J.S., Irwin, L.L. & Boyce, M.S. (1998) Influence of habitat abundance and fragmentation on spotted owls in western Oregon. *Wildlife Monographs* **139**: 1–51.

Moore, H.D.M., Jenkins, N.M. & Wong, C. (1997) Immunocontraception in rodents: a review of the development of a sperm-based immunocontraceptive vaccine for the grey squirrel (*Sciurus carolinensis*). *Reproduction, Fertility and Development* **9**: 125–29.

Rushton, S.P., Hill, D. & Carter S.P. (1994) The abundance of river corridor birds in relation to their habitats – a modeling approach. *Journal of Applied Ecology* **31**: 313–28.

Rushton, S.P., Lurz, P.W., Fuller, R. & Garson, P.J. (1997) Modelling the distribution of the red and grey squirrel at the landscape scale: a combined GIS and population dynamics approach. *Journal of Applied Ecology* **34**: 1137–54.

Rushton, S.P., Lurz, P.W.W., Gurnell, J. & Fuller, R. (2000) Modelling the spatial dynamics of parapoxvirus disease in red and grey squirrels: a cause of the decline in the red squirrel in the United Kingdom. *Journal of Applied Ecology* **37**: 991–1012.

Rushton, S.P., Gurnell, J., Lurz, P.W.W. & Fuller, R. (2002) Modelling impacts of gray squirrel control regimes on the viability of red squirrel populations. *Journal of Wildlife Management* **66**: 683–97.

Schtickzelle, N. & Baguette, M. (2004) Metapopulation viability analysis of the bog fritillary butterfly using RAMAS/GIS. *Oikos* **104**: 277–90.

Starfield, A.M (1997) A pragmatic approach to modelling for wildlife management. *Journal of Wildlife Management* **61**: 261–70.

Sutherland, W.J., Pullin, A.S., Dolman, P.M. & Knight, T.M. (2004) The need for evidence-based conservation. *Trends in Ecology and Evolution* **19**: 305–8.

Vucetich, J.A. & Creel, S. (1999) Ecological interactions, social organization, and extinction risk in African wild dogs. *Conservation Biology* **13**: 1172–82.

Wiens, J.A. (1989) *The Ecology of Bird Communities*, Vol. 1. Cambridge Studies in Ecology, Cambridge University Press, Cambridge.

Yasukawa, K., Boley, R.A., McClure, J.L. & Zannoco, J. (1992) Nest dispersion in the red-winged blackbird. *Condor* **94**: 775–7.

10

Conservation in the tropics: evolving roles for governments, international donors and non-government organizations

Stephen Cobb, Joshua Ginsberg and Jorgen Thomsen

In preparing for battle I have always found that plans are useless, but planning is indispensible.
(Dwight D. Eisenhower (1890–1969), US General and Republican Politician.)

Introduction

Enormous changes have overtaken the world of conservation in the past 30 years. Conservation organizations have proliferated locally, nationally, regionally and globally. The growth and diversification of Non-Government Organizations (NGOs) with international or global focus has changed the way in which they are able to work with, and relate to, other private and government institutions. Yet although they have proliferated and grown in strength the state conservation agencies of most tropical countries have often not prospered.

Little has been written on the underlying causes for the evolution of conservation NGOs. We examine how the role of international conservation organizations has changed, and how this has been linked to changes in the relationships between government and non-government institutions, and the role of international aid. We will examine the issues and the processes that have led to the changes in

these relationships in middle- and lower-income countries, particularly in the tropics. In closing, we speculate on how things might look at the end of the next 30 years.

The role of the State

In the early 1970s the practice of conservation in the tropics was institutionally rather simple: in the majority of tropical countries it was an activity that was conceived of by the State, undertaken on land reserved by the State for the purposes of conserving samples of the nation's natural endowment, and put into practice by agencies of the State.

There are four important facts to note about conservation at that time and preceding the 1970s: civil society, in the form of local NGOs and interest groups, had very little influence on the creation of the conservation estate (the protected area networks of each country); the traditional or indigenous occupants of land set aside for conservation rarely had a say in the creation of parks and reserves and in the uses of them that they might or might not continue to enjoy; the management of these areas was entrusted to central government agencies, which were usually part of the central ministerial apparatus of the State; often, but not always, conservation activities were undertaken as part of a broader land-management effort, usually overseen by departments or ministries concerned with extraction and management of natural resources.

None of this is very surprising. Many of the countries concerned (in the Caribbean, Africa, south-east Asia and the Pacific) were in their first decade of independence, and were concerned to ensure that Government was firmly in control. Because the State was then, and remains to this day, the owner of the majority of rural land in most of these countries, it was not unreasonable to deduce that conservation was the proper business of the State.

In many cases, Governments were clearly keen to show that they were not only in control, but willing to back their commitments with resources: for example, the National Parks systems of at least a dozen East and Southern African countries were rather efficient, relatively well-funded operations. They survived on subventions from the State and in some cases the retention of revenues from tourism, but very little support from the outside world. Aid programmes and international NGOs were scarcely visible.

Elsewhere, most countries, of course, did not have tourist industries, nor much prospect of developing them, until the relatively recent advent of specialist nature and adventure holidays as a segment of the growing global tourist market. Those tropical countries that first developed nature tourism were able to focus their efforts on savannah ecosystems where wildlife is easily visible. To this day, nature tourism dollars flow disproportionately to such areas.

Yet much of the biodiversity in tropical countries is to be found in its forests. Thus, under growing pressure from the international financial institutions and bilateral donors, many forest departments developed, from the late 1980s onwards, more explicitly conservation-oriented agendas. A paradoxical consequence for this has been that in a number of countries, forest reserves in which strong conservation policies were pursued have found themselves elevated to National Park status, thereby changing their institutional affiliation. This has happened in recent years, to the evident disquiet of forestry professionals, in Gabon, Suriname, Thailand and Uganda, amongst others. Curiously, governments that were economically sufficiently savvy to concede forest management on long leases (25 years and more) to private logging companies, have not, in the main, felt able to do the same with protected areas, although Kasanka National Park in Zambia and Malekele National Park in South Africa are places where the management has been conceded to non-government entities.

A further complication has been that Ministries of Forestry have tended, as guardians of a sector that is a net contributor to the economy, to be relatively powerful, politically. The fashion for separation of wildlife conservation activities from forestry has frequently seen them transferred to Ministries of the Environment, which typically are weak and under-resourced entities. Many such environment ministries and secretariats were created through donor pressure in response to the rise in international environmental consciousness in the late 1970s and early 1980s, but they never had political *gravitas*. Paradoxically, therefore, National Park agencies or Environment ministries may have gained international visibility and new areas to protect, but may have done so at the expense of national political and financial support.

The early 1970s were also a time of the first real collective international acknowledgement of the importance of the natural environment to human well-being, a consciousness that took shape at the Stockholm Conference on the human environment in 1972 (Chapter 18). This gave birth to the United Nations

Environment Programme (UNEP) a couple of years later. Nineteen seventy-two was also the year of the second World National Parks conference, held in Yellowstone, USA.

In the years since, many of the countries concerned have seen their human populations double, and more. Economic growth, and government tax revenues in many countries have not kept pace with this expansion of population. Increased debt owed to developed nations has placed a further pressure on the government purse. The consequent increasing demands on government resources, coupled with widespread declines in the quality of government services, has led to an equally widespread reduction in the budgets that these governments have been able to make available for many activities, among them conservation. From the late 1980s onwards, this has become increasingly acute.

In the majority of low- and middle-income countries, there simply is not enough money coming from the State to meet the costs of conservation. Recent analyses, such as those by James et al. (1999a,b; 2001), Balmford et al. (2003), Balmford & Whitten (2003) and Bruner et al. (2004), all point to the unmet funding gap needed to undertake conservation effectively (Chapter 3) and belie the simplistic view that natural ecosystems should be able to look after themselves (and should cost relatively little to manage).

During the same period, however, subtle pressure has been exerted by the international community, notably by the World Commission on Protected Areas of the International Union for the Conservation of Nature (IUCN), that has made a virtue of protecting a target proportion of more than 10% of national land area. By the time of the fifth World Parks Congress in South Africa in 2003, three decades after the meeting in Yellowstone, this target had been exceeded globally, thanks to some large, sparsely populated countries that were able to exceed the 10% target by a substantial margin. All too little attention had been given, however, to planning how to fund this increasing burden of protected land. The global community had advocated for more to conserve, but not adequately provided funds with which to do so; staff and other resources were stretched ever more thinly in many places; the funds available were failing to meet the minimum needs, and this situation has already prevailed for two decades or more in some places (e.g. Leader-Williams & Albon 1988).

All this created empty space, waiting to be filled. Filled by the lawless, profiting from the all-too-evident inability of government wildlife agencies to uphold the wildlife laws (frequently called poaching, but sometimes the exercise of tenure and use-rights pre-existing the establishment of protected areas); filled by the land-hungry, wishing to find new lands, or to reclaim old ones; filled by public international donor agencies, wishing to make their mark in a domain that captured the public mood in Europe and North America in the 1980s and 1990s; and filled by international conservation NGOs, having meanwhile become increasingly robust, well-funded and strategic.

The States' agents

Ministries of Finance have had an enormous impact on conservation. Under strong political pressure to meet the needs of healthcare, education, national security and infrastructure provision, centrally controlled budgets for marginal activities such as nature conservation have suffered constant attrition. External policies of financial assistance and reform, such as budget support (previously called structural adjustment), have tended to accentuate, rather than alleviate, this process. Any national wildlife conservation agency that has been an integral part of a parent ministry (and whose budgets, staff salaries and operating expenses are therefore controlled by the Ministry of Finance's budget management process) will have been a relatively powerless prey item in this hungry world.

A number of countries have had the good fortune to be able to combine their nature conservation policies with the development of tourist industries, which in turn have created flows of revenues capable of contributing to the costs of conservation. Some of these countries have also had the foresight to create partially autonomous agencies, with the right to retain these revenues, and to manage them in pursuit of their overall conservation objectives. Conventional wisdom is that they are more likely to be better managed, more likely to embrace some of the skills and approaches of the private sector, and more likely to create a coherent approach to revenue generation and expenditure management. However, it has not always been a straightforward transition.

Kenya's experience has been as tortuous as any. Until the mid-1970s, it had an efficient parastatal body, Kenya National Parks, operating under the control of a Board of Trustees, and receiving annual support from Government. It did its own fund-raising outside the country, and kept its tourist revenues. The government then decided to merge it in 1976 with the Game Department, an agency responsible for wildlife management outside the parks, making a single department within a parent ministry, enjoying none of the freedoms or motivation that the partial independence of a parastatal agency had engendered. Although cause and effect are difficult to establish, the quality of wildlife conservation in the field declined markedly at this time. These were the years that Kenya's elephant population declined by over 90% and its rhino population by 99%. Government took note of this failure, so a new parastatal agency was created in 1989, the Kenya Wildlife Service (KWS), which was vested with a large measure of freedom and autonomy. This new organization attracted substantial international financial support ($150 million of grants and loans were made available in the period 1990–1995), which itself acted as a lightning conductor for political and public scrutiny. As a consequence, a succession of changes have been made to the way the KWS is managed and governed; changes that have reduced its freedom and influence at every stage. When, in late 2005, Amboseli National Park was degazetted by Government, KWS was not even consulted. Once again, the quality of conservation in Kenya has visibly declined in parallel to the loss of autonomy in management.

In neighbouring Tanzania, a parastatal agency Tanzania National Parks had been created in 1959. Operating under the control of a Board of Trustees, drawn from within Government and from civil society, this body received annual grants from Government but also enjoyed the freedom to retain its revenues from tourist operations. In socialist Tanzania of the 1960s and 1970s, this was counter-revolutionary. But in the early days, it did not generate much money, either. Four decades later, this basic model has hardly changed. What has changed, though, is that tourism in Tanzania has grown and the willingness of tourists to pay higher fees for their experience has been successfully put to the test. Although there have been rocky patches along the way (financial and political interference), the point has now been reached where the Government no longer gives an annual subvention to the National Parks: it taxes their revenues, which are currently running at about $24 million per year.

If there is a conclusion to be drawn from this comparison of two neighbouring countries (which, after all, share many transfrontier wildlife populations, as well as rather similar tourist offerings), it must be that the freedoms and incentives to good management that are offered by the semi-autonomy of the parastatal wildlife agency can be enjoyed to the full only when the quality of management is high and government is prepared not to interfere.

The role of international aid

The volume of international aid available to wildlife conservation grew enormously during the decades under review. Although a systematic and authoritative review of both bilateral and multilateral agency funding is sorely lacking, a report by Lapham & Livermore (2003) provide a clear overview of more recent trends, and of the way in which global donor priorities for development assistance have affected biodiversity-related expenditures.

During the 1970s, modest support in this domain came from the United Nations, particularly from the Food and Agriculture Organization (FAO), which was always interested in funding programmes relating to the utilitarian aspects of wildlife (through hunting, cropping and rearing schemes). The World Bank dipped its toe in these waters, mainly to promote tourism as a means to stimulate wildlife-based economies. Bilateral aid, from the USA, Canada, Britain, Germany, The Netherlands and France, in particular, was always modest in scale (see Annexes in Lapham & Livermore 2003), and much of it was devoted to training and skills transfer, but not to direct operational support to conservation agencies.

A spur to foreign assistance in this domain was the second world conference on the human environment, held in Rio de Janeiro, Brazil, in 1992, 20 years after the Stockholm conference. Not only did this see the creation of the Convention on Biological Diversity (Chapter 18), but it also saw a heightening of global consciousness about the needs of biodiversity conservation. This translated into a greater willingness of international donor agencies to fund programmes supporting biodiversity conservation. The mid-1990s were boom years in this regard; many new partnerships were formed around the world by donors not traditionally known for their interest in biodiversity conservation.

The 1990s also saw the creation of the Global Environment Facility (GEF), an enterprise under the stewardship of the World Bank and the United Nations. Although overall commitments exceed US$6 billion (Lapham & Livermore 2003), grants to biodiversity related projects (historically, one of four areas of funding for the

GEF) average approximately $200 million per year. The World Bank manages a significant portion of GEF biodiversity funding and also makes additional grants and loans in the biodiversity and protected area domains, historically at a rate of about $100 million per year (GEF 1996a). Most of these funds are expended through and by national governments.

Funding by GEF is intended to catalyse important biodiversity projects and investments, not to provide complete financing. When the GEF was created in 1991, donor governments understood that the GEF's resources were quite limited in comparison with the cost of the environmental problems it was asked to tackle. A strategic decision was made to limit GEF financing to 'incremental costs' – those costs required to achieve global environmental benefits over and above national development goals (GEF 1996b). (Incremental costs are determined through negotiation by estimating the difference between the cost of realistic baseline investment in the national interest and the cost of the 'GEF alternative' that provides the added global benefits. In the case of biodiversity projects, the baseline is often estimated from existing development plans and budgets (GEF 1996a).) Thus, GEF projects require significant complementary funding (co-financing). For example, the GEF has invested in a number of successful conservation trust funds (e.g. in Peru, Brazil, Bhutan and Uganda), which help provide long-term financing to support basic management activities in protected areas (GEF 2002). Although the GEF's investments have been critical, significant resources from other donors (public and private) have been necessary to capitalize these funds adequately.

The USA Government, a major contributor to the budgets of the UN, the GEF and the World Bank, is pursuing an increasingly bilateral approach to funding for Official Development Assistance (ODA) generally, and biodiversity assistance specifically. The overall impact of this on global biodiversity funding is unclear. Despite not ratifying the Convention on Biological Diversity (or the Kyoto protocol on climate change), the USA remains the largest single donor to the GEF, providing approximately 20% of its overall funding. The future of this is currently uncertain.

Although it is impossible to make accurate estimates of financial flows from public donors to protected areas worldwide, a recent Organization for Economic Cooperation and Development (OECD) study gives a rough gauge of the level of bilateral investment in conservation. The study asked donors to report all bilateral biodiversity-related assistance, and found that annual average investment from 1998–2000 was approximately $995 million, or roughly 2.7% of total official development assistance (OECD 2002). This figure represents a broadly defined conservation investment, including funding for some of the projects in sectors such as agriculture, water supply and forestry.

Even if all ODA biodiversity funds were applied to protected area management and creation, the $200 million spent annually by the GEF can be placed in the context of projected costs for conservation. For instance, Frazee et al. (2003) have recently estimated that the cost of developing and maintaining a protected areas network in a single biodiversity hotspot (the Cape Floristic Region of South Africa) would be $800 million over a 20 year period, $40 million a year, or 20% of the total annual GEF biodiversity budget. Simply put, international aid is not meeting the needs.

However, attention spans are short, politics are fickle and the public in the donor nations seem to have little success in badgering their politicians to keep the issues to the fore. The environment has dropped down the list of voters' priorities and budgets are tightening. The conference convened in Johannesburg, South Africa, in 2002, to examine 10 years' progress since Rio, was no longer called an environment conference, but the World Summit on Sustainable Development. Along with formulation of the Millennium Development Goals the attention of the world community had shifted perceptibly towards poverty alleviation and other concerns more visibly related to

human welfare (water supply, health and education) and away from the environment. A cursory analysis of this is found in Lapham & Livermore (2003), but the growing impact of changes in donor focus on biodiversity funding requires more detailed analysis.

In sum, the political mood of the world has changed, and there is perceptibly less international aid money available now, in the domain of biodiversity conservation, than there was just a few years ago. Of course, this will surely change again.

The growth of international non-governmental organizations

Increasing budgets

International organizations with a focus on conservation are not a new invention, but have been with us for over a century. For instance, the Royal Society for the Protection of Birds was founded in 1889; the Sierra Club in 1892; the Audubon Society for the Protection of Birds in 1886; the Wildlife Conservation Society (as the New York Zoological Society) in 1895; and Fauna and Flora International (as the Society for the Preservation of the Fauna of the Empire) in 1903. Some of the largest organizations were founded more recently, but with a specific focus on international conservation, including the World Wide Fund for Nature (founded as, and with some of its constituent organizations still operating as, the World Wildlife Fund) in 1961, and Conservation International in 1987.

Although many of these organizations initially had national mandates, or a focus on research when working internationally, they have grown significantly in the past few decades and greatly expanded their conservation efforts on the ground in tropical countries. For instance, the international conservation programmes of the Wildlife Conservation Society (WCS) have grown from just over $3 million in the late 1980s, to nearly $50 million in 2005, nearly half of WCS's overall operating budget. Although the WCS has programmes in North America, all but $5 million of these funds are sent overseas in developing countries (WCS 2004). The Nature Conservancy, founded in 1951 primarily to focus on USA conservation issues, is probably the largest conservation organization in the world, with one million members, and an annual budget (The Nature Conservancy 2004) of over $800 million. Conservation International (CI), with a focus on biodiversity hotspots and tropical wilderness areas, was founded in 1987 and has grown to an organization with an annual budget of over $100 million. One-third of CI's annual budget is shared with other organizations in the form of grants (Conservation International 2005).

The World Wildlife Fund (WWF) was set up to act as a source of funding for the programmes of IUCN, the World Conservation Union (founded in 1948 as the International Union for the Conservation of Nature and Natural Resources). It was also to serve as a fund to which other organizations could apply to undertake their own, small-scale programmes; by the late 1970s, WWF had developed a profile, public recognition and a stature that exceeded that of IUCN from which it separated 20 years ago. Through the 1980s WWF became increasingly disinclined to fund other organizations' projects, relying rather on the growth of its own staff to undertake their own programmes. It is significantly more difficult to estimate the total budget of WWF because the organization acts as a franchise of more than 20 independent NGOs. Because significant monies move from WWF organizations that raise funds in the developed countries to WWF organizations in developing countries, tracking income and expenses for the global institution is difficult. Nonetheless, the WWF network has a budget of the order of $300 million per year.

In many developing countries, individual international NGOs now have larger budgets than the state wildlife or forestry conservation agencies with which they work. This is a reflection both of the relative lack of political importance assigned locally to conservation efforts, and of the increased effectiveness of NGOs in raising and disbursing funds.

That NGOs have such success in securing funds undoubtedly creates tensions, but is it the way of the future? It has been argued recently that major international conservation NGOs have stifled the development of local or indigenous NGOs with similar missions (Chapin, 2004). Responses to this assertion have been many (see a multitude of responses in the *World Watch* magazine of January/February 2005). Whether true or not, we confine ourselves to describing some of the domains in which the international conservation NGOs seem to have had the most positive impact, as conservation practitioners in the tropics.

Priority setting

A key area in which international conservation NGOs have made an imprint is in assisting the global community to establish global priorities. Such approaches have included: those that look at representation of species and habitat types (e.g. Olson et al. 2002); those that focus on species diversity and levels of threats (Myers 1988; Myers et al. 2000); those that focus on areas of higher or lower human impact as a surrogate for scaling threat (Sanderson et al. 2002); those that focus on levels of endemism and evolutionary uniqueness in a particular taxon (e.g. Sattersfield 1998). Although much has been made about the difference between these approaches, a recent synthesis suggests that they are highly complementary and that differences are more apparent than real (Redford et al. 2003)(Chapter 2).

Such strategies have also been instrumental in setting priorities for expenditure of conservation funds, both private and public. The MacArthur Foundation, the World Bank, USAID, and other multilateral and bilateral donors have all used a variety of these exercises to assist in guiding investments in a way that was relatively apolitical and objective.

The success of defining a set of clear global or regional priorities for conservation, and the ability to work and move freely across international frontiers, has allowed international conservation NGOs (e.g. The Nature Conservancy 2005) to adopt strategic approaches, and regional or species-based visions, that often escape conservation managers at national level. Although valuable, this can also engender some disquiet among these managers, who see decision-making and financial capacity being wrested from their grasp.

Acting as intermediaries

International conservation NGOs have also positioned themselves successfully as the intermediaries of choice between the development agencies and recipient countries and organizations. The impact of this far exceeds the actual financial importance of NGOs on the global stage, although for some of the NGOs this has had important budgetary implications. Conservation groups, through the effective use of media campaigns, direct lobbying and development of position papers, can influence national legislation, and international treaty development and implementation. At regular meetings of the parties to major international treaties (Convention on Biological Diversity, Convention on International Trade in Endangered Species, Ramsar Convention on Wetlands), international NGOs frequently have influence by providing technical information and recommendations both to the signatory nations of these treaties, and to the secretariats of the conventions.

These groups are also actively engaged in lobbying international agencies and the development agendas of national governments to adopt more explicitly conservation-oriented policies. With donors focusing more intently on poverty reduction, such activities have become increasingly important in the USA and Europe if biodiversity conservation is to remain on the global agenda.

We should recognize that when looking at applying pressure to the governments of developing countries, within the broader conservation community, NGO pressure often comes, in the main, from a different set of NGOs than those listed above, those that focus on lobbying rather than on the implementation of conservation on the ground. That said, international NGOs that focus on implementation of conservation often generate information that assists in global campaigns, and often campaign and implementation focused organizations assist and collaborate indirectly, if not directly.

Global reach

The global reach of some international NGOs, and their apparent ability to set priorities independent of the political concerns that influence multilateral organizations, allows them certain freedoms.

Non-Government Organizations also can often be the laboratories for innovation and the testing of new ideas. For instance, after setting priorities for conservation we need to work out how to implement conservation and test the effectiveness of conservation actions. The conservation community, like many, can easily become entranced with a particular vision of how to achieve its aims (e.g. IUCN, UNEP & WWF 1991) but critical reviews and evaluations of a particular or different approaches to conservation are rare (but see Wells et al. 1999).

International conservation NGOs, however, have recently started both to systematize the approaches they take to conservation (e.g. Groves 2003; Sanderson et al. 2002) and to test the effectiveness of various approaches (Margoluis & Salafsky 1998; CMP 2005). The global reach of these organizations, their focus and long-term commitment allow them to develop and test such approaches in ways that would be difficult at an international scale for other types of institution. Their reach and scale are extending, their power and influence increasing, their partnerships and joint ventures with public and private institutions strengthening. They have been gaining ground on a broad front of the conservation world. Yet, as the traditional distinguishing characteristics of NGOs blur and as they become more like the establishment, less like the civil societies from which they grew, perhaps they are losing something too.

From simplicity to complexity

We started this chapter by depicting conservation as a relatively simple activity in the early 1970s; there is no escaping the conclusion that it has since become enormously complicated (Chapter 18). In part this is because we know so much more than we did previously about the issues with which conservationists must grapple. This knowledge has taken conservation out beyond the domains of the game warden and the research biologist, to include the social and political aspects of the relations of conservation areas with their neighbours; the partnerships that much be established with other economic interest groups operating in the same area; the political dimensions of land-use, land claims and resource rights; and the thorny issue of the relevant competence to manage conservation areas being demonstrated by different types of institution.

Another feature of the changing world of conservation has been the recognition of the larger scale at which conservation must be approached. This has led conservation NGOs to examine global priority areas, to examine and then work at the scale of ecologically homogeneous regions, to go beyond the boundaries of protected areas to include the landscapes within which they are situated, to work to link such areas together to create biologically meaningful corridors, to link protected areas together across international frontiers. All these new approaches acknowledge reality, and yet make conservation ever more complex.

Perhaps of greatest importance, however, is that while the conservation landscape has become more complex, the tools more sophisticated, and the ability to exploit new funding mechanisms in support of conservation has become more extensive, the problems facing conservation practitioners have exploded. The loss of forests, the legal and illegal overharvest of wildlife both in terrestrial and marine ecosystems, the increase in introduced and invasive species, are all well known to students and conservation practitioners, as are the stresses and underlying drivers of these losses: consumption; population growth; economic growth and the demand for land; more recently globalization and the improved communication and transport networks that drive and are driven by it.

What will happen to tropical conservation institutions over the next three decades?

Without speculating on major global issues and their impact (such as climate change, the growing energy crisis, demographic growth, shifts in global economic power and global peace and security), any of which could have fundamental impacts on biodiversity conservation in the tropics, we will confine ourselves to a few predictions, most of which we see as non-controversial:

1 government departments will wither, while good parastatals thrive; those that survive will tend to be smaller, filling the role of supervisors and controllers of specialist work that is subcontracted to others;
2 parallel national bodies, such as the executive arms of conservation trust funds will increase, both in number and financial muscle;
3 the old model of the central conservation authority will continue to decline in influence (and purchasing power); they will continue, however, to exercise their traditional law-enforcement functions;

4 private business will be increasingly involved in nature management and the provision of support services;

5 stronger local and national NGOs will increase their capability and take on the role of field conservation managers, managing protected areas on contract;

6 international NGOs and their local affiliates will continue to grow and thereby to assume many roles previously the preserve of government agencies; this will face increasing political challenge;

7 illogical cycles of donor support, in pursuit of political correctness, rather than visionary sound sense, will continue to frustrate long-term financial planning in the conservation sector;

8 endowment capital to provide sustained funding of nature conservation will play an increasingly important role, partially off-setting the caprices of official development assistance;

9 fiscal innovation will lead to a mature marketplace for ecosystem services, many of which are provided by natural protected areas.

In conclusion, whatever agencies are at work, be they government departments, parastatal agencies, international donors or big international conservation NGOs, they are all micromanaging the effects of global phenomena at a relatively small scale. What is really needed to support these innumerable worthy conservation efforts is global leadership, global political will and global public consciousness of the imperative need to act to contain those human-made phenomena that are making the long-term aspirations of nature conservation so hard to attain.

The chessboard is the world, the pieces are the phenomena of the universe, the rules of the game are what we call the laws of nature, the player on the other side is hidden from us.

(Thomas Henry Huxley, *A Liberal Education*, 1870.)

References

Balmford, A. & Whitten, T. (2003) Who should pay for tropical conservation, and how could the costs be met? *Oryx* **37**: 238–50.

Balmford, A., Gaston, G.J., Blyth, S., James, A. & Kapes, V. (2003) Global variation in conservation costs, and unmet conservation needs. *Proceedings of the National Academy of Sciences, USA* **100**: 1046–50.

Bruner, A.G., Gullison, R.E. & Balmford, A. (2004) Financial costs and shortfalls of managing and expanding protected-area systems in developing countries. *Bioscience* **54**: 1119–26.

Chapin, M. (2004) A challenge to conservationists. *World Watch* **November/December**.

Conservation International (2005) *Annual Report, 2004*. Conservation International, Washington, DC.

CMP (2005) Conservation Measures Partnership, Bethesda, MD. http://www.conservationmeasures.org/CMP/.

Frazee, S.R., Cowling, R.M., Pressey, R.L., Turpie, J.K. & Lindenberg, N. (2003) Estimating the costs of conserving a biodiversity hotspot: a case-study of the Cape Floristic Region, South Africa. *Biological Conservation* **112**: 275–90.

GEF (1996a) Biological diversity. In *Operational Strategy* (Chapter 2). Global Environment Facility, Washington, DC. Available online: www.gefweb.org/public/opstrat/ch2.htm [accessed 30 May 2003].

GEF (1996b) Incremental costs. *Paper prepared for the 2–4 April Council Meeting*. Global Environment Facility, Washington, DC. Available at *http://www.gefweb.org/meetings/ council7/c7inf5.htm*

GEF (2002) *Biodiversity Matters: GEF's Contribution to Preserving and Sustaining Natural Systems that Sustain our Lives*. Global Environment Facility, Washington, DC. Available online: gefweb.org/Outreach/ outreach- PUblications/GEF_ Biodiversity_CRA. pdf [accessed 30 May 2003].

Groves, C. (2003) *Drafting a Conservation Blueprint: a Practitioner's Guide to Planning for Biodiversity*. Island Press, Washington, DC.

IUCN, UNEP & WWF (1991) *Caring for the Earth: a Strategy for Sustainable Living.* Gland, 228 pp.

James, A.N., Green, M.J.B. & Paine, J.R. (1999a) *Global Review of Protected Area Budgets and Staff.* World Conservation Monitoring Centre, Cambridge.

James, A.N., Gaston, K.J & Balmford, A. (1999b) Balancing the Earth's accounts. *Nature* **401**: 323–4.

James, A.N., Gaston, K.J. & Balmford A. (2001) Can we afford to conserve biodiversity? *BioScience* **51**: 43–52.

Lapham,N. & Livermore,R. (2003) *Striking a Balance.* Conservation International, Washington, DC.

Leader-Williams, N. & Albon, S. (1988) Allocation of resources for conservation. *Nature* **336**: 533–5.

Margoluis, R. & Salafsky, N. (1998) *Measures of Success: Designing, Managing, and Monitoring Conservation and Development Projects.* Island Press, Washington, DC.

Myers, N. (1988): Threatened biotas: 'hotspots' in tropical forests. *Environmentalist* **8**: 187–208.

Myers, N., Mittermeier, R.A., Mittermeier, C.G., et al. (2000) Biodiversity hotspots for conservation priorities. *Nature* **403**: 853–8.

OECD (2002) *Report of the Conference on Financing the Environmental Dimensions of Sustainable Development,* 24–26 April, Organization for Economic Cooperation and Development, Paris.

Olson, D.M., Dinerstein, E., Wikramanayake, E.D., et al. (2002) Terrestrial ecoregions of the world: a new map of life on Earth. *Bioscience* **51**: 933–8.

Redford, K.H., Coppolillo, P., Sanderson, E.W., et al. (2003) Mapping the conservation landscape. *Conservation Biology* **17**: 116–131.

Sanderson, E.W., Jaiteh, M., Levy, M.A., Redford, K.H., Wannebo, A.V. & Woolmer, G. (2002) The human footprint and the last of the wild. *BioScience* **52**: 891–9.

Sattersfield, A.J., Crosby, M.J., Long, A.J. & Wege, D.C. (1998) *Endemic Bird Areas of the World: Priorities for Biodiversity Conservation.* BirdLife International, Cambridge.

The Nature Conservancy (2004) *Conservation that Works: Annual Report of The Nature Conservancy.* Washington, DC, 20 pp.

The Nature Conservancy (2005) *Conservation Action Planning, Developing Strategies, Taking Action, and Measuring Success at Any Scale: Overview of Basic Practices.* The Nature Conservancy, Washington, DC, 20 pp.

WCS (2004) *Annual Report of the Wildlife Conservation Society.* Bronx, New York.

Wells, M., Guggenheim, S., Khan, A., Wardojo, W. & Jepson, P. (1999) *Investing in Biodiversity.* World Bank, Washington, DC.

11

Do parasites matter?
Infectious diseases and the
conservation
of host populations

Philip Riordan, Peter Hudson and Steve Albon

Diseased nature oftentimes breaks forth in strange eruptions ...
(**William Shakespeare, *Henry IV*, Part I, 1598.**)

Introduction

The catalogue of species threatened by parasitic infections is extensive. To name a few, rabies caused dramatic declines in threatened populations of the African wild dog (*Lycaon pictus*) and the Ethiopian wolf (*Canis simiensis*) (Gascoyne et al. 1993; Sillero-Zubiri et al. 1996; Laurenson et al. 1998; Randall et al. 2004). Canine distemper virus hit a broad range of carnivores including lions (*Panthera leo*) in the Serengeti (Roelke-Parker et al. 1996) and seals in Lake Baikal (*Phoca sibirica*: Mamaev et al. 1995) and the Caspian Sea (*Phoca caspica*: Forsyth et al. 1998). Avian malaria devastated endemic bird populations on the Galapagos (Wikelski et al. 2004) and Hawaii (Woodworth et al. 2005) and unrecorded infections such as chytridiomycosis, a fungal infection of frogs, appears to have caused massive declines in Australian rainforest frogs (Berger et al.1998; Daszak et al. 2003; but see also Cleaveland et al. 2002). There have been many noble attempts to mitigate these threats, such as vaccination programmes of domestic dogs (Laurenson et al. 1997; Randall et al.

2004, in press), the treatment of black-footed ferrets (*Mustela nigripes*) with insecticide to remove arthropod vectors of sylvatic plague (Thorne & Williams 1988) and the application of medicated grit to reduce worm infections in grouse (*Lagopus lagopus*) (Hudson 1992, Newborn & Foster 2002). However, despite the obvious disease-induced threats to wildlife species there is only one account of a species being driven completely extinct by a parasite: the last individual snail *Partula turgida* was killed by a microsporidian infection (Cunningham & Daszak 1998).

Conservation biologists only noticed the pervasive nature of parasites, both micro- (e.g. viruses, bacteria, protozoans and fungi) and macro- (e.g. helminths and arthropods) (*sensu* Anderson & May 1979) relatively late in the game. The history of humankind is scattered with descriptions of parasitic organisms causing tragic disease outbreaks which have devastated human populations, provided an opportunity for invading forces to conquer new lands and changed the way human society developed (Cartwright & Biddis 1972). To many of our forefathers, infectious disease was a major hazard resulting in death of infants, debilitating adults and reducing the productivity of their livestock. With the advent of vaccines and antibiotics, the role of infectious diseases in humans was transformed so that by the 1960s there was a general consensus amongst health workers that the disease problems that had ravaged our ancestors had past. Smallpox was eradicated through vaccination; penicillin was an antibiotic able to cure most bacterial infections and workers in the World Health Organization considered it only a matter of time before the insecticide DDT would decimate mosquito populations and lead to the extinction of malaria. Interestingly, the success of these medical developments were reflected in the views of ecologists: ecological texts of the day played scant attention to the effects of disease in wild animal populations, considering outbreaks a consequence of secondary factors, as they assumed that selection should favour the parasites that did not devastate their own hosts.

The bubble burst with the emergence of infections humanity had never experienced before, such as HIV, Ebola and other frightening hemorrhagic diseases. Many of the old infections thought to have been defeated in the past also started to re-emerge: tuberculosis developed resistance to drugs; malaria increased to become a major new killer; and it emerged that strain-specific vaccines against influenza offered little more than partial immunity to other strains. Simultaneously the views of ecologists also changed; Anderson & May (1978) provided a theoretical foundation, showing that parasites could play a major role in regulating the abundance of hosts. Conservation workers started to notice how threatened wildlife species were being affected, particularly by parasitic agents that infected multiple host species.

Against this background, it might seem facetious to address the question 'Do parasites matter?' by considering public and scientific attitudes to threats to the diversity of parasites themselves. There are few who would mourn the loss of any parasitic species and although the biodiversity of parasitic species far exceeds that of the vertebrates it is unlikely that any conservation body would dedicate themselves to their protection. However,

herein lies a couple of philosophical paradoxes. First, should conservationists question the relative value placed (perhaps arbitrarily) on individual species (see Whiteman & Parker 2005)? Indeed since the majority of living organisms are parasitic in one form or another, be they a parasitic worm in a dog, a virus infecting a human or a rust on a prairie grass, they are a major component of biodiversity themselves. What is clear is that global biodiversity is threatened at all taxonomic levels and it is axiomatic that when parasites act as species' executioner they are also threatened with extinction themselves.

The second paradox is another dichotomy between the role of parasites as a factor reducing biodiversity through their impacts on mortality and the appreciation that parasites are probably a major driving force in the evolution of biodiversity (e.g. Nunn et al. 2004). The selective pressure that parasites exert on their hosts drives the variation between hosts in an attempt to fight future infections better. Indeed a creditable hypothesis is that sex evolved primarily as a means of generating diversity and avoiding the ravages of parasitism. As such, does intervention remove a driving force for the future generations of biodiversity when we reduce levels of infection and vaccinate hosts against pathogens.

Parasites also provide a series of other benefits for conservation; they have been used as a biological control agent to reduce or eliminate the effects of pests and exotic invading species (Cleaveland et al. 1999). They have aided conservation efforts directly by keeping people from colonizing parts of the world, for example, infectious and lethal zoonotic infections have kept humans out of some areas of the world such as the Darien on the Panamanian and Colombian border, and consequently this area has been left as wilderness that has benefited wildlife and the conservation of biodiversity. They may also play a key role in ecosystem functioning, although we are only just starting to appreciate how important this may be (Thomas et al. 2005). For example a recent field experiment studying the intertidal soft-bottom substrate assemblage of macro-invertebrates demonstrated that a single type of parasite can have community wide effects solely through the parasites' impact on the behaviour of one species of bivalve (Mouritsen & Poulin 2005). In this system, a trematode parasite infecting the foot tissues of cockles influences the host's ability to move and bury into soft-bottoms. As a result, cockles in heavily infected populations spend more time at the surface, altering the surface environment allowing other species of invertebrates to colonize.

In short, the answer to the question 'Do parasites matter?' may be a simple 'of course they do since they are ubiquitous and they impact all aspects of conservation from causing local extinctions through to influencing ecosystem functioning'. But this superficial answer will not suffice here because it fails to reveal the complexities of parasitic infection, the important dynamical non-linearities between parasite and host and the heterogeneities that can lead to disease emergence. Neither does it address the role that parasites have in conservation. In a nutshell parasitic infections are important to conservation for three reasons. First, they cause epidemics, leading to heavy mortality and reducing abundance to low levels where the hosts become vulnerable to all the vagaries facing small populations. Second, as wildlife is the reservoir for the majority of the world's emerging human diseases, a logical step for health workers can be the decimation of wildlife reservoirs or their habitat in the name of benefiting humans. Third, are the issues of conserving parasites, not only as part of biodiversity but as the drivers of biodiversity. The objective of this essay is to reveal these issues in the context of a dynamic understanding, drawing on examples of parasitic infection in both wildlife and humans.

Why do parasites matter?

The study of parasites contributes a powerful conceptual paradigm, linking on the one hand, ecology, systematics, evolution, biogeography and behaviour, and on the other an array of biological disciplines that link from the molecular, through the response at the organismal level to population dynamics, community structure and ecosystem functioning (Hudson et al. 2002; Begon et al. 2005).

Parasites divert energy resources from their host and thus direct energy away from other consumers, changing energy flow patterns and ecosystem functioning (e.g. Thompson et al. 2005). Indeed one could imagine that they play a pivotal role in influencing energy flow at a fundamental level and so we should expect them to be major players in ecosystem functioning and yet, until recently, their role in ecosystems has been ignored. Classic studies such as those of Polis (1999) considered parasite biomass negligible and yet data emerging from studies of the trematodes on the Carpinteria salt marsh in California indicate the biomass is huge with a massive reproductive rate (Hudson 2005). Given that in all likelihood each species within an ecosystem will have at least one specialist parasite, such an oversight risks missing at least half the species present. Added to this, it is important to consider 'interactive species' (Soulé et al. 2003) and parasites and their hosts are the fundamental example of an intimate coupling of species. Hence the inclusion of parasites with ecological and conservation-based research should be a major goal. The issue is how to obtain this understanding. Should we start from a broad community study that encompasses all species or examine specific systems and then build towards unification? We take the latter approach and start by examining the dynamics of parasite–host relationships and then build on this to reveal the emergent properties at a larger scale and the issues that this evokes for conservation biology.

The population ecology of parasitic infections

The study of disease dynamics has emerged from the application of population ecology to the study of parasitic infection in wildlife, by applying models of host and parasite death and birth processes to the biological features of parasitism (e.g. Hudson et al. 2002). Probably, the key epidemiological parameter that has allowed us to encapsulate the broad features of parasite–host dynamics, and at the same time have a measure of parasite fitness, is the basic reproductive number (R_0), defined as the average number of new infections arising from each infected individual in a population of susceptibles. In many respects this measures the success of transmission and with parasites this is the key to high fitness. For example, if an African woodland worker becomes infected with Ebola then R_0 would be the average number of people he subsequently infects before he dies. This deceptively simple metric determines the nature of an epidemic, whether the parasite has a chance of becoming endemic and if the parasite is even able to invade the host population in the first instance. Where R_0 is < 1, then, by definition the average invading parasite will fail to replace itself and an epidemic will not get started. The effective value of R_0 (R_e) is dependent on the number of susceptible individuals within the population (see Fig. 11.1). For a

parasite to invade a host population R_e therefore needs to exceed one. Epidemics involving parasites with higher R_0 values will tend to spread through the host population rapidly, infect a large proportion of the population and burn through the population of susceptibles so fast that the number of infectious individuals at the end of the epidemic is too small for the parasite to persist in the host population. A lower R_0 value will tend to produce a slower and less severe parasite outbreak that is longer lasting; sometimes long enough for an input of susceptible hosts, through birth or immigration, to allow the parasite to eventually become endemic (Fig. 11.2). Indeed it is important for students of epidemiology to appreciate that not all R_0s of the same value actually generate the same dynamics in the host population. The reproductive number, R_0, is essentially the ratio between the birth rate (transmission) and death rate (period of infectiousness) of the infection. If we keep R_0 the same but increase the

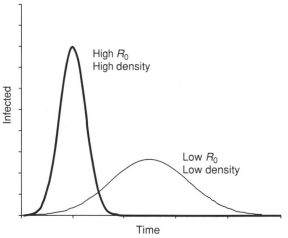

Fig. 11.2 reproductive number, R_0, determines the rate at which the epidemic curve commences: low R_0 results in a longer epidemic, which fades out more slowly; high R_0 gives rise to more rapid epidemics. Similarly, large populations will experience more rapid epidemics than small populations. Note that the longer epidemic curve can allow for the input of new susceptibles through birth and immigration, which can lead to the infection becoming endemic.

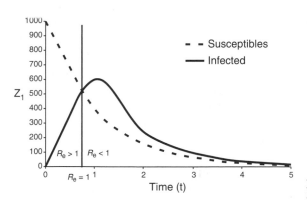

Fig. 11.1 As the number of infected individuals increases, the pool of susceptibles Z_1 becomes depleted. The effective value of the reproductive number, R_0 (termed R_e), is related to the number of susceptible individuals within the population, such that $R_e = R_0 S$. Where $R_e < 1$, the number of infected cases starts to fall and the epidemic fades out.

period of infectiousness then we eliminate the explosive epidemic nature of an infection and lead to improved persistence within the population. Similarly changing host birth rate or the transmission rate has an effect on the instability and nature of subsequent epidemics.

Furthermore, R_0 represents an average case for the whole population and we can expect large variations between individual hosts, both in their susceptibility and their ability to transmit. Ecologists study individual variation and the individual variations in hosts can have important repercussions to the dynamics of infections. After a parasite invades a host population, the first one or two hosts in the chain of transmission will play a major role in influencing the likelihood of an epidemic taking off so that even when R_0 is relatively low, the early infection of a 'superspreader' could initiate an important epidemic. For example, in the 2003 Severe Acute Respiratory Syndrome (SARS) epidemic, most infected people did not infect any susceptible hosts and yet a very small number of infected hosts infected a large number of other hosts. The SARS epidemic in Singapore was initiated by a single superspreader that brought the infection from Hong Kong. There is now growing evidence that susceptibility and transmissibility tends to covary between individual hosts so that a highly susceptible host will tend to infect more susceptibles, this introduces an interesting non-linearity into the epidemic and has a profound effect on the size of R_0, which will be increased about fourfold and result in different dynamics from that predicted by the 'average case'.

To launch an epidemic, an infectious host needs to infect at least one susceptible host. There then needs to be sufficient susceptible hosts for a chain of infection to start, with host density also determining the epidemic curve (Fig. 11.2). The number of susceptible individuals available will vary with host population size, giving rise to a critical threshold in host abundance (N_t), below which an epidemic cannot take hold. The pattern of mixing between infectious and susceptible hosts is also important, and the critical threshold will only exist when there is free mixing such that transmission is effectively density dependent. So, for example, measles is considered a disease with density dependent transmission, we assume there is free mixing between individuals and so the likelihood of becoming infected is simply the probability of a susceptible host coming into contact with an infectious host and then the probability of becoming infected from that host. Contact rate will increase with density so we have density dependent transmission. In contrast, when animals live in tight social groups, or when transmission depends on specific behaviours such as sexual contact between individuals, then transmission is not going to be dependent on host density but on the precise rate of contact. For example, it is possible to live in a high density population of individuals infected with HIV but an individual will only become infected with HIV if they have an intimate relationship with an infectious individual; the more intimate contacts they have the more likely they are to become infected and thus transmission can be considered frequency dependent (see Hudson et al. 2002; Altizer et al. 2003). In reality this is a continuum so that at the small scale, transmission is usually frequency dependent (measles transmission is between class mates at school) but at the larger population scale it appears density dependent and can be modelled as such. As host density is also a predictor of the number and diversity of parasites, because higher densities increase the spread of parasites within host populations (Stanko et al. 2002; Poulin & Mouillot 2004), those at high density may be at risk from a wider range of diseases.

What this all means for conservation is that infectious diseases with density dependent transmission will not drive species to extinction but those with frequency dependent transmission could cause extinctions. Those infections transmitted directly through sneezing and coughing, such as phocine distemper virus, canine distemper virus and rabies will tend to exhibit density dependent transmission and although these diseases will generate epidemics and reduce density dramatically they are unlikely to wipe out their host species, as they reduce host density until the chain of transmission is broken and then the parasite becomes locally extinct before the host, unless there is a disease reservoir. Infections that have frequency dependent transmission on the other hand, do not have the chain of transmission broken by a fall in density and indeed the vectors may act as a reservoir themselves so these diseases can drive populations to very low levels and ultimately extinction. So transmission is the key to understanding parasite dynamics but it is important not to get hung up about whether this is density dependent or not. In all reality it will be a combination of both. The important point is what type of disease are we looking at? Is it directly transmitted? Is there a reservoir?

Parasites as threats

Intuitively, one may suppose the most virulent parasites will be the ones that present the greatest threat to individuals and host populations because these, by definition, have the greatest case mortality rate; however this need not be the case. Extremely virulent parasites, with high levels of host mortality will tend to kill infected hosts before they are able to transmit to the next susceptible host. Ebola is highly virulent and tends to burn through a small proportion of the human population very fast and then be lost when the chain of transmission is broken. Parasites of moderate virulence are predicted to have a greater impact on the host population because they cause some parasite-induced mortality but the hosts live long enough for the chain of transmission to be sustained. However, when the parasite is a generalist, infecting a range of species and the impact it has varies between species, then it may persist within a reservoir species and occasionally spill-over to the more vulnerable species and lead to local populations of the

vulnerable species being wiped out. In this instance, transmission is not dependent on host density but on the occasional spill-over event from a reservoir host. This form of apparent competition is often referred to as parasite mediated competition (Hudson & Greenman 1999), and is immensely important in conservation when the reservoir host is a domestic animal such as a dog that harbours rabies or canine distemper and passes it to threatened carnivore, such as the Ethiopian wolf. Indeed, Macdonald (1993) highlighted how infectious disease in an abundant host (such as rabies in red foxes, *Vulpes vulpes*) might imperil a rare host (such as the Blandford's fox, *Vulpes cana*) living in their midst.

Parasites can regulate a host population when the growth rate of the parasite population is faster than the growth of the host population, such that parasite-induced effects (mortality and reduced fecundity) increase with host density, leading to regulation (Anderson & May 1978). A common misconception is to expect that when a population is being regulated by a parasite then we should observe parasite-induced mortality to be the principal cause of death. Field-workers collect dead bodies, undertake post-mortem analysis and invariably find that a small proportion of hosts have died from infections and then assume parasitism has no part to play in regulating the host population. This need not be the case. When parasites regulate a population, the proportion of the host population dying from an infection in a population at equilibrium, will be proportional to the growth rate of the population. Thus in a host population with a low growth rate and at equilibrium, parasite prevalence and the associated mortalities arising from infection do not have to be large for the parasite to be regulating the population. Anderson (1995) estimated that for a respiratory viral infection in foxes, a prevalence of just 0.18% could be regulatory, when the disease mortality rate is 50%. A survey of such a population would find that fewer that 1% of animals had died of infection and the conclusion could be drawn that the disease was of little significance to population regulation compared with, say, mortality due to road traffic accidents. This conclusion would be wrong, emphasizing that the dominant mortality factor need not be the one controlling the population and this needs to be kept in mind when judging the importance of any potential constraints on endangered populations.

Two detailed studies in the wild have provided evidence to suppose that macroparasites can regulate host populations: red grouse (*Lagopus lagopus*) (Hudson et al. 1992, 1998); and reindeer (*Rangifer tarandus*) (Albon et al. 2002). Interestingly, in both these systems the host population is unstable, showing cyclic fluctuations in abundance, and this instability has allowed the workers to examine how parasite-induced effects vary with host density. We know from the modelling work of May & Anderson (1978) that parasites will, in theory, regulate and generate instability when the parasites have a larger impact on host fecundity than on host mortality, when the parasites are distributed randomly between individual hosts and when there are time delays in the parasite life cycle. In a series of field experiments, Hudson et al. (1992) demonstrated that the parasitic nematode *Trichostrongylus tenuis* reduced the breeding production of the grouse. In a year of population crash a treated parasite-free hen raised nearly five chicks whereas a hen with a natural infection raised, on average, just half a chick. These huge effects of parasitism on breeding production coupled with parasites being weakly aggregated in the host population were sufficient to generate, in a computer model, the population cycles observed (Dobson & Hudson 1992). This nicely demonstrates scaling from individual to population effects and the need for population level experiments to reveal them. Reducing parasitism within the model significantly reduced the instability, demonstrating that the parasites played an important role in driving instability in these populations (Fig. 11.3; Hudson et al. 1998). In many respects this provided a fundamental understanding of how parasites influence host dynamics and the researchers used this to then explore how parasites interact with predators, with pathogens and competed with other vertebrates in the moorland communities of upland Britain (Hudson et al. 2002).

Regulation of a reindeer population was also shown to occur as a consequence of the impact of macroparasitic infection on fecundity (Albon et al. 2002; Stein et al. 2002). Again the host population exhibits unstable, oscillatory

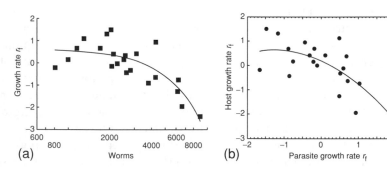

Fig. 11.3 Dynamics of *Trichostrongylus tenuis* in red grouse. (a) Red grouse population growth rate in relation to intensity of infection in adults, illustrating reduced growth rate when intensity of infection was high. (b) Growth rate of grouse population in relation to growth rate of parasite population showing that when parasite population growth rate is high it is associated with a negative growth rate of the host population and the host population growth rate is positive when the parasite population growth rate is negative.

change in abundance and the experimental removal of the gastrointestinal nematode *Ostertagia gruehneri* showed increased fecundity in terms of calf production. In both this study and the grouse study, it is the impact of the parasites on the mother's ability to care for their young that was particularly important in regulating the host population rather than the direct impact the parasites may have on the survival of the young. Similar to the red grouse, if we examine the time course of parasite and host abundance then the peak parasite abundance lagged behind the peak reindeer abundance, in this instance by 2 years (Fig. 11.4). Here is a case study that reveals that a small parasite-mediated reduction in fecundity (5–14%) in reindeer can regulate the host population simply because the average host population growth is inherently low (1–5%) in an environment where winter precipitation (snow/freezing rain) can have a strong, density independent influence on host calf production. Most calves may die as a consequence of the direct and indirect effects of the harsh winters but the critical, density dependent process that

is regulating the reindeer population is actually the parasite induced effects on fecundity.

The critical point is that both the studies of red grouse and reindeer were able to detect the impact of parasites because they conducted the experiments on the same populations over a number of years and could tease out potentially confounding results. In reality, doing long-term manipulative experiments in natural systems is difficult and may explain the paucity of examples demonstrating the role of parasites in the regulation of host populations (Albon et al. 2002). Furthermore both studies focused on an unstable population where the effects of parasite removal could be easily identified in the control populations. These findings help us appreciate how subtle the effects of parasitism can be and yet have a major role to play in regulating populations and of course the parasites may act with other factors to determine abundance; even populations greatly reduced by habitat loss may indeed be regulated by parasites within the remaining area of habitat.

Parasites as units of biodiversity

The contribution made by parasites to global biodiversity is large as we believe the majority of living organisms exhibit a parasitic form of life. It is reasonable to assume that each species of animal or plant will coexist with at least one specialist macroparasite species and at least one specialist microparasite species, although in many cases there is a whole community of parasites associated with each host species (Roberts et al. 2002; Lello et al. 2004). Although many parasites will have multiple hosts, any that are specific to a single host are potentially at risk if the host species is threatened, but few people will rally in the streets in protest at their loss. Invariably it is the charismatic and economically important species, be they revenue earners (e.g. through tourism) or revenue consumers (e.g. though conflict with people) that receive attention and resources. In species recovery programmes for example,

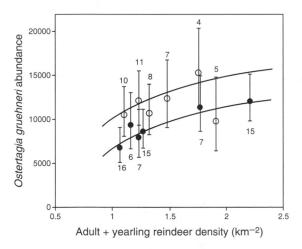

Fig. 11.4 Estimated *Ostertagia gruehneri* abundance in October in relation to adult reindeer numbers in summer 2 years earlier. Open and closed circles denote different sites, with associated regression lines, and error bars give 95% confidence intervals. (Adapted from Albon et al. 2002.)

parasites are undesirable and require control, if not eradication. Part of the immense efforts employed in the recovery programme for the black-footed ferret (*Mustela nigripes*) in North America involved dusting both ferrets and their prey, prairie dogs, with insecticide to reduce flea populations that carry sylvatic plague (Thorne & Williams 1988). There is some evidence that at least two specialist parasites may have been 'eradicated' or 'driven extinct' (the choice of term depending on whether this is judged from an epidemiological or conservationist perspective) in attempts to protect the black-footed ferret (Gompper & Williams 1998).

In considering the contribution to global biodiversity made by parasites we hit a snag. The common currency of biodiversity is the 'species', however, when comparing microorganisms with higher taxa, taxonomic inconsistencies arise that question whether the concept of 'species' represents the same thing across all taxonomic groups (May 1994). The example provided by May illustrates this neatly, by pointing out that the span of genetic diversity found within the bacterial species *Legionella pneumophila* is as great as the genetic distance between mammals and fishes.

Parasites, species diversity and human economic development

There are distinctive patterns in the global distribution of biodiversity, such as species diversity increasing towards the tropics but falling with altitude (Begon et al. 2005). Coupled with this we also see changes globally in human economic development, with the poorer nations within tropical regions where the high biodiversity is also found. There is evidence to suggest that parasites may play a role in both of these phenomena.

Environmental conditions within the tropics tend to suit parasite survival, with warmer, less variable temperatures and higher humidity. It has been argued that the greater parasite pressure within the tropics, compared with temperate and polar regions, has prevented any single species or group dominating ecological communities, permitting greater species coexistence at lower individual densities. Recent economic studies have suggested that this same parasite pressure within the tropics has hindered economic development (Fig. 11.5) relative to temperate regions, due to higher parasite burdens and a greater expression of disease (Bloom &

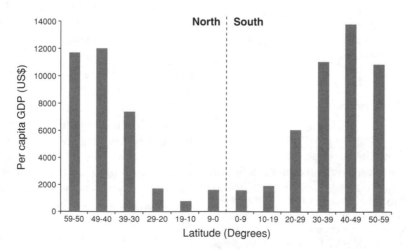

Fig. 11.5 Average gross domestic product (GDP) per capita for countries within latitudinal bands from north to south. (Source: Bloom & Sachs 1998.)

Sachs 1998; Gallup et al. 1999). Economic studies of biodiversity loss based on utilitarianism, regard biodiversity as an asset to be placed within the human societal portfolio. In these terms a driving force behind a species loss will be the choice of human investment in human-made or other natural assets over biodiversity as a biological 'asset' (Swanson 1994). Given the desire for economic growth within those nations containing most of the world's biodiversity, parasites may have unwittingly placed biodiversity in the path of an oncoming train, presenting two paradoxical desires: wanting to conserve biodiversity within developing nations, while also allowing those nations to develop economically and socially.

Conversely, parasitic pressure within the tropics still plays a role in protecting biodiversity, although with inevitable human economic cost. The tsetse fly (*Glossina* spp.) in Africa is a good example. Tsetse flies are vectors for trypanosomes responsible for sleeping sickness in humans and infest an area of Africa approximately 30% larger than the USA. The presence of tsetse flies and the failure of many years of attempted control have prevented the establishment of farming involving livestock, thus preventing human encroachment in large areas of Africa and preserving regional biodiversity (Rogers & Randolph 1988).

Further conservation problems and dilemmas.

The human–biodiversity interface

The majority of occasions when we observe parasitic infections threatening wildlife populations arise at the interface between people and biodiversity (Lafferty & Gerber 2002). This may be as a result of human encroachment on wild areas, with parasites being introduced via livestock or companion animals such as the introduction of rabies with domestic dogs with the Masai, from the consumption of bush-meat

which could lead to outbreaks of Ebola, or from the anthropogenic introduction of non-native species. Additionally, this interface has given rise to zoonotic disease outbreaks, posing direct threats to human life from wildlife (Weiss 2001), such as the recent emergence of SARS in China (Weiss & Mclean 2004), avian influenza virus and the origins of HIV/AIDS from a primate precursor possibly 70 years ago, although not becoming widespread in humans until recently due to the properties of R_0, with variation in contact rates within and between villages limiting spread until a 'critical mass' was reached with overall R_0 rising substantially above unity (May et al. 2001). Indeed the concept of 'viral chatter' involves pathogens spilling over from wild host species to humans regularly and then occasionally they become established and take off as a significant infection. For example, a recent study by Wolfe et al. (2005) showed that human T-lymphocyte viruses regularly spill over from wild primates to humans that hunt and keep primates. People also interact with biodiversity on a number of fronts, and parasites present unique issues in both directions across this multifaceted interface (McMichael 2004; Heymann 2005).

Controlling parasites

Efforts to control potentially devastating parasitic infections in endangered populations have mostly been retrospective crisis management, with epidemics often being well established and probably past their peak of mortality before control measures were initiated. Controlling infections in endangered populations presents real problems, given that time may be short.

First, the parasite must be identified. For example, the rapid decline in the abundance of vultures (*Gyps* spp.) in India showed all the characteristics of an infectious disease and efforts immediately focused on identifying the elusive pathogen. However, subsequent studies have shown that they died after eating an anti-inflammatory compound when scavenging livestock carcasses; cattle are sacred in India and the vultures play a major role in cleaning up cattle carcasses, but now the vultures die after eating the corpses. Large amounts of carrion are left, encouraging scavenging dogs and leading to outbreaks of rabies (Oaks et al. 2004).

Second, the significant sources of risk must be determined. Many parasites affecting endangered species have been introduced by human activities, often via livestock and companion animals. Where generalist parasites are concerned, spill-over host species may become infected, creating a reservoir for possible future epidemics within the endangered species. Examples of this have included vaccinating cattle against rinderpest in East Africa (Plowright 1982) and domestic dogs in the vicinity of endangered populations of Ethiopian wolf or African wild dog against rabies or canine distemper virus (Laurenson et al. 1997).

Given that we are starting to appreciate the importance of individual variations and that some individual hosts may be more important than others for initiating an epidemic (see above), then maybe the future lies in identifying the individuals that could be responsible for much of the epidemics. For example, retrospective studies of the Foot and Mouth epidemic in the UK in 2001 identified the presence of key farms (e.g. with high stock turnover or particular farming practises) that were critical in the chain of transmission, i.e. those where the spatial spread of the epidemic had to pass through, and vaccinating livestock on these would slow down and help with the control of any future epidemic (Keeling et al. 2003). In a similar manner detailed studies found that the transmission of the emerging disease tick borne encephalitis (TBE) in rodents flows mostly through the large body mass, sexually active male mice, indeed they contribute 93% towards the value of R_0, and focused treatment of these individuals would stop the spread of the disease through the reservoir host and then on to humans (Perkins et al. 2003, Hudson 2005).

SOCIAL PERTURBATION

The critical host density threshold (N_t), below which a density dependent epidemic cannot persist, has implications for parasite control. To achieve eradication, the proportion of susceptible individuals only needs to be reduced below N_t, for example, by vaccination or culling in the case of non-threatened reservoir hosts. Classically, density dependence has been regarded across entire populations, assuming that each individual has an equal chance of encountering and transmitting a parasitic infection to every other member of the population in some form of Brownian motion. With social species (and probably most others) this is not the case, where individuals in stable populations form aggregations or groups. Infectious contact rates will be far greater within social groups than between them, particularly where individuals aggregate in shared roosts, colonies or dens (or schools in the case of human childhood disease).

An early observation from red fox (*Vulpes vulpes*) culling within Europe for rabies control was the effect of population reduction on the social structure of the targeted populations. Group formation within culled populations was less apparent, prompting suggestions that this may consequently affect disease transmission (Macdonald 1995). Social perturbation to Eurasian badger (*Meles meles*) populations as a result of culling has also been suggested as a reason for this failure to control bovine tuberculosis (bTB) in the UK (Tuyttens & Macdonald 2000; Donnelly et al. 2003; Macdonald et al. in press). Badgers are a wildlife host for bTB, thought to constitute a reservoir of infection for cattle. Despite culling since the early 1970s, bTB prevalence in both badger and cattle populations has continually risen (Macdonald et al. in press). Within the UK badgers form stable, mixed-sex groups of up to 30 individuals, sharing a common range and one or more communal setts (Kruuk 1978). Badger culling operations have tended to target social groups associated with farms suffering a bTB outbreak, initially with gassing and later with cage trapping (Krebs et al. 1997). Whether the goal of these removals has been to locally reduce population density below the critical density N_t or remove infected animals has never really been spelled out. The absence of an effective test for bTB in live badgers (Woodroffe et al. 1999) has precluded the latter option, resulting in the often incomplete removal of social groups irrespective of disease status (Macdonald et al. in press). The complete or partial removal of social groups obviously leaves gaps within the population, creating new opportunities for social flux.

Culled badger populations have disrupted social organization, with group ranges overlapping significantly more after removals (Tuyttens et al. 2000). Such perturbation increases the levels of movement and contact between individuals. Thus, although population density may be locally reduced, social changes within the population effectively reduce N_t and thereby increase the effective R_0 (Macdonald et al. in press). An additional complication is that the social disruption associated with culling may increase stress within the surviving populations, as seen for example in elephants (Bradshaw 2005). Stress can have an immunosuppressive effect, increasing susceptibility and further enhancing the spread of disease in groups in the vicinity of culls.

VACCINATION

Vaccination has become a significant weapon for controlling disease in wild animals. As with culling, the principle is to reduce the density of susceptible individuals below the critical threshold, N_t, thus preventing the parasite from persisting within the population. The proportion of animals required to be vaccinated can be estimated to be greater than $1 - 1/R_0$. However, as detailed epidemiological information about natural infections in animal populations is often unavailable, determining the level

and spatial extent of vaccination necessary to achieve local eradication is difficult. Furthermore, where a reservoir host coexists with the endangered species, effective management may depend on the decision of which species to vaccinate.

Lessons may be learnt from the many rabies vaccination programmes that have been implemented (Macdonald 1980; Bacon 1985), directed at both wildlife populations and domestic species, usually dogs. Vaccination programmes have varied in their approach and their success, with occasional controversy. Vaccination of African wild dogs against rabies in the Serengeti prompted criticism that latent disease was being activated as a result of handling stress, thereby causing increased mortality (Burrows et al. 1994; East & Hofer 1996), a view questioned by others (Creel 1992; Macdonald 1992; Ginsberg et al. 1995). A precautionary upshot of this argument was to vaccinate domestic dogs, rather than risk endangered populations. However, concerns have been raised that the vaccination of domestic dogs may lead to increases in their numbers by removing rabies as a significant source of mortality (Cleaveland et al. 2002).

Despite these potential problems vaccination is considered an important tool in species conservation. The success of a recent vaccination of Ethiopian wolf against rabies in controlling disease spread within the wolf population (Randall et al. in press) highlights the advantage of vaccination as part of an effective strategy. The alternative of culling reservoir host species may be either unacceptable, for example, few Masai herders would want their dogs shot, or ineffective (see above). As an example the eradication of brucellosis from bison (*Bison bison*) in Yellowstone would require reducing the population to an unacceptably low level. Even then it is not clear if this would result in eradication because the elk (*Cervus elaphus*) are also infected and transmission effectively becomes frequency dependent as the animals are social (Dobson & Meagher 1996).

Small populations, metapopulations and introduced pests

For parasites transmitted in a frequency-dependent manner (e.g. via sexual transmission or arthropod vectors) there is no threshold host density for invasion, so these parasites have the potential to become established and even to wipe out endangered populations. Density-dependent parasite persistence, however, does require N_t to be exceeded and so it is possible that small, endangered populations may lose their density-dependent parasites. Indeed, this has been suggested as a potential mechanism to explain the success of invasive species, whereby the sizes of founding populations are too small to sustain parasitic infection. Once free from their parasites, these exotic species are able to be more successful in their new location (Torchin et al. 2003) (see Chapter 13).

Density-dependence is inevitably linked to the spatial scale at which the host population is examined (Hess et al. 2002). Spatial structuring within populations, given environmental heterogeneities, will lead to regions of relatively high and low densities, with density-dependent parasite transmission being unequal between patches (Macdonald et al. in press). Where hosts exist as metapopulations, the extent of density-dependent transmission will depend on patch connectivity, with parasites failing to persist in smaller and more isolated patches (Hess et al. 2002). The use of wildlife corridors connecting small populations effectively increases N_t and thus potentially permits parasites to invade.

Small isolated populations may also suffer from inbreeding with close relatives, thereby reducing genetic diversity and possibly reducing fitness due to an associated higher susceptibility or vulnerability amongst relatively homozygous individuals (Coltman et al. 1999) (see Chapter 4). The mechanism implicated is the major histocompatibility complex (MHC), responsible for immune function in vertebrates, which may increase susceptibility to parasitic infections (Paterson et al. 1998; Grenfell et al. 2002). Inbred populations of lion (*Panthera leo*) have been shown to have greater parasite loads than outbred populations; the Florida panther (*Puma concolor coryi*), which was reduced to fewer than 50 individuals, appears to be more sensitive to microparasite infection than neighbouring subspecies of puma that have not undergone such population bottlenecks, and genetic impoverishment in the cheetah (*Acinonyx jubatus*) is thought to have placed this endangered felid at even greater risk from parasites (O'Brien et al. 1985; O'Brien & Yuhki 1999).

Future directions

The question of whether parasites matter to biodiversity conservation can be answered only in the affirmative, but it is less obvious how to balance the positive and negative effects of parasites and their control in biodiversity action planning (Lafferty & Gerber 2002). In 2005 a workshop on this topic (Macdonald & Laurenson, in press) highlighted the importance of screening programmes in relation to both the protection of threatened populations and the emergence of zoonotic diseases, however, inadequate funding is almost inevitably an impediment. That said, predicting future risk is a difficult proposition and may possibly offer false hope. For example, few people would have predicted the emergence of an unknown coronavirus such as SARS (Weiss & Mclean 2004), irrespective of the amount of screening. Similarly, wildlife populations may have lived with pathogens of potential concern for many generations (e.g. rabies). In the case of natural diseases, it is perhaps arguable that conservation practitioners should adopt a non-interventionist strategy for managing disease outbreaks in wildlife populations. Such an approach is possibly an anathema in a risk averse society and requires a robust philosophical concept of 'naturalness' against which to judge the appropriateness of human interventions.

Fear of unknown risk was marvellously illustrated by US Defence Secretary Donald Rumsfeld in a speech about global terrorism in 2003: '...as we know, there are known knowns; there are things we know we know. We also know there are known unknowns; that is to say we know there are some things we do not know. But there are also unknown unknowns – the ones we don't know we don't know'. Precautionary principals arising out of such uncertainly may lead to the conclusion that the risks associated with letting natural diseases run their course are too great. For example, such an approach may be considered foolhardy for small, threatened populations such as the Ethiopian wolf (Haydon et al. 2002), where the risk is species extinction.

We have highlighted some of the subtleties of the interaction between parasites, wildlife and people and the consequences for conservation. Great progress has been made, but contradictions, paradoxes and confusion remain. In drawing attention to these we hope to contribute to a platform from which difficulties can be tackled, although it is doubtful we will ever 'know' all of our 'unknowns'.

Acknowledgements

We are indebted to Claudio Sillero, Merryl Gelling and anonymous referees for their valuable comments on earlier drafts of this essay.

At the beginning of a pestilence and when it ends, there's always a propensity for rhetoric.

(**Albert Camus, *The Plague*, 1946.**)

References

Albon, S.D., Stein, A., Irvine, R.J., Langvatn, R., Ropstad, E. & Halvorsen, O. (2002) The role of parasites in the dynamics of a reindeer population *Proceedings of the Royal Society of London Series B – Biological Sciences* **269**: 1625–32.

Altizcr, S., Nunn, C.L., Thrall, P.II., ct al. (2003) Social Organisation and Parasite Risk in Mammals: Integrating Theory and Empirical Studies. *Annual Reviews of Ecology and Systematics* **34**: 517–47.

Anderson, R.M. (1995) Evolutionary pressures in the spread and persistence of infectious agents in vertebrate populations. *Parasitology* **111**: S15–S31.

Anderson, R.M. & May, R.M. (1978) Regulation and stability of host-parasite interactions I: regulatory processes. *Journal of Animal Ecology* **47**: 219–47.

Anderson, R.M. & May, R.M. (1979) Population biology of infectious diseases: Part I. *Nature* **280**: 361–7.

Bacon, P.J. (1985) *Population Dynamics of Rabies in Wildlife*. Academic Press, London.

Begon, M., Townsend, R.C. & Harper, J.L. (2005) *Ecology: from Individuals to Ecosystems*. Blackwell Publishing, Oxford.

Berger, L., Speare, R., Daszak, P., et al. (1998) Chytidiomycosis causes amphibian mortality associated with population eclines in the rainforests of Australia and Central America. *Proceedings of the National Academy of Science, USA* **95**. 9031–9036.

Bloom, D.E. & Sachs, J.D. (1998) Geography, demography and economic growth in Africa. *Brookings Papers on Economic Activity* **2**: 207–95.

Bradshaw, G.A., Schore, A.N., Brown, J.L., Poole, J.H. & Moss, C.J. (2005) Elephant breakdown. *Nature* **433**: 807.

Burrows, R., Hofer, H. & East, M.L. (1994) Demography, extinction and intervention in a small population: the case of the Serengeti wild dogs. *Proceedings of the Royal Society of London, Series B* **256**: 281–92.

Cartwright, F.F. & Biddis, M. (1972) *Disease and History*. Rupert Hart Davis, London.

Cleaveland, S., Thirgood, S. & Laurenson, K. (1999) Pathogens as allies in island conservation? *Trends in Ecology and Evolution* **14**: 83–4.

Cleaveland, S., Hess, G.R., Dobson, A.P., et al (2002) The role of pathogens in biological conservation. In *The Ecology of Wildlife Diseases* (Eds P.J. Hudson, A. Rizzoli, B.T. Grenfell, H. Heesterbeek & A.P. Dobson), pp. 139–150. Oxford University Press, Oxford.

Coltman, D.W., Pilkington, J.G., Smith, J.A. & Pemberton, J.M. (1999) Parasite-mediated selection against in-bred Soay sheep in a free-living island population. *Evolution* **53**: 1259–67.

Creel, S. (1992) Cause of wild dog deaths. *Nature* **360**: 633.

Cunningham, A.A. & Daszak, P. (1998) Extinction of a species of land snail due to infection with a Microsporidian parasite. *Conservation Biology* **12**: 1139–41.

Daszak, P., Cunningham, A.A. & Hyatt, A.D. (2003) Infectious disease and emphibian population declines. *Diversity and Distributions* **9**: 141–50.

Dobson, A.P. & Hudson, P.J. (1992) Regulation and stability of a free-living host parasite system – *Trichostrongylus tenuis* in Red Grouse. II. Population models. *Journal of Animal Ecology* **61**: 487–98.

Dobson, A. & Meagher, M. (1996) The population dynamics of Brucellosis in the Yellowstone National Park. *Ecology* **77**: 1026–36.

Donnelly, C.A., Woodroffe, R., Cox, D.R., et al. (2003) Impact of localized badger culling on tuberculosis incidence in British cattle. *Nature* **426**: 834–7.

East, M.L. & Hofer, H. (1996) Wild dogs in the Serengeti ecosystem: what really happened. *Trends in Ecology and Evolution* **11**: 509.

Forsyth, M.A., Kennedy, S., Wilson, S., Eybatov, T. & Barrett, T. (1998) Canine distemper virus in a Caspian seal. *Veterinary Record* **143**: 662–4.

Gallup, J.L., Sachs, J.D. & Mellinger, A.D. (1999) *Geography and Economic Development*. CID Working Paper No. 1, Centre for International Development, Harvard University.

Gascoyne, S.C., Laurenson, M.K., Lelo, S. & Borner, M. (1993) Rabies in African Wild Dogs (*Lycaon pictus*) in the Serengeti Region, Tanzania. *Journal of Wildlife Diseases* **29**: 396–402.

Ginsberg, J.R., Mace, G.M. & Albon, S. (1995) Local extinction in a small and declining population: wild dogs in the Serengeti. *Proceedings of the Royal Society of London, Series B* **262**: 221–228.

Gompper, M.E. & Williams, E.S. (1998) Parasite conservation and the black-footed ferret recovery program. *Conservation Biology* **12**: 730–732.

Grenfell, B.T., Amos, W.A., Arneberg, P., et al. (2002) Visions for future research in wildlife epidemiology. In *The Ecology of Wildlife Diseases* (Eds P.J. Hudson, A. Rizzoli, B.T. Grenfell, H. Heesterbeek & A.P. Dobson), pp. 151–164. Oxford University Press, Oxford.

Haydon, D.T., Laurenson, M.K. & Sillero-Zubiri, C. (2002) Integrating epidemiology into population viability analysis: managing the risk posed by rabies and canine distemper to the Ethiopian wolf. *Conservation Biology* **16**: 1372–1385.

Hess, G.R., Randolph, S.E., Arneberg, P., et al. (2002) Spatial aspects of disease dynamics. In *The Ecology of Wildlife Diseases* (Eds P.J. Hudson, A. Rizzoli, B.T. Grenfell, H. Heesterbeek & A.P. Dobson), pp. 102–18 Oxford University Press, Oxford.

Heymann, D.L. (2005) Social, behavioural and environmental factors and their impact on infectious disease outbreaks. *Journal of Public Health Policy* **26**: 133–9.

Hudson, P.J. (1992) *Grouse in Space and Time*. Game Conservancy Trust, Fordingbridge.

Hudson, P.J. (2005) Parasites, diversity and the ecosystem. In *Parasitism and Ecosystems* (Eds F. Thomas, F. Renaud & J. Guegan), pp. 1–12. Oxford University Press, Oxford.

Hudson, P.J. & Greenman, J.V. (1998) Parasite mediated competition. Biological and theoretical progress. *Trends in Ecology and Evolution*. **13**: 387–90.

Hudson, P.J., Newborn, D. & Dobson, A.P. (1992) Regulation and stability of a free-living host parasite system – *Trichostrongylus tenuis* in red grouse. I. Monitoring and parasite reduction experiments *Journal of Animal Ecology* **61**: 477–86.

Hudson, P.J., Dobson, A.P. & Newborn, D. (1998) Prevention of population cycles by parasite removal. *Science* **282**: 2256–8.

Hudson, P.J., Rizzoli, A., Grenfell, B.T., Heesterbeek, H. & Dobson, A.P. (Eds) (2002) *The Ecology of Wildlife Diseases*. Oxford University Press, Oxford.

Keeling, M.J., Woolhouse, M.E.J., May, R.M., Davies, G. & Grenfell, B.T. (2003 Modelling vaccination strategies against foot-and-mouth disease. *Nature* **421**: 136142.

Krebs, J.R., Anderson, R., Clutton-Brock, T., et al. (1997) *Bovine Tuberculosis in Cattle and Badgers*. The Ministry of Agriculture, Fisheries and Food Publications, London.

Kruuk, H. (1978) Spatial organisation and territorial behaviour of the European badger (*Meles meles*). *Journal of Zoology* **184**: 1–19.

Lafferty, K.D. & Gerber, G.H. (2002) Good medicine for conservation biology: the intersection of epidemiology and conservation theory. *Conservation Biology* **16**: 593–604.

Laurenson, K., Shiferaw, F. & Sillero-Zubiri, C (1997) The Ethiopian wolf: status survey and conservation action plan. In *Disease, Domestic Dogs and the Ethiopian Wolf: the Current Situation* (Eds C. Sillero-Zubiri & D.W. Macdonald), pp. 32–42.

International Union for the Conservation of Nature and Natural Resources, Gland, Switzerland.

Laurenson, K., Sillero-Zubiri, C., Thompson, H., Shiefraw, F., Thergood, S. & Malcolm, J. (1998) Disease as a threat to endangered species: Ethiopian wolves, domestic dogs and canine pathogens. *Animal Conservation* **1**: 273–80.

Lello, J., Boag, B., Fenton, A., Stevenson, I.R. & Hudson, P.J. (2004) Competition and mutualism among the gut helminths of a mammalian host. *Nature* **428**: 840–4.

McMichael, A.J. (2004) Environmental and social influences on emerging infectious diseases: past, present and future. *Philosophical Transactions of the Royal Society of London, Series B* **359**: 1049–58.

Macdonald, D.W. (1980) *Rabies and Wildlife: a Biologist's Perspective*. Oxford University Press, Oxford.

Macdonald, D.W. (1992) Cause of wild dog deaths. *Nature* **360**: 633–4.

Macdonald, D.W. (1993) Rabies and wildlife: a conservation problem? *Onderstepoort Journal of Veterinary Research* **60**(4): 351–5.

Macdonald, D.W. (1995) Wildlife rabies: the implications for Britain – unresolved questions for the control of wildlife rabies: social perturbation and interspecific interactions. In *Rabies in a Changing World*, pp. 33–48. Proceedings of the British Small Animal Veterinary Association, Cheltenham.

Macdonald, D.W. & Laurenson, K. (Eds)(In press) *Infectious Disease and Mammalian Conservation. Biological Conservation* (Special Issue).

Macdonald, D.W., Riordan, P. & Mathews, F. (In press) Biological hurdles to the control of TB in cattle: a test of two hypotheses to explain the failure of control. In *Infectious Disease and Mammalian Conservation* (Eds D. W. Macdonald & K. Laurenson). *Biological Conservation* (Special Issue).

Mamaev, L.V., Denikina, N.N. Belikov, S.I., et al. (1995) Characteristics of morbilliviruses isolated from Lake Baikal seals (*Phoca sibirica*). *Veterinary Microbiology* **40**: 251–9.

May, R.M. (1994) Conceptual aspects of the quantification of the extent of biological diversity. *Philosophical Transactions of the Royal Society of London, Series B* **345**: 13–20.

May, R.M. & Anderson, R.M. (1978) Regulation and stability of host-parasite population interactions II: destabilizing processes. *Journal of Animal Ecology* **47**: 249–67.

May, R.M., Gupta, S. & Mclean, A.R. (2001) Infectious disease dynamics: what characterizes a successful invader? *Philosophical Transactions of the Royal Society of London, Series B* **356**: 901–10.

Mouritsen, K.N. & Poulin, R. (2005) Parasites boosts biodiversity and changes animal community structure by trait-mediated indirect effects. *Oikos* **108**: 344–50.

Newborn, D. & Foster, R. (2002) Control of parasite burdens in wild red grouse *Lagopus lagopus scoticus* through the indirect application of anthelmintics. *Journal of Applied Ecology* **39**: 909–14.

Nunn, C.l., Altizer, S., Sechrest, W., Jones, K.E., Barton, R.A. & Gittleman, J.L. (2004) Parasites and the evolutionary diversification of primate clades. *American Naturalist* **164**: S90–S103.

O'Brien, S.J. & Yuhki, N (1999) Comparative genome organization of the major histocompatibility complex: lessons from the Felidae. *Immunol Rev.* **167**: 133–44.

O'Brien, S.J., Roelke, M.E., Marker, L., et al. (1985) Genetic basis for species vulnerability in the cheetah. *Science* **227**: 1428–34.

Oaks, J.L., Gilbert, M., Virani, M.Z., et al. (2004) Diclofenac residues as the cause of vulture population decline in Pakistan. *Nature* **427**: 630–3.

Paterson, S., Wilson, K. & Pemberton, J.M. (1998) Major histocompatability complex variation associated with juvenile survival and parasite resistance in a large unmanaged ungulate population. *Proceedings of the National Academy of Sciences, USA* **95**: 3714–9.

Perkins, S.E., Cattadori, I.M., Taglieppetta, V., Rizzoli, A.P. & Hudson, P.J. (2003) Empirical evidence for key hosts in persistence of a tick borne disease. *International Journal of Parasitology* **33**: 909–17.

Plowright, W. (1982) The effects of rinderpest and rinderpest control on wildlife in Africa. *Symposium of the Zoological Society of London* **50**: 1–28.

Polis, G.A. (1999) Why are parts of the world green? Multiple factors control productivity and the distribution of biomass. *Oikos* **86**: 3–15.

Poulin, R. & Mouillot, D. (2004) The evolution of taxonomic diversity in helminth assemblages of mammalian hosts. *Evolutionary Ecology* **18**: 231–47.

Randall, D.A., Williams, S.D., Kuzmin, I.V., et al. (2004) Rabies in endangered Ethiopian wolves. *Emerging Infectious Diseases* **10**: 2214–7.

Randall, D.A., Marino, J., Haydon, D., et al. (In press) Impact and management of rabies in Ethiopian wolves. In *Infectious Disease and Mammalian Conser-*

vation (Eds D. W. Macdonald & K. Laurenson) *Biological Conservation* (Special Issue).

Roberts, M.G., Dobson, A.P., Arneberg, P., et al. (2002) Parasite community ecology and biodiversity. In *The Ecology of Wildlife Diseases* (Eds P.J. Hudson, A. Rizzoli, B.T. Grenfell, H. Heesterbeek & A.P. Dobson), pp. 63–82. Oxford University Press, Oxford.

Roelke-Parker, M.E., Munson, L., Packer, C., et al. (1996) A canine distemper virus epidemic in Serengeti lions (*Panthera leo*). *Nature* **381**: 172.

Rogers, D.J. & Randolph, S.E. (1988) Tsetse flies in Africa: bane or boon? *Conservation Biology* **2**: 57–65.

Sillero-Zubiri, C., King, A.A. & Macdonald, D.W. (1996) Rabies and mortality in Ethiopian wolves (*Canis simensis*). *Journal of Wildlife Diseases* **32**: 80–96.

Soulé, M.E., Estes, J., Berger, J. & Martinez del Rio, C. (2003). Ecological effectiveness: conservation goals for interactive species. *Conservation Biology* **17**: 1238–50.

Stanko, M., Miklisova, D., de Bellocq, J.G. & Morand, S. (2002) Mammal density and patterns of ectoparasite species richness and abundance. *Oecologia* **131**: 289–95.

Stein, A., Irvine, R.J., Ropstad, E., Halvorsen, O., Langvatn, R. & Albon, S.D. (2002) The impact of gastrointestinal nematodes on wild reindeer: experimental and cross-sectional studies. *Journal of Animal Ecology* **71**: 937–45.

Swanson, T.M. (1994) The economics of extinction revisited and revised: a generalised framework for the analysis of the problems of endangered species and biodiversity losses. *Oxford Economic Papers* **46**: 800–21.

Thomas, F., Renaud, F. & Guegan, J. (2005) *Parasitism and Ecosystems*. Oxford University Press, Oxford.

Thompson, R.M., Mouritsen, K.N. & Poulin, R. (2005) Importance of parasites and their life cycle characteristics in determining the structure of a large marine food web. *Journal of Animal Ecology* **74**: 77–85.

Thorne, E.T. & Williams, E.S. (1988) Disease and endangered species: the black-footed ferret as a recent example. *Conservation Biology* **2**: 66–74.

Torchin, M.E., Lafferty, K.D., Dobson, A.P., McKenzie, V.J. & Kuris, A.M. (2003) Introduced species and their missing parasites. *Nature* **421**: 628–30.

Tuyttens, F.A.M. & Macdonald, D.W. (2000) Consequences of social perturbation for wildlife management and conservation. *Behaviour and Conservation* (Eds M. Gosling & W. Sutherland), pp. 315–329. Cambridge University Press, Cambridge.

Tuyttens, F.A.M., Delahay, R.J., Macdonald, D.W., Cheeseman, C.L., Long, B. & Donnelly, C.A. (2000) Spatial perturbation caused by a badger (*Meles meles*) culling operation: implications for the function of territoriality and the control of bovine tuberculosis (*Mycobacterium bovis*). *Journal of Animal Ecology* **69**: 815–28.

Weiss, R.A. (2001 The Leeuwenhoek Lecture (2001) Animal origins of human infectious disease. *Philosophical Transactions of the Royal Society of London, Series B* **356**: 957–77.

Weiss, R.A. & Mclean, A.R. (2004) What have we learnt from SARS? *Philosophical Transactions of the Royal Society of London, Series B* **359**: 1137–40.

Whiteman, N.K. & Parker, P.G. (2005) Using parasites to infer host population history: a new rationale for parasite conservation. *Animal Conservation* **8**: 175–81.

Wikelski, M., J. Foufopoulos, H. Vargas & H. Snell. (2004) Galápagos birds and diseases: invasive pathogens as threats for island species. *Ecology and Society* **9**(1): 5. [Online at URL: http://www.e-cologyandsociety.org/vol9/ iss1/art5]

Wolfe, N.D., Heneine, W., Carr, J.K., et al. (2005) Emergence of unique primate T-lymphotropic viruses among central African bushmeat hunters. *Proceedings of the National Academy of Sciences* **102**: 7994–9.

Woodroffe, R., Frost, S.D.W. and Clifton-Hadley, R.S. (1999) Attempts to control tuberculosis in cattle by removing infected badgers: constraints imposed by live test sensitivity. *Journal of Applied Ecology* **36**: 494–501.

Woodworth, B.L., Atkinson, C.T., LaPointe, D.A., et al. (2005) Host population persistence in the face of introduce vector-borne diseases: Hawaii amakihi and avian malaria. *Proceedings of the National Academy of Sciences, USA* **102**: 1531–6.

12

The nature of the beast: using biological processes in vertebrate pest management

Sandra Baker, Grant Singleton and Rob Smith

The affair runs always along a similar course. Voles multiply. Destruction reigns. There is dismay, followed by outcry, and demands to Authority. Authority remembers its experts and appoints some: they ought to know. The experts advise a cure. The cure can be almost anything: golden mice, holy water from Mecca, a Government Commission, . . . prayers denunciatory or tactful, a new God, a trap, a Pied Piper. The Cures have only one thing in common: with a little patience they always work. They have never been known to fail. Likewise they have never been known to prevent the next outbreak.

(Charles Elton,*Voles, Mice and Lemmings: Problems in Population Dynamics*, Oxford University Press, 1942.)

Introduction

Vertebrate pests conflict with the economic, health and recreational interests of people worldwide, and many pests have adverse impacts on wildlife conservation, e.g. introduced rat species have caused extinction of many island populations of endemic bird species. The traditional approach to reducing conflict is to kill animals believed to be responsible. Lethal control tackles human–wildlife conflict directly, but is rarely straightforward in practice, e.g. culling is often perceived as undesirable in terms of welfare, ethics or conservation. These are valid philosophical issues, but a more important practical point is that culling might also prove ineffective, or even counterproductive, in terms of reducing problems (Baker & Macdonald 1999).

Conservation is concerned with maintaining biodiversity and viable populations and the effects of some pests on conservation values are severe. Vertebrate pests together with land clearing contributed to the extinction of more than 20 mammal species endemic to Australia in less than 200 years, and a further 43 mammals are endangered or vulnerable. The economic impact in Australia of 11 vertebrate pest species was recently estimated at US$500 million each year (McLeod 2004). An important goal for conservation biologists is to develop management methods that are both sustainable and acceptable (environmentally and socially).

Conservation biologists frequently attempt to control pest animals, such as rats, in order to protect other species, and a deep understanding of the ecology of the pest species is necessary to achieve this goal. It is often essential in particular to understand the foraging behaviour (Macdonald et al. 1999), population structure (Smith 1999) and breeding ecology (Singleton et al. 2001) of the targeted pest in order to develop effective management strategies. It also can be important to understand wider aspects of the behaviour and biology of the target animal. Complex behaviour can render lethal control ineffective, even counterproductive, whereas understanding and thus using such complexities can make non-lethal control surprisingly effective. We will use examples of both ineffective control and effective, biologically based management of both native and introduced pest species (primarily rodents and canids) in this essay in order to illustrate the importance of these principles to conservation management.

Efficacy problems associated with culling

Density-dependent compensation

Culling includes lethal management programmes aimed either at keeping a population within a desired band of densities (harvesting), or below critical threshold levels, or at eliminating animals from certain areas (control) (Caughley & Sinclair 1994). Although culling temporarily reduces population numbers and can occasionally eradicate a species (e.g. Tasmanian Tiger: Guiler 1985), removing individuals from a population does not generally lead to a lasting reduction in population size after culling stops (e.g. Reynolds et al. 1993).

The aim of effective wildlife management is usually to limit, rather than to exterminate, a population. A population that has grown to the maximum that can be supported by a limiting resource is said to be at **carrying capacity**. Changes in the size of a population depend on the number of animals joining it, through birth or immigration, and leaving, through death or emigration. Birth, death, immigration and emigration rates are not, however, constant – one or more of them often changes with population density in a predictable way known as **density dependence**. Birth/immigration and death/emigration are opposing processes influenced by factors that include competition, predation, disease and weather as well as management.

As populations increase towards their carrying capacity, density-dependent compensation may come into force, acting to reduce the population growth rate. Density-dependent processes include reduction in birth or immigration rates, increase in mortality or dispersal, or some combination of these. If population size is reduced by an external influence, e.g. a bout of disease or culling, density-dependent constraints are relaxed and the population growth rate increases (Smith 1999). Even irruptive species, which include many rodent pests that are characterised by irregular outbreaks, are subject to intrinsic density-dependent compensation, which often interacts with extrinsic environmental factors, including culling.

A key goal of management is therefore to increase the intensity of a density-dependent force or to add a density independent force (e.g. fertility control) that is sufficient to reduce the population, and will hold it below the natural equilibrium density (Krebs 2001). See Fig. 12.1. This is not always easy; culling constitutes little more than a harvest sustained by compensatory production or survival. This phenomenon is exploited, for example, for venison production, in which carefully planned harvesting aims to ensure that deer are removed without reducing productivity in future years (Macdonald et al. 2000). Harvesting and pest control both involve the same biological principles; only the desired outcomes are different and achieving these outcomes can be difficult for practitioners – if their management models are wrong, they could reduce something they were trying to preserve (as has happened in fisheries management), or fail to reduce something they were trying to control. Sustained population decline is achieved only when cull rate exceeds the maximum rate of growth the population can achieve. In addition, culling itself is density-dependent; it becomes more difficult to kill each additional animal as population density decreases. Culling needs to be conducted regularly to have a lasting effect. In territorial species, culling should be targeted at breeding females, because animals killed

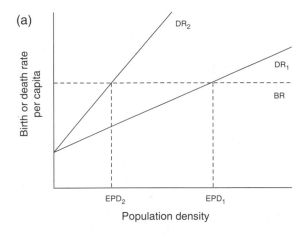

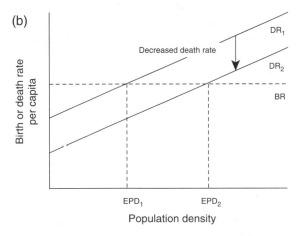

Fig. 12.1 Simple graphical model to illustrate how two populations may differ in average abundance. Population density comes to equilibrium when the per capita birth rate (BR) equals the per capita death rate (DR), shown by the dotted lines. Only the death rate is density dependent here (assumed linear for simplicity). (a) The density dependence is stronger in DR_2 than in DR_1, and the equilibrium population density (EPD) is lowered. (b) The density-independent component of the death rate is decreased from DR_1 to DR_2, and this increases the equilibrium population density. Similar changes could result from shifts in the birth rate, which is illustrated here as a constant for simplicity. (Modified from Krebs 2001.)

non-selectively, especially between breeding seasons, are more likely to be dispersing juveniles (Reynolds et al. 1993) from the 'doomed surplus', destined to die anyway when resources are limited, e.g. over winter (Errington

1946). As well as causing reductions in the effectiveness of lethal control, density-dependent responses may sometimes even render culling counterproductive. Intensive removal of territorial red foxes (*Vulpes vulpes*) in order to protect partridges in southern England led to their swift replacement by immigrants, thereby producing only local and short-term reductions in partridge predation (Reynolds et al. 1993). Non-lethal techniques have the potential to prove more efficient than lethal control by avoiding the density-dependent consequences that can follow culling (Baker & Macdonald 1999), in addition to increasing acceptability in welfare and ethical terms. Examples include successful (non-lethal) oral vaccination campaigns against rabies in foxes (see Macdonald 1988) and social perturbation resulting from lethal control (see Chapter 11).

Why is it so difficult to control rats? Neophobia, bait aversion and resistance

Rats defy conservation efforts across the world because they are adaptable, aggressively competitive/predatory and notoriously difficult to control. Introduced rat species (primarily *Rattus norvegicus* and *Rattus rattus*) threaten many island populations of seabirds and have driven other island populations of many endemic bird species to extinction. Rats feature conspicuously among the mammals that have most affected the course of human history, costing billions of dollars annually in terms of prophylactic or remedial control, disease transmission and damage to crops and stored food. In Laos, Kenya and Tanzania, rodent damage is episodic and often patchily distributed (Singleton et al. 1999) such that some families commonly lose over 70% of their crop to rodents; such high losses in consecutive years can be catastrophic. In Indonesia, rats consume enough rice each year to feed 25 million people (Stenseth et al. 2003), and more land has to be cultivated to make good these losses, with predictably adverse effects on natural habitats and biodiversity. Rat-control is attempted primarily through the use of poisoned bait, and rats have two main defences against poisoning: the highly evolved behavioural adaptations described below and, in some areas, physiological resistance to the most effective (anticoagulant) poisons. As a result, poisoning efforts constitute a series of chemical attacks and evolutionary defences, in what amounts to a billion-pound experimental demonstration of evolution in action (Macdonald et al. 1999).

Thompson demonstrated in 1948 that rats react to novel stimuli with extreme caution. Such **neophobia** assists rats in avoiding poison, as does their ability to associate a food with the metabolic consequences long after it was ingested (learned or conditioned taste aversion (CTA) leading to **bait shyness**, which can persist for eight months), and their capacity to interpret food-safety cues from other group members (the 'demonstrator' effect). Bait shyness is the reason why fast-acting poisons are generally ineffective at controlling rats. The introduction in the 1950s of delayed-action, anticoagulant poisons based on warfarin marked a breakthrough in rat control. Anticoagulant poisons act slowly such that animals generally ingest a lethal dose before the symptoms of toxicosis develop; this overcomes the problem of aversion learning (see 'learned food aversions' below).

Warfarin binds an enzyme in the vitamin K cycle and indirectly reduces levels of blood-clotting factors, resulting in poor blood coagulation and ultimately death through many small internal haemorrhages. Control failure owing to warfarin resistance was reported in 12 areas of the UK less than two decades after its introduction. The altered enzyme present in warfarin-resistant individuals is less affected by the poison, but is also less efficient at recycling vitamin K, and resistant rats need a higher vitamin K intake to retain normal clotting activity, at least in one (Welsh) population. Homozygous resistant Welsh rats are most dramatically affected, suffering reduced growth, status and reproduction (Smith et al. 1991), as well as an increased mortality rate. Balanced polymorph-

ism in wild populations regularly controlled with warfarin suggests the existence of a selective advantage to heterozygotes. Susceptible rats suffer reduced fitness via the poison, homozygous resistant rats have reduced fitness (*via* vitamin K deficiency) and heterozygotes enjoy an efficient compromise between warfarin resistance and vitamin K deficiency. The lower fitness of both resistant genotypes in the absence of poison advocates a temporary relaxation of warfarin use to reduce the frequency of resistant individuals (Smith & Greaves 1987). More recent work in central southern England, however, suggests that some resistant genotypes have high fitness even in the absence of poison (Smith et al. 1993).

Warfarin resistance was overcome by the development of more toxic, 'second generation' anticoagulants, such as brodifacoum, in the 1970s. There is now evidence that rats are resistant to two of the four second-generation anticoagulants in parts of the UK. Lethal control of rodent pests using anticoagulants is, however, still effective as a short-term culling measure in most of the UK and the rest of the world. Lethal control using anticoagulants has been effective as a longer-term measure for conserving native flora and fauna on islands (1–1970 ha) in New Zealand through successfully removing rodent species that recently invaded the islands. This is a special case because the bait formulation is unattractive to ground-dwelling birds and there are no native mammals present, hence there are no important concerns about poisoning non-target species.

Indirect effects of pest control on conservation: secondary poisoning of non-target species

The main route of pesticide exposure to terrestrial mammals is through accumulation of pesticide residues via feeding on contaminated food items. Rodenticides are, by necessity, highly toxic to vertebrates and they pose a direct (primary poisoning) threat to non-target species (birds and mammals) that consume poisoned bait. They also present considerable risk of secondary poisoning to scavenging or predatory species feeding both on contaminated rats and mice and on non-target species that consume bait, e.g. bank voles and wood mice (Harrington & Macdonald 2002; Brakes & Smith 2005). Resistance to some first-generation anticoagulants instigated their replacement largely with second-generation anticoagulants, which are 100–1000 times more toxic to mammals. This increased toxicity can provide a lethal dose in one meal, but the delayed action of all anticoagulants allows a poisoned animal to continue feeding on poisoned bait after consuming a lethal dose, thereby accumulating higher residue levels (Smith 1999). This delayed action increases the problem of secondary poisoning, as does poison resistance in rats.

Poisoned rodents have been known to survive for up to two weeks before dying, during which time pre-lethal effects of anticoagulants cause rats to spend more time in the open, and mice above ground, thereby increasing the risk of predation. Rodenticide-resistant rats may carry residue levels over three times greater than non-resistant rats, effectively becoming mobile packages of poison to potential predators (Helen MacVicker's data, referred to by Smith 1999), e.g. mustelids such as European polecats (*Mustela putorius*), stoats (*M. erminea*) and weasels (*M. nivalis*). Post-mortem examination of two polecats, found dead in a Welsh farmyard, first revealed signs of secondary, anticoagulant poisoning in the late 1960s. Similar incidents have occurred since and two recent surveys suggested extensive, widespread exposure to secondary rodenticide poisoning among polecats (Shore et al. 1996), stoats and weasels (McDonald et al. 1998). Rodenticides were detected in the livers of stoats (22.5%) and weasels (30%) killed by gamekeepers on shooting estates, but not in stoats from an estate where rodenticides were not used. Contamination was detected in polecats (26%) from three of five UK counties tested. There is insufficient toxicological information to know whether the concentrations observed would have caused mortality or sublethal effects. Secondary poisoning may be underestimated because of the improbability of finding dead or dying animals. Polecats forage and rest in farmyards mainly in winter when most rodenticide is used. In both McDonald et al.'s (1998) and Shore et al.'s (1996) studies, all contaminated animals were killed in winter or spring. The lack of residues in animals killed between May and October suggests that rodenticides accumulating over winter were metabolized and eliminated by that time, or that contaminated animals had died earlier. In Britain, rodenticide use is greater in eastern arable parts of Britain than in the pastoral west. As the polecat recovers its range eastwards it might therefore encounter increasing levels of exposure. Furthermore, in areas where intensive farming restricts other prey, polecats may be forced to feed even more heavily on rats, further increasing the chances of secondary poisoning.

Secondary poisoning was used to advantage in New Zealand where introduced mustelids (stoats, weasels and ferrets) are major predators of the native parrots and robins. Of these predators, the European stoat has the greatest impact. Stoats shift between eating black rats (*Rattus rattus*) and native birds, depending upon the abundance of rats. Second generation anticoagulants proved more appropriate than poisons such as 1080 (sodium fluoroacetate) in controlling these rodent species, because anticoagulants killed stoats through secondary poisoning (as well as the target rats), while 1080 simply killed rats and encouraged predators to switch to birds for food (Murphy et al. 1998).

Exploiting behaviour in non-lethal control

Learned food aversions

Conditioned taste aversion (CTA) is a type of learned food aversion with potential for use as a wildlife-management tool. Conditioned taste aversion is a natural phenomenon that evolved to prevent poisoning, and it is widespread across the taxa, from molluscs to humans. To form a CTA, an animal must ingest a toxic food item and correlate the resulting post-ingestional effects (illness, nausea, malaise) with the taste of that food after a single or small number of exposures to it (see Baker et al. in press). Conditioned taste aversion retains an association between the conditioned and unconditioned stimulus, i.e. food taste and illness respectively (Cowan et al. 2000). The animal forms an association between the taste of that food and the post-ingestional effects, and subsequently avoids the taste of that particular food. Such aversions are said to be acquired 'subconsciously'. Stimulus of a CTA depends on the severity and nature of the illness, the time between the conditioned and unconditioned stimuli, the salience of the conditioned stimulus, and whether it is novel or familiar to the target species. Both the development and duration of CTAs are highly complex, and influenced by social and environmental factors, sex, individual variation, prey-recognition cues, stimulus reinforcement, length of training period and choice of aversive agent.

A CTA can be caused deliberately in a pest species by administering an emetic to the target food commodity; this should induce vomiting or nausea, and subsequently aversion to the target food, in the pest species. Where learned food aversions are created in a resident population of a problem territorial species, managers should benefit, because food preferences in the target species are altered, while other ecological relationships are left intact, including continued exclusion of untrained conspecifics through territorial defence (Reynolds 1999). In other words, the 'poacher' becomes the 'gamekeeper'. In 1974, Gustavson and colleagues conditioned captive coyotes not to attack live sheep while retaining their appetite for alternative prey. Researchers quickly moved to ambitious, large-scale field trials that attempted to use CTA to deter wild coyotes from attacking sheep, but trials were poorly designed and re-

sults inconclusive (see Reynolds 1999). In the early 1980s, Lowell Nicolaus and colleagues established two important principles. First, Nicolaus et al. (1982) showed unequivocally (using raccoons and chickens) that it was possible to inhibit the killing of live prey under field conditions. Second, Nicolaus et al. (1983) demonstrated that conditioned aversion using model baits can cause extensive reductions in predation on mimics; wild American crows (*Corvus brachyrhynchos*) that ate green-painted eggs containing an emetic subsequently avoided green eggs whether or not they were treated. Cox et al. (2004) used CTA to train captive carrion crows (*Corvus corone*) to delay their attack on a previously favoured egg colour, even when this was subsequently untreated. They also established that crows did not generalize their aversion to eggs of a different colour, a potentially important consideration when designing realistic wildlife-management strategies. Nicolaus and co-authors have now demonstrated the capacity of a variety of free-ranging egg predators to learn to avoid consuming eggs through CTA. These include ravens, crows, raccoons, mongooses and guilds of mammalian predators. More recently Baker et al. (2005, in press) demonstrated that badgers could be conditioned, via CTA, to avoid foods on the basis of an odour cue. This most likely occurred through second-order conditioning, a two-stage process in which first the taste of a food becomes aversive, and then an odour cue becomes associated with the aversive taste.

Cowan et al. (2000) concluded that research on learned food aversions needs to progress on two fronts: (i) fundamental studies on learned food aversion; and (ii) field research on target animals' behaviour and field logistics, which may ultimately limit or prevent the efficient exploitation of CTA in wildlife management.

Scaring devices

Scaring devices were recommended for controlling bird pests as long ago as 1668 (Crocker

1984). Since then, a wide range of visual and acoustic devices has been used for scaring animals. These devices include scarecrows, lanterns, flashing lights, reflectors, flags, agricultural gas-guns, bangers, broadcasts of warning or distress calls, as well as model aeroplanes, predators or con-specifics. However, the results of some scaring methods are inconsistent, and many tests have proved unsuccessful. Draulans (1987) concluded that any effect of many scarers tended to be short-term, both audio and visual deterrents being prone to habituation. For example, audio deterrents failed to prevent gulls from roosting, and coyotes from preying on sheep. Ultrasonic devices produced no demonstrable deterrent effects on rodents. Some ultrasonic devices, and a compressed air alarm, each elicited aversive responses in some dogs, whereas a flashing light and other ultrasonic devices did not. The responses of raccoons (*Procyon lotor*) and opossums (*Didelphis marsupialis*) to starling (*Sturnus vulgaris*) distress calls varied widely as they attacked a caged starling; some individuals rejected the bird, while others attacked more. Birds habituated to visual deterrents, lanterns, bangers, and hanging sacks. Flashing lights had no consistent effect against fox predation on pheasants, and lanterns did not prevent fox predation on little terns (*Sterna albifrons*).

Scarers incorporating elements of biological significance to the target animal might have a greater chance of success. Scarers that feature movement or unpredictability, or that are reinforced naturally in the target animal's daily life (e.g. through real predation attempts by hawks in the case of a hawk-kite), might enhance or prolong scaring effects (Conover 1984; Crocker 1984). Conover (1984) cost-effectively reduced damage to corn by red-winged blackbirds (*Agelaius phoeniceus*) by 77%, using exploders, and 83%, using hawk-kites. A kite reduced wood-pigeon damage to spring cabbage for over 3 months, with no sign of habituation. Indeed, predator models are often used in an attempt to reduce bird damage to crops, although birds tend to habituate to inanimate models because the models' spatial context does not change and they never attack (Conover 1985). Traditional scarecrows can be similarly ineffective (Conover unpublished data), or have no long-term effect, e.g. on predation by kestrels (*Falco tinnunculus*) on little terns at Rye Harbour RSPB Reserve, and stationary gull corpses have been ineffective at deterring gulls from roosting or feeding.

Combinations of aversive stimuli may be more repellent than the individual components (e.g. Avery & Mason 1997). Scarers integrating different related stimuli can increase the effective area of a scarer, e.g. a stuffed crow with crow's call had a larger effective area than the call alone. Crocker (1984) promoted combining stimuli in the design of scarers, such that, rather than needing artificial reinforcement, e.g. by shooting, they are reinforced naturally in the target species' daily lives, e.g. by mimicking features of natural predators, or incorporating signals given by alarmed conspecifics. Goose models that 'head flagged' as if to take flight (as they would when frightened) (Crocker 1984), and open-winged woodpigeon models exposing white wing marks, were more effective than static, and closed-winged models respectively. Conover (1985) mounted a model of an owl grasping a crow on to a weather vane, such that it rotated, and its wings moved in the wind. Vegetable plots with moving models suffered 81% less damage by American crows (*Corvus brachyrhynchos*) than controls, and 83% lower damage than plots with static models. A person scaring birds was more effective at reducing goose damage than conventional scarecrows (Vickery & Summers 1992).

Scarers could disturb a prey species under protection, and therefore might be inappropriate for the protection of live prey. Predators' hunting success could also be disrupted, however, if prey became more vigilant as a consequence. Owners of domestic cats often exploit this phenomenon, by fitting their pets with bells to reduce predation on wild birds and mammals.

Guardian animals

Using one species of domesticated animal to protect another from predation is an ancient concept (Linnell et al. 1996). Dog and sheep remains have been found together in archaeological excavations since 3685 BC (Olsen 1985; cited in Coppinger & Coppinger 1993). There are reports that animals as diverse as dogs, donkeys, baboons, cattle, goats, ostriches and llamas can be trained to act as livestock guardians, and guardians have been used with varying success to protect a range of livestock including sheep, goats, cattle and poultry from a variety of canids, felids, bears and baboons. Protecting sheep from predatory canids with another type of canid is common in parts of Europe, but until recently was virtually unknown in the USA. Disadvantages of guarding dogs are that they sometimes harass livestock, humans or wildlife, might not guard the flock, and are expensive and labour intensive to train. Advantages, besides reduced predation, include reduction in the human labour required to tend the livestock, and opportunities to make more effective use of available grazing land.

The guardian species used will depend on the type of livestock being defended, the predator species, the intensity of predation, the grazing habitat of the livestock, and the management system employed by the producer (Linnell et al. 1996). The relative size and hunting style of the predator will determine the likely effectiveness of a particular guard animal, and the numbers required, e.g. a pack of wolves is more likely to be deterred by two dogs than by one (Coppinger 1992). Guardians need to demonstrate attentiveness and protectiveness towards their charges, as well as trustworthiness. Dog strains vary widely in attentiveness and guard animals could be bred selectively if such qualities are genetically based (Coppinger et al. 1983).

Guard dogs are quite distinct from herding dogs. Herding dogs are actively bonded with humans from a young age, whereas guard dogs often work independently of the herder, needing to be more strongly bonded with their flock. Guard dogs are long removed from the predatory end of the canid behaviour spectrum, having been genetically selected to retain adolescent behavioural traits. Care and training are important factors in the successful establishment of a guardian, a dog needing correction if it leaves its charges or indulges in negative play behaviour. Young guard dogs should ideally be placed with the appropriate species at a few weeks of age (Linnell et al. 1996).

In a survey of Namibian livestock farmers participating in a guarding dog trial, 73% reported a large decline in losses to cheetah and leopard since acquisition of the dog. Before placement of the dog, 71% of farmers reported a loss of 10 or more livestock per annum, whereas afterwards 65% reported none (Marker et al. 2005). In another survey, 82% of livestock producers who used guard dogs believed them to be an 'economic asset', and the vast majority of producers who grazed either pasture or open ranges recommended guard dogs (Linnell et al. 1996). Using dogs as guardians seems to have proved particularly successful, suggesting that there might be scope not only for developing further the breeding and training methods used with dogs, but also for employing all sorts of other species for this purpose.

Integrating other aspects of biology in pest-control programmes

Ecologically based management of rodent pests

An increase in the intensity of agricultural practices (a change from one or two rice crops per year, to two or three, plus subsidiary crops) has contributed to the re-emergence of rodents as major agricultural pests in Asia and Africa. Increasing need for environmentally sensitive wildlife management has stimulated a reassessment of approaches to rodent management that were often loosely described as integrated pest management. The concept of ecologically based rodent management (EBRM) was introduced as a formal framework for developing integrated management strategies, for rodent pests, based on sound ecological understanding (Singleton et al. 1999). This approach was first tested in the field for strategic management of mouse plagues in south-eastern Australia and has since proven successful in the protection of rice fields in Indonesia and Vietnam. We consider Indonesia briefly as a case study in Box 12.1.

Ecologically based management can be strengthened if farmers are provided with predictions of years when rodent numbers are likely to be high, and if they have access to advice on the likely economic benefits, given that they need to outlay money to conduct control early in the cropping season. Progress has begun along these lines for particular rodent species in Australia, Africa, Asia and Chile (Stenseth et al. 2003).

Biological control

The success of using predators and disease agents (e.g. fungi and viruses) for managing invertebrate populations leads people to question whether biological control could be helpful in controlling vertebrate pests. Unfortunately, despite 60 years of research, there are only two major success stories, the use of myxoma virus and rabbit haemorrhagic disease (RHD) on wild rabbit (*Oryctolagus cuniculus*) populations in Australia. Myxoma virus was successfully released in 1950 and it was estimated that mortality was often around 99% in rabbit populations that became infected (Cooke & Fenner 2002). The success of myxoma virus was confined to areas (mainly in the temperate and semi-arid zones) where there was good survival of the vectors of the virus – rabbit fleas and mosquitoes. Rabbit populations bounced back within 5 years as they became immune to various strains of the virus. The

Box 12.1 Application of ecologically based rodent management in Indonesia

After a 4-year ecological study in West Java of the main rodent pest (the rice-field rat, *Rattus argentiventer*), ecologically based management methods were tested in a replicated experimental study at a village level (100 ha and about 120 families) for three years. Management actions adopted and applied by farmers included better synchrony of cropping (which reduces the breeding season of rats), reduced width of water-retaining banks or 'bunds' (to minimize availability of nest sites), a new technology known as the community trap-barrier system, and community campaigns to catch rats at a time of year when they are aggregated in key habitats and prior to the onset of breeding. At the end of the study there was a reduction of 49% in chemical usage, a mean yield increase of 6% and reduced costs of management actions (Singleton et al. 2005). The benefit-to-cost ratio for all years (five crop seasons) averaged 25:1 but varied considerably, from a low of 2:1 to a high of 63:1.

myxoma virus also changed genetically with selection favouring a virus that was more persistent, but which had lower mortality rates, and this highlights a potential risk; deliberate introductions might evolve, jump species and become alien invaders (see Chapter 13).

In late 1995, RHD was accidentally introduced to the Australian rabbit populations after it escaped from an island where it was being tested under quarantine conditions, and was subsequently unlawfully released into New Zealand. Cooke & Fenner (2002) contributed to a special volume of *Wildlife Research* that deals with the impact of RHD in rabbit populations in Australia and New Zealand, and its interaction with myxoma virus. Rabbit haemorrhagic disease has been particularly successful in managing rabbit populations in low rainfall areas through annual epizootics in both Australia and New Zealand. This has led to significant environmental and economic benefits. There is limited evidence of interaction between myxomatosis and RHD; myxomatosis epizootics are more common in autumn since the introduction of RHD, whereas spring epizootics are delayed. A later paper reported that the rabbit population in southern Queensland was estimated to have declined by 90% some 30 months after RHD had spread to that region (Story et al. 2004).

Fertility control

Fertility control as an alternative approach to lethal control has generated interest in recent years, with special editions of journals covering some of the developments (*Reproduction, Fertility and Development* 6, 1994; *Reproduction* **60** (Supplement), 2002). Gao & Short (1993) reviewed the potential of chemosterilants for rodent control and Chambers et al. (1999) reviewed the relative advantages and disadvantages of reducing fertility versus increasing mortality for managing populations of rodent pests.

A relatively new concept is the delivery of a reproductive protein that generates an immune response, with the antibodies blocking fertilization in the female host. This approach is known as immuno-contraception – it has been used successfully on island populations of horses and white-tailed deer (Kirkpatrick et al. 1997) and holds potential for controlling fox, mouse and rabbit populations. The real challenge for immuno-contraception is the delivery mechanism. In Kirkpatrick et al.'s study, horses were vaccinated individually, which is impractical for most vertebrate pest populations. The immuno-contraceptive potentially could be delivered by oral baits or using a virus as a carrier. Laboratory studies on the house mouse (*Mus domesticus*) have confirmed proof of the concept for viral delivery of an immuno-contraceptive antigen. Furthermore, studies of wild mouse populations and the epidemiology of mouse cytomegalovirus (MCMV) in enclosures and cereal fields indicate that MCMV would be an effective vector for an immuno-contraceptive vaccine of wild mice (Singleton et al. 2002). The use of a virus as a carrier of a sterility agent is controversial, even

in the case of MCMV, which appears to be species-specific and has relatively benign effects on wild mouse populations. The main issue revolves around an international debate on the safety of genetically modified organisms; risks need to be adequately assessed and weighed against the benefits of reduced impacts and reduced suffering of animals compared with other methods for managing wildlife (Oogjes 1997).

Fertility control is strongly favoured by many animal-welfare groups over current mortality methods (Oogjes 1997). Anti-fertility agents could, however, have adverse effects on the target animal's well-being through altered behavioural effects. For example, there are concerns over sterilized females experiencing an abnormal number of oestrous cycles and thus expending more energy, and increased aggression between males in their battle for access to cycling females (e.g. the rutting season in white-tailed deer is extended if females continue to cycle after being immuno-sterilized). In addition, loss of libido and any associated reduction in territorial defence might affect the social structure of treated populations, for example by loss of status, and reduce the efficacy of the treatment by allowing immigration of non-sterile individuals. Interestingly, a study of 18 immuno-sterilized female African elephants in Kruger National Park found no changes in their general behaviour patterns including bull–cow interactions (Delsink et al. 2002).

Conclusions

We do not suggest that lethal control is ineffective, nor are we claiming that non-lethal methods are always effective or more humane. Rather, we aim to illustrate that managing vertebrate pests is complicated and can be upset by many biological factors, which include evolution of resistance, behavioural habituation and density-dependent compensation. Sometimes the most efficient strategy might involve combining lethal and non-lethal management techniques. An example, using foxes, could involve shooting breeding vixens in spring (raising difficult issues about starving their cubs to death, or sending terriers in to kill them), and a combination of habitat management and learned food aversions to reduce predation on the eggs of populations of rare ground-nesting birds during the brief period for which they are vulnerable. Much research remains to be done on non-lethal control methods, including repellency, fertility control, diversionary feeding and fencing, as well as the other methods described above. The most important route for increasing efficacy and public acceptability of wildlife management is to develop integrated control strategies based on a solid understanding of the population ecology of the pest that needs to be managed (Singleton et al. 1999).

Another issue that needs more detailed examination is the relative cost-effectiveness of various approaches. There are recent, illuminating examples of the economic value of different strategies for wildlife management developed from long-term population data sets (Stenseth et al. 2003). Costs include more than the dollars lost through damage and spent on control. There are ethical issues concerning suffering of pests, loss of habitats and extinctions, which will almost certainly (and understandably) be viewed differently in different countries; in the UK, we can afford to be more concerned about these ethical issues because our children are not starving and there is a relatively low risk of rat-borne diseases, which suggests that ethics are flexible. People certainly apply their principles inconsistently, objecting to shooting pigeons in towns while accepting that rats may be killed by a variety of painful and stressful methods and moles asphyxiated by an excruciatingly painful poison (strychnine) in order to maintain the appearance of ornamental gardens and lawns. Conservation is concerned with maintaining biodiversity and viable populations, and it is hard to balance the population-level outcomes of conservation management against increased suffering and death of individual animals that may be caused along the way. These are real ethical dilemmas that require value judgements and we offer no answers, but it is clear that socio-economics of wildlife management will become an integral part of ecologically based integrated management in future years.

In summary, we are led to two overall conclusions. The first seems obvious: understanding more fully the biology of pest species and the ecosystems within which they exist will help to achieve more effective wildlife management. The second is more difficult: strategic development of wildlife management must take account of a great number of biological processes that are not only intrinsically non-linear (e.g. density dependence) but also interact with each other and with non-biological processes – interdisciplinarity is essential and difficult judgements about ethical trade-offs are inevitable.

He who wants to catch foxes must hunt with geese.

(Danish Proverb.)

References

Avery, M.L. & Mason, J.R. (1997) Feeding responses of red-winged blackbirds to multisensory repellents. *Crop Protection* **16**: 159–64.

Baker, S.E. & Macdonald, D.W. (1999) Non-lethal predator control: exploring the options. *Advances in Vertebrate Pest Management* (Eds P.D. Cowan & C.J. Feare), pp. 251–66. Filander Verlag, Furth.

Baker, S.E., Ellwood, S.A., Watkins, R.W. & Macdonald, D.W. (2005) Non-lethal control of wildlife: using chemical repellents as feeding deterrents for the European badger. *Journal of Applied Ecology* **42 (5)**: 921–31.

Baker, S.E., Johnson, P.J., Slater, D., Watkins, R.W. & Macdonald, D.W. (In press) Learned food aversion with and without an odour cue for protecting untreated baits from wild mammal foraging. In *Conservation Enrichment and Animal Behaviour* (Ed. R. Swaisgood) *Applied Animal Behaviour Science* special edition.

Brakes, C.R. & Smith, R.H. (2005) Exposure of non-target small mammals to rodenticides: short-term effects, recovery and implications for secondary poisoning. *Journal of Applied Ecology* **42**(5): 118–28.

Caughley, G. & Sinclair, A.R.E. (1994) Wildlife Ecology and Management. Blackwell Science, Massachusett.

Chambers, L.K., Lawson, M.A. & Hinds, L.A. (1999) Biological control of rodents – the case for fertility control using immunocontraception. In *Ecologically-based Management of Rodent Pests* (Eds G.R Singleton, L.A. Hinds, H. Leirs & Z. Zhang), pp. 215–42. Australian Centre for International Agricultural Research, Canberra.

Conover, M.R. (1984) Comparative effectiveness of Avitrol, exploders, and hawk-kites in decreasing blackbird damage to corn. *Journal of Wildlife Management* **48**: 109–16.

Conover, M.R. (1985) Protecting vegetables from crows using an animated crow-killing owl model. *Journal of Wildlife Management* **49**(3): 643–5.

Cooke, B.D. & Fenner, F. (2002) Rabbit haemorrhagic disease and the biological control of wild rabbits, *Oryctolagus cuniculus*, in Australia and New Zealand. *Wildlife Research* **29**: 689–706.

Coppinger, R. (1992) Can dogs protect livestock against wolves in North America? *The Livestock Guard Dog Association DogLog* **3**(2): 2–4.

Coppinger, L. & Coppinger, R. (1993) Dogs for herding and guarding livestock. In *Livestock Handling and Transport* (Ed. T. Grandin), pp. 179–96. CAB International, Wallingford.

Coppinger, R., Lorenz, J., Glendinning, J. & Pinardi, P. (1983) Attentiveness of guarding dogs for reducing predation on domestic sheep. *Journal of Range Management* **36**(3): 275–9.

Cowan, D.P., Reynolds, J.C. & Gill, E.L. (2000) Reducing predation through conditioned aversion. In *Behaviour and Conservation* (Eds L.M. Gosling & W.J. Sutherland), Vol. 2, pp. 281–99. Cambridge University Press, Cambridge.

Cox, R., Baker, S., Macdonald, D.W. & Berdoy, M. (2004) Protecting egg prey from Carrion Crows: the potential of aversive conditioning. *Applied Animal Behaviour Science* **87**: 325–42.

Crocker, J. (1984) How to build a better scarecrow. *New Scientist* **101**(1403): 10–12.

Delsink, A.K., van Altena, J.J., Kirkpatrick, J., Grobler, D. & Fayrer-Hosken, R.A. (1982) Field applications of immunocontraception in African elephants (*Loxodonta africana*). *Reproduction Supplement* **60**: 117–24.

Draulans, D. (1987) The effectiveness of attempts to reduce predation by fish-eating birds. A review. *Biological Conservation* **41**: 219–32.

Errington, P.L. (1946) Predation and vertebrate populations. *Quarterly Review of Biology* **21**: 144–77, 221–45.

Gao, Y. & Short, R.V. (1993) The control of rodent populations. *Oxford Reviews of Biology* **15**: 265–310.

Guiler, E. (1985) *Thylacine: The Tragedy of the Tasmanian Tiger*. Oxford University Press, Oxford.

Gustavson, C.R., Garcia, J., Hankins, W.G. & Rusiniak, K.W. 1974. Coyote predation control by aversive conditioning. *Science* **184**: 581–3.

Harrington, L.A. & Macdonald, D.W. (2002) *A Review of the Effects of Pesticides on Wild Terrestrial Mammals in Britain*. Report to the Royal Society for the Prevention of Cruelty to Animals. Wildlife Conservation Research Unit, Oxford.

Kirkpatrick, J.F., Turner, J.W. Jr, Liu, I.K.M., Fayrer-Hosken, R. & Rutberg, A.T. (1997) Case studies in wildlife immunocontraception: wild and feral equids and white-tailed deer. *Reproduction, Fertility and Development* **9**: 105–110.

Krebs, C.J. (2001) *Ecology: the Experimental Analysis of Distribution and Abundance*, 5th edn. Addison Wesley Longman, San Francisco, 695 pp.

Linnell, J.D.C., Smith, M.E., Odden, J., Kaczensky, P. & Swenson, J.E. (1996) Carnivores and sheep farming in Norway 4. Strategies for the reduction of carnivore–livestock–conflicts: a review. *NINA Oppdragsmelding* **443**: 1–116.

Macdonald, D.W. (1988) Rabies and foxes: the social life of a solitary carnivore. In *Vaccination to Control Rabies in Foxes*, pp. 5–13. Office for Official Publications of the European Communities, EEC Symposium (Agriculture), Brussels.

Macdonald, D.W., Mathews, F. & Berdoy, M. (1999) The behaviour and ecology of *Rattus norvegicus*: from opportunism to Kamikaze tendencies. In *Ecologically-based Rodent Management* (Eds G. Singleton, L. Hinds, H. Leirs & Z. Zhang), pp. 49–80. Monograph 59, Australian Centre for International Agricultural Research, Canberra.

Macdonald, D.W., Tattersall, F.H., Johnson, P.J., et al. (2000) *Managing British Mammals: Case Studies from the Hunting Debate*. Wildlife Conservation Research Unit, Oxford.

Marker, L.L., Dickman, A.J. & Macdonald, D.W. (2005) Perceived effectiveness of livestock guarding dogs placed on Namibian farms. *Rangeland Ecology and Management* **58**(4): 329–36.

McDonald, R.A., Harris, S., Turnbull, G., Brown, P. & Fletcher, M. (1998) Anticoagulant rodenticides in stoats (*Mustela erminea*) and weasels (*Mustela nivalis*) in England. *Environmental Pollution* **103**: 17–23.

McLeod, R. (2004) *Counting the cost: impacts of invasive animals in Australia*. Cooperative Research Centre for Pest Animal Control, Canberra, 70 pp.

Murphy, E.C., Clapperton, B.K., Bradfield, P.M.F. & Speed, H. (1998) Effects of rat-poisoning operations on abundance and diet of mustelids in New Zealand podocarp forests. *New Zealand Journal of Zoology* **25**(4): 315–28.

Nicolaus, L.K., Hoffman, T.E. & Gustavson, C.R. (1982) Taste aversion conditioning in free-ranging raccoons, *Procyon lotor*. *Northwest Science* **56**: 165–9.

Nicolaus, L.K., Cassel, J.F., Carlson, R.B. & Gustavson, C.R. (1983) Taste aversion conditioning of crows to control predation on eggs. *Science* **220**: 212–4.

Oogjes, G. (1997) Ethical aspects and dilemmas of fertility control of unwanted wildlife: an animal welfarist's perspective. *Reproduction, Fertility and Development* **9**: 163–7.

Reynolds, J.C. (1999) The potential for exploiting conditioned taste aversion (CTA) in wildlife management. In *Advances in Vertebrate Pest Management* (Eds D.P. Cowan & C.J. Feare), pp. 267–82. Filander Verlag, Furth.

Reynolds, J.C., Goddard, H.N. & Brockless, M.H. (1993) The impact of local fox (*Vulpes vulpes*) removal on fox populations at two sites in southern England. *Gibier Faune Sauvage* **10**: 319–34.

Shore, R.F., Birks, J.D.S., Freestone, P. & Kitchener, A.C. (1996) Second generation rodenticides in polecats (*Mustela putorius*) in Britain. *Environmental Pollution* **91**: 279–82.

Singleton, G.R., Leirs, H., Hinds, L.A. & Zhang, Z. (1999) Ecologically-based management of rodent pests – re-evaluating our approach to an old problem. In *Ecologically-based Rodent Management* (Eds G. Singleton, L. Hinds, H. Leirs & Z. Zhang) pp. 17–29. Monograph 59, Australian Centre for International Agricultural Research, Canberra.

Singleton, G., Krebs, C., Davis, S., Chambers, L. & Brown, P. (2001) Reproductive changes in fluctuating house mouse populations in Southeastern Australia. *Proceedings of the Royal Society of London Series B – Biological Sciences* **268**: 1741–8.

Singleton, G., Farroway, L., Chambers, L., Lawson, M., Smith, A. & Hinds, L. (2002) Ecological basis for fertility control of house mice using immunocontraceptive vaccines. *Reproduction Supplement* **60**: 31–9.

Singleton, G.R., Sudarmaji, Jacob, J. & Krebs, C.J. (2005) An analysis of the effectiveness of integrated management of rodents in reducing damage to lowland rice crops in Indonesia. *Agriculture, Ecosystems and Environment* **107**: 75–82.

Smith, P., Townsend, M.G. & Smith, R.H. (1991) A cost of resistance in the brown rat: reduced growth rate in warfarin resistant lines. *Functional Ecology* **5**: 441–447.

Smith, P., Berdoy, M.L., Smith, R.H. & Macdonald, D.W. (1993) A new aspect of warfarin resistance in wild rats: benefits in the absence of poison. *Functional Ecology* **7**: 190–4.

Smith, R.H. (1999) Population biology and non-target effects of rodenticides: trying to put the eco into ecotoxicology. In *Advances in Vertebrate Pest Management* (Eds D.P. Cowan & C.J. Feare), pp. 331–346. Filander-Verlag, Fürth.

Smith, R.H. & Greaves, J.H. (1987) Resistance to anticoagulant rodenticides: the problem and its management. In *Proceedings of the 4th International Conference on Stored-Product Protection* (Eds E. Donahaye & S. Navarro), pp. 302–15, Tel Aviv, Israel, September 1986. Agricultural Research Organization, Bet Dagon, Israel.

Stenseth, N.C., Herwig, L., Skonhoft, A., et al. (2003) Mice, rats, and people: the bio-economics of agricultural rodent pests. *Frontiers in Ecology and the Environment* **1**: 367–75.

Story, G., Berman, D., Palmer, R. & Scanlan, J. (2004) The impact of rabbit haemorrhagic disease on wild rabbit (*Oryctolagus cuniculus*) populations in Queensland. *Wildlife Research* **31**: 183–93.

Thompson, H.V. (1948) studies of the behaviour of the common brown rat: 1. Watching marked rats taking plain and poisoned bait. *Bulletin of Animal Behaviour* **6**: 2–40.

Vickery, J.A. & Summers, R.W. (1992) Cost-effectiveness of scaring brent geese *Branta bernicla* from fields of arable crops by a human bird scarer. *Crop Protection* **11**(5): 480–4.

13

Introduced species and the line between biodiversity conservation and naturalistic eugenics

David W. Macdonald, Carolyn M. King and Rob Strachan

There must have been plenty of them about, growing up quietly and inoffensively, with nobody taking any particular notice of them ... and so the one in our garden continued its growth peacefully, as did thousands like it in neglected spots all over the world ... it was some little time later that the first one picked up its roots and walked.

(John Wyndham, *The Day of the Triffids*, 1951.)

Introduction

Are introduced species – those transported by people beyond their natural geographical range – different from other species in terms of ecological process, and is that what makes them a hot topic in biodiversity conservation? Consider the grey squirrel, *Sciurus carolinensis*, deliberately transported from the USA to England and Wales between 1876 and 1929, and from Canada to Scotland between 1892 and 1920 (Corbet & Harris 1991). Now hugely abundant, conservationists regard expatriate grey squirrels as a 'bad thing', and cheap spin-doctoring labels them American tree rats. But those same bushy-tailed visitors are judged a 'good thing' by the millions of British families enchanted by their acrobatics in garden or park. The damage done by grey squirrels in bark-stripping trees makes them pests of forestry, costing the British timber industry around £10 million per rotation. Since the now lamented red Squirrel Nutkin (*Sciurus vulgaris*) had also been a forestry pest, ethicists might struggle to see why it is meretricious to poison grey individuals but criminal to poison red ones; but the issues go deeper than the problems of replacing one pest of forestry with another. Conservation is inescapably multidisciplinary, and involves both technical and cultural judgements (Lawton 1997).

Although it is obvious that importing species which become economic pests or threaten public health, or which destroy native biota, is undesirable, many issues provoked by introductions (past and present) are far from obvious. The reasons why conservationists deplore such assisted passages, although proximately to do with maintaining local biodiversity and/or community composition, are ultimately to do with a philosophical preference for allowing natural processes to run their own course without human interference. Fulfilling this preference is, however, increasingly tricky when even the passage of the seasons has a human taint (see Chapter 16), and the distinction between natural and unnatural movements of species is increasingly opaque. There is nothing unusual about one species abetting another in extending its geographical range: doubtless the first *Smilodon* to trot south across the Panamanian land-bridge took a community of fleas with it. Furthermore, it is not unnatural for a species arriving in a new land to disrupt gravely, or even to extinguish, the lives of its new compatriots, as did *Smilodon* and its northern contemporaries when they encountered the previously isolated South American biota (Simpson 1980). No, what conservation biologists worry about is that the perturbations that might always have arisen when an immigrant arrived have, recently, happened very often because of modern people, and are continuing at an ever-increasing rate.

Some assisted passages are deplored, others rejoiced in, or at least accepted. Few would treasure less the endemic and now endangered foxes (*Urocyon littoralis*) of the Californian Channel Islands because they exist thanks only to the pre-Columbian canoeists who transported their ancestors to the islands from the mainland between 2200 and 5200 years ago (Roemer 2004). The understandable abhorrence of Australian conservationists for a nineteenth century imported canid – the red fox, *Vulpes vulpes* – does not extend to the dingo, *Canis lupus dingo*, which, some 5000 years earlier, was not merely a transported canid but also a domestic one. Kiore (*Rattus exulans*) are regarded as tribal treasures by some Maori, whose ancestors brought kiore to New Zealand in their double-hulled canoes from eastern Polynesia about 700 years ago, but as just another introduced rat by European conservationists. One of the four remaining endemic Galapagos rice rats, the Santa Fe rice rat (*Oryzomys bauri*), which is believed to have reached the archipelago aboard aboriginal boats 700–1000 years ago (Patton & Hafner 1983) is chromosomally identical to its Peruvian ancestors, *O. xanthelous*, and, some

argue, should not be given priority for conservation while a natural stock exists in Peru. Clearly, the consistent treatment of introduced species is a challenge.

The pivotal question of when a species should be considered naturalized (and treated as native) is a cultural matter, rife with 'speciesism' and illogicality, and invites unhappy although often mistaken parallels with the treatment of human immigrants. For example, in Britain, the fallow deer, *Dama dama*, introduced in the tenth century, is widely accepted whereas the muntjac, *Muntiacus reevsii*, a late nineteenth century arrival, is deplored. The possibility that the rabbit, *Oryctolagus cuniculus*, brought in by the Romans, might succumb to haemorrhagic fever (a recent invasive from China) was greeted as a disaster by conservationists who appreciate the rabbit as the engine of natural grassland maintenance, and the staple of the native predatory community. Furthermore, the rabbit, an expensive and populous introduced consumer of British agriculture, is rare and valued in its native Spain (Thompson & King 1994), illustrating that pestilential immigrants may be coveted, and even imperilled, in their native land. The wallabies (*Macropus eugenii eugenii* and *Petrogale penicillata*) that are unwelcome invasives on New Zealand's Kawau Island are being repatriated to Australia where they are valued (King, 2005), and eastern England's Chinese water deer *Hydropotes inermis* represent over 10% of the species' global population. In many matters associated with introductions, it may well be that the quest for consistency is both hopeless and incapacitating, and that case-by-case pragmatism is more helpful.

The lexicon of this topic is as colourful – aliens, invasives, non-natives, immigrants – as it is ambiguous. Our usage of the term **introduced species**, as defined above, implies no judgement of their impact, and we use **invasiveness** as a measure of the extent to which an introduced species is successful in colonizing large areas or reaching high numbers. Others interpret **invasive** as automatically implying a capacity for trouble-making, but in fact introduced species may be both invasive and (seemingly) neutral (e.g. the collared dove (*Streptopelia decaocto*) in the UK). Some may even be benign (Canadian golden-rod *Solidgo canadensis* may be an important nectar source for some British insects), and the rare native noble chafer (*Gnorimus nobilis*) survives only in old decaying fruit trees in traditional planted orchards. We distinguish those that are both successful and demonstrably damaging as **malign invasive**, and a precautionary generality with introduced species would be to treat them as guilty until proven innocent. Of course, categorizing a species as benign or malign (with respect to its impact on human interests) involves value judgements that are neither easily quantified nor widely agreed, and may need to be changed as evidence unfolds.

Do introductions matter? Extinctions are currently running at 100–1000 times background rates, very often as consequences of humans transporting organisms beyond their natural range (Soulè 1990; Williamson 1999). In the UK, 23% and 12% of Biodiversity Action Plans for habitats and species, respectively, cite nonnatives as problematic. Introduced species may 'weaken' an ecosystem, and tip native biodiversity over a cliff edge to which it has already been brought by population fragmentation, habitat degradation or loss, or even earlier invasive onslaughts. Introduced species may therefore be convenient scapegoats misdirecting attention from more fundamental problems such as habitat loss. One such debate is whether the greatest risk of extinction to the native freshwater mussels in the Great Lakes is the explosively invasive (and originally Caspian) zebra mussel, *Dreissena polymorpha*, or the insidiously toxic pollutant that poisons the mud in which they live (Gurevitch & Padilla, 2004). Nonetheless, of the 941 vertebrate taxa in danger of extinction, 18.4% face threats from invasives (Macdonald et al. 1989) and people have introduced c.400,000 species of plants, vertebrates, invertebrates and microbes worldwide (IUCN 1997) – the largest violation of biogeographical

boundaries since the Great American Faunal Interchange ended the isolation of South America 5–8 Ma (Simpson 1980).

An earlier generation thought it interesting to transport species widely. In Australia, where the devastating effects of sheep, cats, etc., on native species was especially clear, Gerard Krefft (1866) was among the first to raise the alarm, followed by Elton (1958). Since then, ecologists have feared the impact on native species, communities and ecosystems (Lodge 1993a,b; McNeely et al. 2001). The collapse of New Zealand's fauna (Wilson 2004), in the face of human hunters and 32 resident species of introduced mammals, ranks high in the litany of disastrous invasions (Parkes & Murphy 2003); 50% of New Zealand's breeding birds have been lost, and a further 10% are on their way out (Holdaway 1999). Half of all mammal species extinguished in the past 200 years were lost from Australia, largely due to predation by invasives (Smith & Quin 1996). In the USA, 958 species are endangered, 400 primarily because of invasives (Nature Conservancy 1996). A catchy, if puzzling, rule of thumb is that 10% of introduced species establish successfully, depending mainly on the size of the colonizing group (Forsyth & Duncan 2001), and about 10% of these become pests (Williamson 1993). Of 12,500 plant species imported into Britain since the 1850s, over 650 have established in the wild and 14 have so far become pests – two, Giant Hogweed (*Heracleum mantegazzianum*) and Japanese Knotweed (*Fallopia japonica*), are so noxious that legislation now makes it illegal to distribute them outside their natural range (Holland 2001).

The ten-ten rule raises the question of why so many introductions, including some that turn out to be successful or even invasive, appear harmless. One possible answer is that natural ecosystems have plenty of vacant niche space – this would be a theoretical revelation. Another is that the impacts of introductions have been inadequately measured. Extinction is a conspicuous end-point, and may distract attention from widespread, but less newsworthy, damage to species composition, distribution, abundance, behaviour, or evolution (Herbold & Moyle 1986), and homogenization (the spread of hardy species that can live anywhere, at the cost of sensitive, endemic species) (Atkinson 1996; Mack et al. 2000).

The effects of invasives are most severe on oceanic islands, where many populations of endemic species combine rarity with vulnerability (Vitousek et al. 1997; Chapin et al. 2000; McNeely et al. 2001; Veitch & Clout 2002). Evidence linking introductions to extinctions on larger land masses or islands close to them is, surprisingly, often circumstantial (Ebenhard 1988).

The distinction between introductions and natural colonists is becoming increasingly blurred (Hulme 2003). Little egrets (*Egretta garzetta*), Nathusius pipistrelles (*Pipistrellus narthusius*), red-veined darters (*Sympetrum fonscolumbei*) and dozens of other species from elsewhere in Europe are now colonizing Britain without direct human help, but as a result of human-induced climate change, thereby eroding the traditional distinction based on direct human assistance. Likewise, a dozen Australian bird species have colonized New Zealand unassisted, mostly taking advantage of the new habitats opened up by human-induced deforestation. Some (e.g. the silvereye *Zosterops lateralis*) are benign invasives that are now very widespread, and accepted as natives; other species, from the same source, but transported by people, are listed as introduced and have become malign invasives, such as the Australian magpie *Gymnorhina tibicen* (Heather & Robertson 1996). These examples raise two questions – to which we return below – about the relevance of the direct intentionality of human involvement, neither of which may affect their biological consequences, but the two might be argued to differ morally.

Are invasive species different from any others? Characteristics common to successful colonists (natural or transported) across taxa include *r*-selected life histories (use of pioneer habitat, short generation time, high fecundity, high growth, environmental tolerance and diet-

ary plasticity, large gene pool and the ability to shift between *r*- and *K*-selected strategies; Kolar & Lodge 2001). Early successional and disturbed habitats, or environments climatically similar to that from which the colonist originates, foster successful invasions, as does a low diversity of native species (Diamond & Case 1986; Lodge 1993b). Colonists tend to fail when facing a new climate, disturbance, competition or predation from native species and diseases (Moyle 1986; Newsome & Noble 1986). Conversely, introduced species live in the absence of competitors and parasites with which they have evolved: for example, Australian brush-tailed possums *Trichosurus vulpecula* carry many fewer parasites in New Zealand than they do in Australia (King 2005). In addition to all these, among the few consistent indicators of colonizing success is propagule size (the number of individuals released together) (Forsyth & Duncan 2001; Forsyth et al. 2004).

Ecological effects of invasive species

Introduced species commonly induce complex ecosystem effects in native communities, catalysing ecological chain reactions with unpredictable consequences (Towns et al. 1997), which are compounded when more than one species is introduced, especially when introduced to degraded or fragmented habitats (Smith & Quin 1996; Macdonald & Strachan 1999). For example, the declines of the bellbird (*Anthornis melanura*) and the lesser short-tailed bat (*Mystacina tuberculata*) due to predation by rats, stoats *(Mustela erminea)* and Australian brush-tail possums (*Trichosurus vulpecula*), plus widespread deforestation, have removed the main pollinators of New Zealand's endemic beech mistletoes (*Peraxilla* sp. and *Alepis flavida*; Robertson et al. 1999) and the woodrose (*Dactylanthus taylorii*; Ecroyd 1996). Even non-native genotypes can be problematic: the flower structure of some imported cultivars of red clover (*Trifolium pratense*) differs sufficiently from the native genotype that British bumblebees cannot feed from them. Darwin foresaw the intricacy of such ecosystem effects when he posited an indirect positive association between house cats and flowers in England. His reasoning was that cats, by keeping mouse numbers in check, indirectly assisted bumble bee numbers which in turn enhanced pollination.

Predation and herbivory

Some 19% of introduced mammal species are carnivores (which comprise 5% of mammalian species), and predation is the cause of about one-third of the documented problems caused by invasive mammals (Macdonald & Thom 2001). As with predator–prey interactions among native species (Macdonald et al. 1999) the effects of invasive predators vary case-by-case, from negligible to catastrophic. Thus, one breeding colony of Hutton's shearwaters (*Puffinus huttoni*) in New Zealand has survived in the presence of stoats for >100 years, mainly because the colony is at high altitude and is so large that stoats can reduce its annual productivity by <1% a year (Cuthbert & Davis 2002); many smaller, lower altitude colonies have been destroyed by stoats and rats (Holdaway 1999). Shearwater eggs and chicks are available for only part of the year, and food at high altitude is scarce at other times, so stoat numbers cannot increase sufficiently to affect this colony as much as they can other colonies at lower altitude. On Lord Howe Island, the arrival of ship-wrecked black rats *(Rattus rattus)* in 1918 led to the extinction of half the 16 endemic bird species within a few years (Hindwood 1940). Between these extremes, fragile native prey may disappear whereas resilient ones co-exist with a new predator; these different outcomes are dictated more by the vulnerability of the prey, than by the predatory prowess of the invaders (King 1984). For example, in New Zealand the opposite population trends of two related native birds, the slow-breeding, endemic, endangered takahe (*Porphyrio mantelli*) and its recently-arrived and very successful relative, the opportunist pu-

keko (*Porphyrio porphyrio*) (Heather & Robertson 1996), show how their different breeding behaviour, productivity and reactions towards the introduced stoat interact to make the difference between survival and extinction (Bunin & Jamieson 1995). New Zealand's national icon, the brown kiwi (*Apteryx australis*) is one of several endemics where survival on the mainland is unlikely without a breakthrough in stoat control (McLennan et al. 1996; Basse et al. 1999). Different predators are differently damaging; some tree-dwelling birds in New Zealand that survived the Polynesian hunters and kiore which, between 1250–1850 eradicated 30 of their ground-dwelling contemporaries, then succumbed when tree-climbing black rats and stoats arrived after the 1880s (Holdaway 1999). Domestic cats are also potent agents of extinction, starkly emphasised by the contrasting surviving faunas of matched pairs of cat-infested and cat-free islands (e.g. Raoul Island and neighbouring Meyer Islets, in the Kermadec group)(Merton 1970); Fitzgerald & Turner (2000) attribute at least 38 cases of population-reducing predation to domestic cats.

Predatory communities are themselves complex, and mesopredator effects (competition between predators, introduced or not, including intraguild predation by large hunters on smaller ones), can dampen their impacts on certain prey. The extinction of the endemic Macquarie Island parakeet (*Cyanoramphus novazealandiae erythrotis*) is an example of the hyperpredation effect (Courchamp et al. 1999). Before the introduction of rabbits, parakeets had coexisted with feral cats for 60 years, but abundant rabbits facilitated an increase in feral cats sufficient to exert intolerable predation on the parakeets (Taylor 1979).

The effects of herbivory on vegetation and soil stability create about a quarter of the problems attributed to mammalian invasives (Ebenhard 1988). Feral goats (*Capra hircus*) have all but obliterated the food of the Galapagos giant tortoise (*Geochelone elephantopus*) (Desender et al. 1999). Comparison of aquatic plant communities before and after the arrival of the

North American muskrat, *Ondatra zibethicus,* on Valaam Island in north-east Russia in the 1970s, revealed the rise to dominance of those species most resistant to muskrat grazing (Smirnov & Tretyakov 1998). Muskrats have similarly affected shoreline vegetation in Finland (Danell 1996), where the dominant species changed from *Equisetum* and *Schoenoplectus* to *Phragmites* and *Typha* (Nummi 1996). A few centuries of seed predation by the kiore has left cohort-gaps in the forest canopy on otherwise undisturbed islands off New Zealand (Campbell & Atkinson 1999). On South Georgia Island, grazing by introduced reindeer (*Rangifer tarandus*) affected the species composition and structure of grass communities, and the indigenous perimylopid beetle (*Hydromedion sparsatum*) declined. Explanations were clouded because of the coincidental arrival of an invasive carnivorous carabid beetle (*Trechisibus antarticus*). Which of the invasives – reindeer or beetle – was the main driver of change? Reindeer browsing caused palatable *Parodiochola flabellata* and *Acaena magellanica* to be replaced by unpalatable *Poa annua* (Leader-Williams et al. 1987), and this had a greater impact on the native beetle (by slowing the growth of its larvae) than did the invasive predatory beetle (Chown & Block 1997).

Effects of competition

Examples of extinctions infallibly attributable to competition with an invasive are few (Mooney & Cleland 2001; Sax et al. 2002), except on islands (MacArthur 1972). However, some aggressive alien (often ornamental) plants are blamed for the loss of entire plant communities (e.g. Crawley et al. 1996; Vitousek et al. 1997). In the UK, the capacity of New Zealand pygmy weed (*Crassula helmsii*) (imported to decorate aquaria) to grow year-round, and thus to steal a competitive march against native species which are mostly dormant in winter, has enabled it to colonize over 10,000 sites in 40 years, forming dense blankets to the detriment

of native invertebrates, amphibians and fish (Dawson & Warman 1987).

Lack (1947) deduced that invasives were least likely to penetrate communities already populated by a similar native, but this generalization begs the question 'how similar does it have to be?'. Furthermore, it seems that the great similarity (morphological and behavioural) between invasive American mink *Mustela vison* and the European mink, *M. lutreola*, was the cause of *lutreola*'s displacement (see Box 13.1) (Macdonald et al. 2002). By contrast, in Finland the introduced racoon dogs (*Nycetereutes procyonoides*) and badgers (*Meles meles)* are also fairly similar, and share many common foods and den sites, apparently harmlessly (Kauhala et al. 1993, 1998). The endemic Galapagos rice rat (*Nesoryzomys swarthi),* and the highly aggressive introduced ship rat apparently coexist in unstable equilibrium in a small homogeneous locality on Isla Santiago (see Box 13.2); the rice rats' survival hinges, paradoxically, on drought years – during which they, unlike the ship rats, can obtain water from cacti (Harris et al. submitted). One difficulty in interpreting these cases is that overlap in use of a particular resource does not constitute evidence for competition (Macdonald & Thom 2001), because the resource may not be limiting (Sale 1994) and/or the percentage of overlap may not equate with the intensity of competition (Colwell & Futuyma 1971). Other factors often confuse simple comparisons: for example, climate change is increasing carrying capacity for badgers (Macdonald & Newman 2002), so Finnish badgers have been able to extend their distribution northwards: it may be that the raccoon dog's arrival in Finland happened to coincide with climate change lifting the local carrying capacity for small carnivores.

INTERFERENCE VERSUS EXPLOITATION

Competition operates through two broadly defined mechanisms: directly (interference

Box 13.1 The American mink abroad

Their expensive coats caused American mink (*Mustela vison*) to be transported to Europe in the 1920s, where they were generally kept in fur farms. Inevitably, they soon escaped to establish feral populations, and in Russia by 1971, 20,400 of them had been deliberately released to bolster fur-trapping revenue. American mink abroad have been wondrously successful invasives, and Macdonald & Harrington (2003) review the cost in damage to species from eider ducks in Iceland, through terns in Scotland to rodents in Patagonia. In Britain, American mink delivered the *coup de grace* to water voles (*Arvicola terrestris*), whose lowland habitat had been reduced by agriculture to narrow waterside ribbons. Linear habitat enhanced their susceptibility to the mink, against which they have no effective defence (they evolved to escape stoats by taking to the water, and otters (*Lutra lutra*) by hiding in burrows, but neither refuge is safe from mink). Water voles remain only where mink are scarce, and their populations may now be so fragmented as to be unviable even if mink were eliminated (Rushton et al. 2000; Bonesi et al. 2002).

Considering that water voles are abundant elsewhere in Europe, does their plight in the UK justify killing mink there? The probable answer is yes and no. Yes, killing mink locally, humanely and methodically is arguably justifiable where it is effective in restoring water voles. Furthermore, the European abundance of water voles is small consolation in Britain, where water voles have been isolated for over 8000 years. No, the aim of national eradication is probably not economically or logistically feasible (nor ethically appropriate where their predation does not threaten native prey). Conservation relies on acting locally even when thinking globally.

Historically, the European mink, *Mustela lutreola*, was widespread throughout Europe, but now survives only in enclaves of France, Spain, Belarus and Russia. The decline in their distribution is caused by intraguild hostility from the American mink (Macdonald et al. 2002). American mink are larger (males: 1310.2 g versus 976.6 g), have larger litters (mean 5.8 versus 2.4), have a competitive advantage, and attack their smaller congener (Sidorovich et al. 1999). After the arrival of American mink, European mink became largely confined to small brooks, the habitat least used by the invaders, but where food may be insufficient to support breeding. The only hope for the European mink may be island sanctuaries, and one such has been established in the Baltic.

Paradoxically, intraguild competition may facilitate the recovery of water voles in Britain. American mink spread in Britain because otters had been wiped out by agrochemicals; now otters are recovering, and seemingly driving a decline in American mink (Bonesi et al. 2004). If otters can hold American mink down to a level that permits water voles to survive, should Britain accept the remaining mink as naturalized?

competition), through face-to-face conflict (perhaps, but not necessarily, over resources), or indirectly (exploitation competition, or more subtle forms of aggression via odours or other signals or remote interactions, as in allelopathy between plants), where competitors race for resources but do not necessarily meet, or both. Interference is often aggressive, whereas the outcome of exploitation competition is decided when the winner usurps the loser by depleting their mutual resources. Whether competition is between natives, or involves invasive species, the processes are the same, but the outcomes have different consequences for local biodiversity. Macdonald et al. (2002) conclude that competition between incoming herbivores and native members of their guild is generally by exploitation. For example, red deer, *Cervus elaphus*, have invaded the last remaining habitat of

the takahe, an endangered flightless bird in New Zealand, and compete with them for the subalpine tussock grass, which is degraded by deer grazing (Lee & Jamieson 2001).

Interference competition – typically bullying or killing of smaller species by larger ones, without reference to resources – is commonplace amongst carnivores. Hersteinsson & Macdonald (1992) argued that the southern limit of the Arctic fox's (*Alopex lagopus*) range has been determined by interference competition with red foxes. Arctic foxes, introduced to over 450 islands during the early nineteenth century, flourished in the absence of red foxes, but disappeared where red foxes were present, or when sterile red foxes were introduced to oust them (West & Rudd 1983; Bailey 1992). Following the natural invasion of red foxes to Dutch sand-dunes between 1968 and 1977, formerly

---

Box 13.2 Galápagos invasion

Over 800 introduced terrestrial species have bombarded the 120 islands of the Galápagos archipelago – feral goats, pigs, dogs and cats are driving endemics, including the Galápagos tortoises (*Geochelone elephantopus*), to extinction (Mauchamp et al. 1998). Goats are arguably the worst of the invaders. A unique international effort, costing $8.5 million over 6 years and involving marksmen in helicopters, hunters on the ground and hounds with specially developed Teflon bootees to enable them to hunt on the sharp-edged lava, has eradicated a population of > 100,000 goats from the 585 km^2 of Santiago Island, before turning their attention to the half million more on Isabella Island. Ingeniously, the last goats are mopped up by luring them to radio-collared, compulsively sociable Judas females sterilized, implanted with hormones and alluringly in permanent oestrus – which lead hunters to the last wild survivors (Campbell 2002; Project Isabela 2004). The aim is to eradicate the goats before they complete their destruction of the native ecosystems en route to starvation.

Second to the goats are the black rats (Key & Muñoz 1994; Bensted-Smith et al. 2000). *Rattus rattus* have contributed to the loss of one race of the giant tortoise, whose hatchlings it kills (Macfarland et al. 1974), and rat predation on eggs and young caused 70% mortality of the Galápagos dark-rumped petrel, *Pterodroma phaeopygia* (Cruz & Cruz 1996) and also (probably) the probable disappearance of the Floreana mockingbird, *Nesomimus trifasciatus* (Grant et al. 2000). Amongst the most dramatic but unsung catastrophes has been the decline – more serious than that of any other vertebrate taxa – from 12 to four species of endemic rice rats. All members of the 'giant' rat genus *Megaoryzomys* have gone, along with most species of the genera *Nesoryzomys* and *Oryzomys* (Dowler et al. 2000). The Santiago rice rat, *Nesoryzomys swarthi*, thought extinct since 1907, was rediscovered in 1997 in one arid cove on Santiago, where it apparently coexists with black rats (and house mice, *Mus domesticus*) in what may be a state of competitive coexistence mediated by its superior adaptations to drought. Harris & Macdonald (unpublished) speculate that the survival of the rice rat is imperilled by climate change: El Niño years are wet and warm, and increase vegetation growth, creating conditions seemingly advantageous to black rats. A run of El Niños – which may become more frequent, owing to human impacts on climate – may destroy the unstable equilibrium that has been the salvation of the rice rat. As an aside, the plight of the rice rats illustrates a different, but pervasive, conservation issue, namely the importance of branding. Despite their colossal interest to science, and biodiversity importance, the Galápagos rice rats have been paid scant attention. Raising conservation support for a creature with the surname 'rat' is challenging; it might be beneficial if this rediscovered species were rebranded the Santiago cactus nibbler!

---

abundant stoats disappeared during the late 1980s from all reserves except the only one where foxes were shot on sight (Mulder 1990).

Effects of disease

The perils to humanity of transporting pathogens, or their reservoirs, beyond their native range are obvious. In the fourteenth century black rats hitch-hiked from Asia to Europe, carrying with them the plague and typhus diseases that have caused 25,000,000 human deaths and changed the course of civilization (Macdonald 1995). Commensal rodents raise an interesting definitional problem – they have evolved to travel with people, and to maintain natural populations on ships. The Norway rats (*Rattus norvegicus*) that reached New Zealand wherever European ships were brought close inshore were simply dispersing as animals do everywhere, so they were not invasives in quite the usual sense. Human actions are hence less directly responsible for the fact that their descendants help maintain leptospirosis and salmonellosis in New Zealand (King 2005), than for the deliberate transport of non-commensal muskrats (*Ondatra zibethicus*) to Germany from the USA, which have became a European reservoir for puumala-like hantavirus strains (Vahlenkamp et al. 1998). Other, deliberately transported invasives have carried diseases to domestic stock. The Australian brush-tailed possum (*Trichosurus vulpecula*), for example, is the main wildlife reservoir for bovine tuberculosis in New Zealand's cattle (Montague 2000).

For almost 100 years the extent to which imported grey squirrels were responsible for the decline of the British red squirrel, and if so by what mechanism, has been debated. Latterly, attention has focused on exploitation competition (Gurnell et al. 2002; Wauters et al. 2002a,b), and on intriguing discoveries about differential capacity to detoxify acorns (Kenward 1985) and differences in energy consumption (Bryce et al. 2001); the crucial revelation was that the invasive grey was host to a parapox virus fatal to the red (Sainsbury & Ward 1996). Reds may persist in Highland Scotland – due to their advantage over greys in (introduced) sitka spruce (Bryce et al. 2002) – so the otherwise reasonable plan to restore native oaks in invasive conifer woods could foster the spread of invasive, poxy greys and hasten the end of native reds! Without the vaccination of reds (Tompkins et al. 2003) or the unfeasibly large-scale eradication of greys, the local extinction of red squirrels south of the Highlands seems inevitable. A parallel case is the devastation of British white-clawed crayfish, *Austropotamobius pallipes*, by the fungus *Aphanomyces astaci* on North American signal crayfish, *Pacifastacus leniusculus*, imported for human consumption by aquaculturalists under a 1976 Government food initiative – although this unhappy outcome had already been witnessed elsewhere in Europe, imports were not adequately screened for disease. Likewise, avian malaria was introduced to Hawaii with exotic birds, but had no vector until 1826 when *Culex quiquefasciatus* arrived aboard ships carrying water barrels. The mosquitoes caused a wave of extinction among the endemic native birds.

Effects of hybridization

How much does it matter that contemporary red foxes in the eastern USA mix native genes with those inherited from their British subspecific cousins, imported by George Washington and other early settlers to bolster opportunities for hunting (Kamler & Ballard 2002)? Or that in Britain some roe deer (*Capreolus capreolus*) descend from native stock and others from contin-

ental reintroductions? Technical considerations include the risk that the mixing of long-separate genotypes reduces taxonomic diversity in the short term, and that hybrid progeny may be less well adapted to local conditions (Rhymer & Simberloff 1996). European representatives of the globally threatened white-headed duck (*Oxyura leucocephala*), expensively restored from only 22 in 1977 to 1300 by 1998, breed in Spain where they face oblivion through producing fertile hybrids with American ruddy ducks (*Oxyura jamaicensis*). Ruddy ducks have been escaping since 1965 from British ornamental collections, and by 1998 numbered 4000 individuals in 19 countries (increasing at 21% per year between 1976 and 1996) (Hughes 2001). Although there is a commitment to eradicate ruddy ducks from France, Portugal and Spain, it seems logistically, economically and ethically dubious to pursue this goal unless there is (i) a unified commitment to eradicate them throughout the Western Palaearctic and (ii) prevention of further escapes from ornamental collections (which probably means either a ban on keeping them, or compulsory sterilization of all captive individuals). In Britain, trials suggest that a 6-year programme costing £3.6–5.4 million might eradicate them, but reinvasion is always possible. In such cases, cultural as well as scientific judgement is necessary to chart a course between innovative management and reactionary naturalistic eugenics.

Perhaps the most provocative example of ethical dilemmas in the context of interbreeding is provided by the langurs of Bhutan. The endemic golden langur (*Trachypithecus geei*) was previously kept apart from the more cosmopolitan capped langur (*T. pileata*) by the Chamkhar–Mangde–Manas River system, but five suspension bridges constructed in the 1980s across the Chamkhar enabled the capped langurs to cross, and cross-bred offspring now abound (Wangchuk 2005). Although further liaisons are partly thwarted by bridge police, the damage has been done, and the golden endemic may be doomed unless the cross-breds are shot – but should they be? After all, their parents crossed of their own volition. On the other hand, would that argument impress if, say, rabies reached the UK through the Channel Tunnel?

The special case of domestics

As with commensals, one might debate whether domestic animals are invasives, but each of the foregoing ecological impacts is illustrated, often catastrophically, by feral domestics. Predation by domestic cats on wildlife is well documented (Woods et al. 2003). Water voles, *Arvicola terrestris,* have suffered exploitation competition from sheep in Britain since the Iron Age (starting about 2700 years ago), such that 99% of their habitat was gone by AD 1900 (Jefferies 2003). Rabies spread by domestic dogs threatens endangered canids, such as Ethiopian wolves, *Canis simensis* (Sillero-Zubiri et al. 1996; Sillero-Zubiri & Macdonald, 1997), and African wild dogs, *Lycaon pictus* (Macdonald 1993; Perry 1993; Woodroffe et al. 1997), and fuels periodic epidemics in more abundant canids such as jackals (Rhodes et al. 1998). Domestic dogs were the reservoir of canine distemper virus that decimated lion, *Patherus leo,* populations in the Serengeti (Roelke-Parker et al. 1996).

Domestics also raise problems of hybridization (such as dogs with Ethiopian wolves, Sillero-Zubiri et al. 1994). The case of hybridization between wildcats (*Felis silvestris*) and feral domestic cats has mind-bending biological, legal and ethical ramifications (Macdonald et al. 2004). Wildcats, isolated in Britain after the end of the last ice age approximately 10,000 years ago, were joined there by domestic cats 2000 years ago or more, and the two have been interbreeding ever since. Scottish wildcats are protected by law, whereas feral domestic cats are regarded as a pest (particularly of grouse). There is debate as to whether they are different species (domestic cats descended perhaps only 4000 years ago from the African subspecies of wildcats), and they are still less genetically distinct than are most breeds of domestic dogs from each other. Fur-

ther, the law gives no protection to hybrids, yet a significant proportion of remaining wildcat genes may be packaged in hybrid bodies, and there are of course no specimens of pre-Iron Age Scottish wildcats from which to judge the original natural variation. Two linked questions thus emerge – first, how to define Scottish wildcats so that they may enjoy the protection of the law that was framed in 1988 to safeguard them indefinitely, and second, how to conserve and restore their populations. These questions themselves provoke another – does it matter if contemporary wild-living cats in Scotland carry domestic genes, or is this a trivial, and perhaps unbecoming, form of naturalistic eugenics?

Variation amongst wild-living cats includes a spectrum of characteristics which are, at one end, inseparable from domestic cats, and at the other, remote from domestic cats (Daniels et al. 1998; Beaumont et al. 2001). Individual Scottish wildcats might be defined on pelage characters that seem likely to have typified the pure form, but these individuals may be scattered adrift in an ocean of hybrids and feral domestics. Further, paradoxically, saving wildcat genes may necessitate conserving some cats in which wildcat and domestic cat genes are packaged together. Domestic cat genes do not necessarily diminish the gene pool of wildcat – they may add to it – but they do alter the genomes of the individuals in whose existence we delight. The case of the Scottish wildcat illustrates difficult issues of how conservation emphasis should be apportioned between genes and the bodies they build. To protect the individuals we must protect the genes, because bodies are built from genes much as musicians build a concert from the score, but that is not to say, by analogy, that a musical score is more important than the performance.

Management options for invasive species

Macdonald & Thom (2001) championed the notion that taking great care over the control of invasive species was important both because of the potential outcome (good or bad) for biodiversity conservation (Zavaleta et al. 2001) and because invasives offer powerful 'experiments' whereby to study fundamental ecological processes. The natural invasion of the collared dove into western Europe (Rocha-Camarero & De Trucios 2002) has been benign and offers no biodiversity case for managing (i.e. removing) them. By contrast, although the arrival of the pied stilt (*Himantopus himantopus*) in New Zealand was natural, the consequences are not benign: pied stilts are now very abundant, and threaten to swamp the endangered endemic black stilt (*H. novaezelandiae*) through hybridization (Pierce 1984). This case raises the awkward question of the priority that conservation should attach to preserving species that are naturally becoming obsolete.

Against this unpromising history, can science help protect the future in what is being dubbed the Homogocene era? The obvious need is for a risk analysis that directs attention to the species that pose the greatest threats as invaders, and to the communities that are most susceptible (yet another irony is that exactly the same predictions are required for selecting biological control agents – which are invaders but with a different name). The central algebra is simple: likelihood × consequence = risk. A combination of life history theory and biogeographical extrapolation offers hope of a predictive, decision support system to identify risk and thus priorities for preventive (and remedial) action. On the other hand, however good the theory may get, the practice will be daunting in a world in which natural boundaries are undermined by the needs of free trade and human mobility, and already in flux through climate change. Furthermore, it is a world full of contradictory imperatives: while conservation agencies decry the transportation of species, development agencies have promoted it through exotic trees and aquaculture. Nonetheless, amongst the 40 or more international instruments relating to non-native species, sig-

natories to the Bern Convention are obliged, by Article 11, Paragraph 2b, 'to strictly control the introduction of non-native species'; Article 8(h) of the Convention on Biological Diversity requires Parties to 'Prevent the introduction of, control or eradicate those alien species which threaten ecosystems, habitats, or species'.

Species susceptible to the ravages of invaders may be identified by their rarity, specialization and life histories – for example, fish are especially at risk of hybridization because of their external fertilization and weakly developed isolating mechanisms. The most dangerous invaders are ecological generalists that are introduced in numbers enabling them quickly to escape the vortex of rarity to new habitats roughly matching those of their native land. Life history analysis revealed that sexually monomorphic birds were successful invaders – a correlation ingeniously, but probably erroneously, interpreted as arising because intensely sexually selected birds may be more extinction prone. More prosaically, the drab species have probably been introduced in greater numbers (Cassey et al. 2004). Because success is predicted by the force of invasion, it could be useful to analyse the life history traits that are associated with frequent transportation – at least for fish, the key predictors of invasion success may differ at different stages of the process (Kolar & Lodge 2002). Communities at risk may also be predictable through comparative analysis. For example, low functional diversity makes grasslands susceptible to invasion (Dukes 2002). Indeed, invasion may be a risk factor for further invasions, in so far as imbalances caused in a community by an earlier invasion may liberate resources that enable subsequent invaders to gain a foothold (Davies et al. 2000). Furthermore, amongst the cascade of effects following invasion may be homogenization of the whole community: the arrival of rainbow trout (*Oncorhynchus mykiss*) in North American lakes led to a simplification of communities of both fish and zooplankton (Beisner at al. 2003). The characteristics of candidate invaders, and of the communities that receive them, can work together in intricate ways. Not only may an invader's success be predicted by the extent of its pre-adaptation to its new circumstances, but these circumstances may select for rapid evolutionary changes in the newcomer that nurture its success. Leaving behind the natural enemies that have evolved to specialize upon it, the invader may find itself less challenged by predators or pathogens, or at least facing a threat that has shifted from specialist to generalist enemies. Under these circumstances the invader may evolve increased competitive ability by channelling more resources, previously required for defence, into growth and reproduction – ideas that have been refined by Muller-Scharer et al. (2004), who show that the allocation of different categories of plant defence (e.g. alkaloids versus lignins) and different life histories (monocarpic versus polycarpic) vary between native and invasive representatives of the same species. An exactly parallel argument, elaborated by Lee & Klasing (2004), suggests that invasive animals, freed from specialist pathogens, need invest less in expensive immunological inflammatory responses. Because they rely instead on cheaper, antibody-mediated immunity, they are freed to spend their savings on growth and reproduction. Indeed, species (e.g. voles) and varieties (e.g. chickens) differ in their relative allocation to humoral versus inflammatory immune defences (Klein & Nelson 1998; Koenen et al. 2002), so perhaps species emphasizing the latter can be predicted to be good invaders.

The goal to predict, and prevent, future invasions, is daunting. The task of stamping out invaders before they take hold is likely to be arduous, especially as, like American mink in England, invaders often escape detection until they are well established. However, the importance of both is great, and the sizes of areas that are being cleared of invasive mammals for conservation purposes is increasing (e.g. Norway rats (*Rattus norvegicus*) were recently cleared off the 110 km^2 Campbell Island, and goats

from the 585 km^2 Santiago Island). Impressive achievements on a mainland include the winkling of the coypu *Myocaster coypus* out of 6000 km^2 of eastern England (Gosling & Baker 1989), and the Western Shield programme in Western Australia which, by the use of broad-scale 1080 poison baiting, has reduced fox numbers so successfully that there have been extraordinary recoveries of many native mammal and bird species.

The qualities that make an invader successful are also likely to make it robust in the face of control (Boitani, 2001). High reproductive rate and dispersal capacity make invaders (and any other pest) resilient to artificially increased mortality. Even the introduced disease feline panleucopaenia, which reduced the feral cat population of Marion Island (South Indian Ocean) from an estimated 3045 in 1977 to 620 in 1982, eventually stimulated an upsurge of antibodies such that the last cats had to be mopped up by shooting and trapping (Bloomer & Bester 1992). Myxomatosis had an even more devastating effect on rabbits in the UK when it was deliberately (and illegally) introduced in 1953, but the development of genetic resistance has permitted British rabbit numbers to recover (Thompson & King, 1994). On Raoul Island, killing goats for several years reduced their numbers, and permitted vegetation recovery, but this led to a density-dependent doubling (from 0.96 to 1.70 kids female^{-1} year^{-1}) of reproduction so the goats bounded back (Parkes & Murphy 2003). The lesson is that eradication schemes should be hard, fast, well planned and irreversible. The problems of controlling abundant species are the same whether it is an invasive alien or a native (Courchamp et al. 1999; Tuyttens & Macdonald 2000; Zavaleta et al. 2001), and solutions generally require attention to human attitudes as well as biology. For example, the successful eradication of invasive coypu from England required an ingenious incentive scheme to ensure that trappers benefited from putting themselves out of work (Gosling & Baker

1989). The eradication cost £2.5 million – for comparison, the estimated cost to eradicate Japanese knotweed from Britain is £1.5 billion.

There has been debate over whether the goal should be to eradicate introduced species (Towns & Broome 2003), contain or exclude them (Moors et al. 1989), live with them (Rosenzweig 2003) or make use of them (Hutton 2003). The answer may often be that whatever can be done most effectively is probably best. Calculations of a cost/benefit ratio may justify an extreme effort to evict a localized alien such as the coypu in England, but the equivalent campaign against the more widely distributed American mink was abandoned as hopeless in 1970. Distinguishing between the alternatives is an awkward matter of judgement involving, as conservation so often does, balancing incommensurable factors, for which no ready rule can be universally valid. The only certainty is that the costs of dealing with ineradicable invasives are very high. For example, in one year (2002–3) New Zealand spent NZ$80 million on pest management ($60 million for possums, $10 million for feral ferrets, $6 million for goats, $2 million for stoats and $1 million for rabbits) (Parkes & Murphy 2003). The best ways to minimize such massive expenditure in future will be swift action against new arrivals, a scheme for prioritizing highest risks and constant monitoring. Set against the costs of attempted eviction, the direct costs to society (agriculture, water supply, etc.) of invasive alien species is estimated by the Global Invasive Species Programme (GISP) at in excess of tens of billions of Euros annually. Who should pay the costs of tackling introduced species? The notion that the polluter pays is generally appealing but, in this case, complicated by the fact that managing most invasives seems to be an interminable task.

Money is an inadequate common denominator for the political, cultural and ethical currencies involved in such debates (Bomford & O'Brien 1995; Eggleston et al. 2003). For example, hedgehogs, *Erinaceus europaeus*, are not native to the Scottish island of South Uist, where four individuals were released in 1974. Their descendents (estimated now at 5000) have spread to neighbouring islands, and their predation threatens internationally important populations of wading birds that have almost halved in numbers since the mid-1980s. As hedgehogs are iconically popular with the British public, but at the same time the Government has responsibility to protect the waders, controversy erupted when in 2002 the statutory agency concluded that suffering would be minimized if hedgehogs were killed rather than translocated or held in captivity – both latter options being unfeasible, and arguably immorally expensive. A parallel case involves plans to remove kiore (*Rattus exulans*) from Little Barrier Island (near Auckland, where these rats killed annually 90% of the chicks of endangered Cook's petrels, *Pterodroma cookii*) against the wishes of the local Maori (Imber et al. 2003). The political flammability of attempting to eradicate invasives is illustrated by the case of the grey squirrels introduced to red squirrel country in the Piedmont (north-west Italy) in 1948, where they smouldered until 1970 before exploding to 2500–6400 by May 1997. Then an eradication campaign started but, one month later, animal rights campaigners won a court action to stop the trial, and subsequent years of legal wrangling have prevented further control and ensured the continued spread of the grey squirrels. It is widely said that education is an essential prerequisite to winning public acceptance of alien eradication, and although this is true, it is also potentially condescending in so far as it implies that if people only knew more they would share the view of the cognoscenti: a more troubling reality is that the more one knows about introductions the more complex the judgements become.

Conclusions

The record of human redistribution of species is an extended tragedy of errors. Historical mixtures of unavoidable accident, misplaced whimsy, bungling and irresponsibility, compounded by careless or prejudiced logic, have created problems that may often prove unassailable. Some may be soluble, but always at great cost in several currencies. It is worth guarding against the verbal sloppiness that 'blames' or vilifies introduced species for the damage they cause – it was the people who moved their ancestors, not the descendents of the émigrés, who precipitated the problems we now face. Further, arguments developed for biodiversity conservation and the ethical treatment of individual animals are each complicated, and it may be impossible either to identify the logical supremacy of one over the other or to find a harmonious reconciliation; ultimately, there may be no dodging a stark choice between contradictory values. Where the native species and the invader cannot coexist, people must decide whether to stand by and let modified nature take its new course, or to act to prevent a foreseeable extinction (often by killing individuals of one species so that populations of another can prosper). In Europe the choice is between the American and the European mink; in New Zealand it is between the stoat and the brown kiwi. As it is humanity's fault that such dilemmas have arisen, the decision to remove an invader puts us in an ethically tawdry position (Fulton & Ford 2001; Mason & Littin 2003; Morris & Weaver 2003), and our only consolation may be at least to make sure the invaders are removed humanely.

We must start from where we are, not from where we might wish to be, and different mixes of incommensurable values can and should lead to different judgements. It would have been better if grey squirrels had never come to Britain, but now they are here, they give pleasure to some and cause losses to others, and they are sentient individuals whose deaths should not be commissioned carelessly. Where killing grey squirrels, or at least, preventing them from breeding, has a realistic chance of preventing the extinction of native reds, so be it. Where the prospect of restoring reds is close to nil, the moral argument for killing greys on biodiversity (as distinct from forestry) grounds is corrupted – from a biodiversity, societal and even moral standpoint, the pragmatic conclusion may be that there are parts of the UK where grey squirrels are here to stay and, indeed, better than no squirrels. It would also have been better if muntjac and Chinese water deer had not established in Britain, but they have, and there is no realistic chance of removing them. So long as their numbers are kept below what causes unacceptable damage to ground flora (a proviso which itself provides sport, meat and revenue), then in a country with an unnaturally impoverished fauna we might as well take the pragmatic view that it is rather pleasing to see them. While acknowledging that there comes a point when it is useless to try shutting the stable door after the proverbial horse has bolted, we should learn the lesson, and strive harder to prevent further invasions. Prevention is easier than cure, but even prevention is proving difficult. The history of introductions is far from over, and the creation of genetically modified organisms may herald a new chapter in this old story.

So, are introduced species special? They can be especially devastating to biodiversity conservation (and to the human enterprise) but, no, they are not biologically special – none of the ecological phenomena associated with an introduced species differs from those associated with natural colonists. 'Invasive species' are a subset of species recently transported by people – some judged benign, others malign – and, with man-made environmental change escalating biogeographical movements, even this definition is evermore arbitrary. This blurring of the boundaries between intended and natural effects does not lessen the danger posed by malign invasives to biodiversity conservation (Hulme 2003), but it does worsen the ethical and logical stresses of deciding upon solutions. Although the distinction between an invader and a guest may sometimes be hazy, there is nonetheless abundant evidence that the consequences of introductions can be dire. When it comes to looking back from the future, this history of transportation is likely to be judged an unhappy one.

Acknowledgements

We warmly acknowledge the help of Kerry Kilshaw, together with the numerous referees who helped us improve this essay.

The wildness of birds . . . is not dependent on any general degree of caution . . . [and] is not acquired by individual birds in a short time, even when much persecuted . . . .there is no way of accounting for it, except as an inherited habit: comparatively few birds, in any one year, have been injured by man in England, yet almost all, even nestlings, are afraid of him; many individuals, on the other hand, both at the Galapagos and the Falklands, have been pursued and injured by man, but yet have not learned a salutory dread of him. We may infer from these facts, what havoc the introduction of any new beast of prey must cause in a country, before the instincts of the indigenous inhabitants have become adapted to the stranger's craft or power.

(Charles Darwin, *The Voyage of HMS Beagle*, 3rd edn, 1860, pp. 399–400. Reprinted by the Folio Society, London, 2003.)

References

Atkinson, I.A.E. (1996) Introductions of wildlife as a cause of species extinctions. *Wildlife Biology* **2**: 135–41.

Bailey, E.P. (1992) Red foxes, *Vulpes vulpes*, as biological control agents for introduced Arctic foxes, *Alopex lagopus*, on Alaskan islands. *Canadian Field-Naturalist* **106**: 200–5.

Basse, B., McLennan, J.A. & Wake, G.C. (1999) Analysis of the impact of stoats, *Mustela erminea*, on northern brown kiwi, *Apteryx mantelli*, in New Zealand. *Wildlife Research* **26**: 227–37.

Beaumont, M.A., Barratt, E.M., Gottelli, D., et al. (2001) Genetic diversity and introgression in the Scottish wildcat. *Molecular Biology* **10**: 319–36.

Beisner, B.E., Ives, A.R. & Carpenter, S.R. (2003) The effects of an exotic fish invasion on the prey communities of two lakes. *Journal of Animal Ecology* **72**: 331–42.

Bensted-Smith, R., Cruz, E. & Valverde, F. (2000) The strategy for conservation of terrestrial biodiversity in Galápagos. *Proceedings of the Symposium Science for Conservation in Galápagos. Bulletin de l'Institut Royal des Sciences Naturelles de Belgique* (Supplement) **70**: 65–70.

Bloomer, J.P. & Bester, M.N. (1992) Control of feral cats on sub-Antarctic Marion Island, Indian Ocean. *Biological Conservation* **60**: 211–9.

Boitani, L. (2001) Carnivore introductions and invasions: their success and management options. In *Carnivore Conservation* (Eds J.L. Gittleman, S.M. Funk, D.W. Macdonald & R.K. Wayne), pp. 123–44. Cambridge University Press, Cambridge.

Bomford, M. & O'Brien, P. (1995) Eradication or control for vertebrate pests? *Wildlife Society Bulletin* **23**: 249–55.

Bonesi, L., Rushton, S.P. & Macdonald, D.W. (2002) The combined effect of environmental factors and neighbouring populations on the distribution and abundance of *Arvicola terrestris*. An approach using rule-based models. *Oikos* **99**: 220–30.

Bonesi, L., Chanin, T. & Macdonald, D.W (in press) Competition between Eurasian otters, *Lutra lutra*, and American mink, *Mustela vison*, probed by niche shift. *Oikos* **106**: 19–26.

Bryce, J.M., Speakman, J.R., Johnson, P.J. & Macdonald, D.W. (2001) Competition between Eurasian red and introduced Eastern grey squirrels: the energetic significance of body-mass differences. *Proceedings of the Royal Society of London Series B – Biological Sciences* **268**: 1731–6.

Bryce, J., Johnson, P.J. & Macdonald, D.W. (2002) Can niche use in red and grey squirrels offer clues for their apparent coexistence? *Journal of Applied Ecology* **39**: 875–87.

Bunin, J.S. & Jamieson, I.G. (1995) New Approaches Toward a Better Understanding of the Decline of Takahe (*Porphyrio mantelli*) in New Zealand. *Conservation Biology* **9**(1): 100–6.

Campbell, K. (2002) Advances in Judas Goat methodology in the Galápagos Islands: manipulating the animals. *Judas Workshop 2002 Proceedings*. Volume 15, Department of Conservation, Otago Conservancy, Dunedin.

Campbell, D.J. & Atkinson, I.A.E. (1999). Effects of kiore (*Rattus exulans* Peale) on recruitment of indigenous coastal trees on northern offshore islands of New Zealand. *Journal of the Royal Society of New Zealand* **29**: 265–90.

Cassey, P., Blackburn, T.M., Sol, D., Duncan, R. & Lockwood, J. (2004) Global patterns of introduction effort and establishment success in birds. *Alphagalileo and Biology Letters* **2004**: 19.

Chapin, F.S., Zavaleta, E.S., Eviner, V.T., et al. (2000) Consequences of changing biodiversity. *Nature* **405**: 234–42.

Chown, S.L. & Block, W. (1997) Comparative nutritional ecology of grass feeding in a sub-Antarctic beetle: the impact of introduced species on *Hydromedion sparsatum* from South Georgia. *Oecologia* **111**: 216–24.

Colwell, R.K. & Futuyma, D.J. (1971) On the measurement of niche breadth and overlap. *Ecology* **52**: 567–76.

Corbet, G.B. & Harris, S. (1991) *The Handbook of British Mammals*, 3rd edn. Blackwell, Oxford, pp. 185–7.

Courchamp, F., Langlais, M. & Sugihara, G. (1999) Cats protecting birds: modelling the mesopredator release effect. *Journal of Animal Ecology* **68**: 282–92.

Crawley, M.J., Harvey, P.H. & Purvis, A. (1996) Comparative ecology of the native and alien floras of the British Isles. *Philosophical Transactions of the Royal Society of London* **351**: 1251–9.

Cruz, J. B. & Cruz, F. (1996) Conservation of the dark-rumped petrel *Pterodroma phaeopygia* of the Galápagos Islands. *Bird Conservation International* **6**: 23–32.

Cuthbert, R. & Davis, L.S. (2002) The impact of predation by introduced stoats on Hutton's shearwaters, New Zealand. *Biological Conservation* **108**: 79–92.

Danell, K. (1996) Introductions of aquatic rodents: lessons of the muskrat *Ondatra zibethicus* invasion. *Wildlife Biology* **2**(3): 213–19.

Daniels, M.J., Balharry, D., Hirst, D, Kitchener, A.C. & Aspinall, R.J. (1998) Morphological and pelage characteristics of wild living cats in Scotland: implications for defining the 'wildcat'. *Journal of Zoology, London* **244**: 231–47.

Davies, M.A., Grime, J.P. & Thompson, K. (2000) Fluctuating resources in plant communities: a general theory of invisibility. *Journal of Ecology* **88**: 528–34.

Dawson, F.B. & Warman, D.A (1987) *Crassula helmsii* (T. Kirk) cockayne: is it an aggressive alien Aquatic plant in Britain. *Biological Conservation* **42**(4): 247–72.

Desender.K., Baert.L., Maelfair, J.P & Verdyck, P (1999) Conservation on Volcan Alcedo (Galápagos): terrestrial invertebrates and the impact of introduced feral goats. *Biological Conservation* **87**(3): 303–10.

Diamond, J.M. & Case, T.J. (Eds) (1986) *Community Ecology*. Harper and Row, New York.

Dowler, R.C., Carroll, D.S. & Edwards, C.W. (2000) Rediscovery of rodents (Genus *Nesoryzomys*) considered extinct in the Galápagos Islands. *Oryx* **34**: 109–17.

Dukes, J.S. (2002) Species composition and diversity affect grassland susceptibility and response to invasion. *Ecological Applications* **12**: 602–17.

Ebenhard, T. (1988) Introduced birds and mammals and their ecological effects. *Viltrevy* **13**: 5–107.

Ecroyd, C.E. (1996) The ecology of *Dactylanthus taylorii* and threats to its survival. *New Zealand Journal of Ecology* **20**: 81–100.

Eggleston, J.E., Rixecker, S.S. & Hickling, G.J. (2003) The role of ethics in the management of New Zealand's wild mammals. *New Zealand Journal of Zoology* **30**: 361–76.

Elton, C.S. (1958) *The Ecology of Invasion by Animals and Plants*. Methuen, London.

Fitzgerald, B.M. & Turner, D. (2000) Hunting behaviour of domestic cats and their impact on prey populations. In *The Domestic Cat: the Biology of its Behaviour* (Eds D.C. Turner & P. Bateson), pp. 152–75. Cambridge University Press, Cambridge.

Forsyth, D.M. & Duncan, R.P. (2001) Propagule size and the relative success of exotic ungulate and bird introductions in New Zealand. *American Naturalist* **157**: 583–95.

Forsyth, D.M., Duncan, R.P., Bomford, M. & Moore, G. (2004). Climatic suitability, life-history traits, introduction effort, and the establishment and spread of introduced mammals in Australia. *Conservation Biology* **18**: 557–69.

Fulton, G.R. & Ford, G.R. (2001) The conflict between animal welfare and conservation. *Pacific Conservation Biology* **7**: 152–3.

Gosling, L.M. & Baker, S.J. (1989) The eradication of muskrats and coypus from Britain. *Linnean Society Biological Journal* **38**: 39–51.

Grant, P.R., Curry, R.L. & Grant, B.R. (2000) A remnant population of the Floreana mockingbird on Champion Island, Galápagos. *Biological Conservation* **92**: 285–90.

Gurevitch, J. & Padilla, D (2004) Assessing species invasions as a cause of extinction. *Trends in Ecology and Evolution* **19**: 620.

Gurnell, J., Wauters, L.A., Lurz, P.W.W. & Tosi, G. (2002) Alien species and interspecific competition: effects of introduced eastern grey squirrels on red squirrel population dynamics. *Journal of Animal Ecology* **73**: 26–35.

Heather, B. & Robertson, H. (1996) *The Field Guide to the birds of New Zealand*. Viking (Penguin Books), Auckland.

Herbold, B. & Moyle, P.B. (1986) Introduced species and vacant niches. *The American Naturalist* **128**: 751–60.

Hersteinsson, P. & Macdonald, D.W. (1992) Interspecific competition and the geographical distribution of Red and Arctic foxes *Vulpes vulpes* and *Alopex lagopus*. *Oikos* **64**: 505–15.

Hindwood, K.A. (1940) The birds of Lord Howe Island. *The Emu* **40**: 1–77.

Holdaway, R.N. (1999) Introduced predators and avifaunal extinction in New Zealand. In *Extinctions in Near Time* (Ed. R.D.E. MacPhee), pp. 189–238. Kluwer Academic/Plenum Publishers, New York.

Holland, D.G. (2001) Giant hogweed and Japanese knotweed. In *Exotic and Invasive Species: Should we be Concerned? Proceedings of the 11th Conference of the Institute of Ecology and Environmental Management*, April 2000, Winchester.

Hughes, B. (2001) Ruddy duck and threat to white-headed duck in Europe. In *Exotic and Invasive Species: Should we be Concerned? Proceedings of the 11th Conference of the Institute of Ecology and Environmental Management*, April 2000, Winchester.

Hulme, P.E. (2003) Biological invasions: winning the science battle but losing the conservation war? *Oryx* **37**: 178–93.

Hutton, J. (2003) Sustainable use and incentive-driven conservation: realigning human and conservation interests. *Oryx* **37**: 215–26.

Imber, M.J., West, J.A. & Cooper, W.J. (2003). Cook's petrel (*Pterodroma cookii*): historic distribution, breeding biology and effects of predators. *Notornis* **50**: 221–30.

IUCN (1997) Conserving vitality and diversity. In *Proceedings of the World Conservation Congress Workshop on Alien Invasive Species* (Compiled by C.D.A. Rubec & G.O. Lee). Species Survival Commission, International Union for the Conservation of Nature and Natural Resources, Gland, Switzerland; North American Wetlands Conservation Council, Ottawa, Canada; Canadian Wildlife Service, Ottawa, Canada.

Jefferies, D.J. (Ed.) (2003) *The Water Vole and Mink Survey of Britain 1996–1998 with a History of the Long Term Changes in the Status of both Species and their Causes*. The Vincent Wildlife Trust, Ledbury.

Kamler, J. F. & Ballard, W. B. (2002) A review of native and nonnative red foxes in North America. *Wildlife Society Bulletin* **30**:370–79.

Kauhala, K., Kaunisto, M. & Helle, E. (1993) Diet of the raccoon dog, *Nyctereutes procyonoides*, in Finland. *Zeitschrift fur Saugetierkunde* **58**: 129–36.

Kauhala, K., Laukkanen, P. & von Rége, I. (1998) Summer food composition and food niche overlap of the raccoon dog, red fox and badger in Finland. *Ecography* **21**: 457–63.

Kenward, R.E. (1985) Ranging behaviour and population dynamics in grey squirrels. In *Behavioural Ecology* (Eds R.M. Sibly & R. Smith), pp. 319–30. Blackwell, Oxford.

Key, G. & Muñoz, E. (1994) Distribution and current status of rodents in the Galápagos. *Noticias de Galápagos* **53**: 21–5.

King, C.M. (1984) *Immigrant Killers: Introduced Predators and the Conservation of Birds in New Zealand*. Oxford University Press, Auckland, 224 pp.

King, C M. (Ed.) (2005). *The Handbook of New Zealand Mammals*, 2nd edn. Oxford University Press, Melbourne.

Klein, S.L. & Nelson, R.J. (1998) Adaptive immune responses are linked to the mating system of arvicoline rodents. *American Naturalist* **151**: 59–67.

Koenen, M.E., Boonstra-Blom, A.G. & Jeurissen, S.H.M. (2002) Immunological differences between layer and broiler-type chickens. *Veterinary Immunology and Immunopathology* **89**: 47–56.

Kolar, C.S. & Lodge, D.M. (2001) Progress in invasion biology: predicting invaders. *Trends in Ecology and Evolution* **16**: 199–204.

Kolar, C.S. & Lodge, D.M. (2002) Ecological predictions and risk assessment for alien fishes in North America. *Science* **298**: 1233–6.

Krefft, G. (1866). On the vertebrated animals of the lower Murray and Darling, their habits, economy, and geographical distribution. *Transactions of the Philosophical Society of New South Wales* **1862–1865**: 1–33.

Lack, D. (1947) *Darwin's Finches*. Cambridge University Press. Cambridge.

Lawton, J. (1997). The science and non-science of conservation biology. *New Zealand Journal of Ecology* **21**: 117–20.

Leader-Williams, N., Lewis-Smith, R.I. & Rothery, P. (1987) Influence of an introduced reindeer population upon the vegetation of South Georgia: results from a long-term exclusion experiment. *Journal of Applied Ecology* **24**: 801–22.

Lee, K.A. & Klasing, K.C. (2004) A role for immunology in invasion biology. *Trends in Ecology and Invasion* **19**: 524–9.

Lee, W.G. & Jamieson, I.G. (Eds) (2001) *The Takahe: Fifty Years of Conservation Management and Research*. University of Otago Press, Dunedin, 132 pp.

Lodge, D.M. (1993a) Biological invasions: lessons for ecology. *Trends in Ecology and Evolution* **8**: 133–7.

Lodge, D.M. (1993b) Species invasions and deletions: community effects and responses to climate and habitat change. In: *Biotic Interactions and Global Change* (Eds P.M. Kareiva, J.G. Kingsolver & R.B. Huey). Sinnauer Associates, Sunderland, MA, 559 pp.

MacArthur, R.M. (1972) *Geographical Ecology: Patterns in the Distribution of Species*. Harper & Row, New York.

Macdonald, D.W. (1993) Rabies and wildlife: a conservation problem? *Onderstepoort Journal of Veterinary Research* **60**: 351–5.

Macdonald, D.W. (1995) *European Mammals: Evolution and Behaviour*. Harper Collins, London.

Macdonald, D. W. & Harrington, L.A. (2003). The American mink: the triumph and tragedy of adaptation out of context. *New Zealand Journal of Zoology* **30**: 421–41.

Macdonald, D.W. & Newman, C. (2002). Population dynamics of badgers (*Meles meles*) in Oxfordshire, UK: numbers, density and cohort life histories, and a possible role of climate change in population growth. *Journal of Zoology* **256**: 121–38.

Macdonald, D.W & Strachan, R. (1999) *The Mink and the Water Vole: Analyses for Conservation*. Wildlife Conservation Research Unit and the Environment Agency, Oxford.

Macdonald, D.W. & Thom, M.D. (2001). Alien carnivores: unwelcome experiments in ecological theory. In *Carnivore Conservation* (Eds J.L. Gittleman, S.M. Funk, D.W. Macdonald & R.K. Wayne), pp. 93–122. Cambridge University Press, Cambridge.

Macdonald, I.A.W, Loope, L.L, Usher, M.B. & Hamann, O. (1989) Wildlife conservation and the invasion of nature reserves by introduced species: a global perspective. In *Biological Invasions: a Global Perspective* (Eds J.A. Drake & H.A. Mooney), pp. 215–54. Scientific Committee on Problems of the Envirornment, International Council of Scientific Unions. Wiley, Chichester.

Macdonald, D.W., Mace, G.M. & Barreto, G.R. (1999) The effects of predators on fragmented prey populations: a case study for the conservation of endangered prey. *Journal of Zoology, London* **247**: 487–506.

Macdonald, D.W., Sidorovich, V.E., Maran, T. & Kruuk, H. (2002) *European Mink, Mustela lutreola: Analyses for Conservation*. Wildlife Conservation Research Unit, Oxford.

Macdonald, D.W., Daniels, M.J., Driscoll, C., Kitchener, A. & Yamaguchi, N. (2004) The Scottish Wildcat: analyses for conservation and an action plan. *Conservation Action Plan: the Wildcat in Scotland*. Wildlife Conservation Research Unit, Oxford.

Macfarland, T., Villa, J. & Toro, B. (1974) The Galápagos giant tortoises (*Geochelone elephantopus*). Part I. Status of the surviving populations. *Biological Conservation* **6**: 118–33.

Mack, R.N., Simberloff, D., Lonsdale, W.M., Evans, H., Clout, M. & Bazzaz, F.A. (2000) Biotic invasions: causes, epidemiology, global consequences, and control. *Ecological Applications* **10**: 689–710.

Mason, G. & Littin, K.E. (2003) The humaneness of rodent pest control. *Animal Welfare* **12**: 1–37.

Mauchamp, A., Aldaz, I., Ortiz, E. & Valdebenito, H. (1998) Threatened species, a re-evaluation of the status of eight endemic plants of the Galápagos. *Biodiversity and Conservation* **7**: 97–107.

McLennan, J.A., Potter, M.A., Robertson, H.A., et al. (1996) Role of predation in the decline of kiwi, *Apteryx* spp., in New Zealand. *New Zealand Journal of Ecology* **20**: 27–35.

McNeely, J.A., Mooney, H.A., Neville, L.E., Schei, P. & Waage, J.K. (Eds) (2001) *A Global Strategy on Invasive Alien Species*. International Union for the Conservation of Nature and Natural Resources, Gland, Switzerland; in collaboration with the Global Invasive Species Programme.

Merton, D.V. (1970) Kermadec Islands expeditions reports: a general account of birdlife. *Notornis* **17**: 147–99.

Montague, T.L. (Ed.) (2000) *The Brushtail Possum: Biology, Impact and Management of an Introduced Marsupial.* Manaaki Whenua Press, Lincoln, New Zealand, 292 pp.

Mooney, H.A. & Cleland, E.E. (2001) The evolutionary impact of invasive species. *Proceedings of the National Academy of Sciences of the USA* **98**(10): 5446–51.

Moors, P.J., Atkinson, I.A.E. & Sherley, G.H. (1989) *Prohibited Immigrants: the Rat Threat to Island conservation.* World Wide Fund for Nature – New Zealand, Wellington, New Zealand.

Morris, M.C. & Weaver, S.A. (2003) Minimising harm in possum control operations and experiments in New Zealand. *Journal of Agricultural and Environmental Ethics* **16**: 367–85.

Moyle, P.B. (1986) Fish introductions into North America: Patterns and ecological impact. In *Ecology of Biological Invasions of North America and Hawaii* (Eds H.A. Mooney & J.A. Drake), pp. 27–43. Springer-Verlag, New York.

Mulder, J.L. (1990) The stoat *Mustela erminea* in the Dutch dune region, its local extinction, and a possible cause: the arrival of the fox *Vulpes vulpes. Lutra* **33**: 1–21.

Muller-Scharer, H., Shaffner, U. & Steinger, T. (2004) Evolution in invasive plants: implications for biological control. *Trends in Ecology and Evolution* **19**: 417–22.

Nature Conservancy (1996) *America's Least Wanted: Alien Species Invasions of the U.S. Ecosystems.* The Nature Conservancy, Arlington, VA.

Newsome, A.E. & Noble, I.R. (1986) Ecological and physiological characters of invading species. In *Ecology of Biological Invasions* (Eds R.H. Groves & J.J. Burdon), pp. 1–20.Cambridge University Press, Cambridge.

Nummi, P. (1996) Wildlife introductions to mammal deficient areas: the Nordic countries. *Wildlife Biology* **2**(3): 221–6.

Parkes, J. & Murphy, E.C. (2003) Management of introduced mammals in New Zealand. *New Zealand Journal of Zoology* **30**: 335–59.

Patton, J. L. & Hafner, M. S. (1983). Biosystematics of the native rodents of the Galápagos archipelago, Ecuador. In: *Patterns of Evolution in Galápagos Organisms* (Eds R.I. Bowman, M. Berson & A.E. Leviton), pp. 539–68. AAAS Pacific Division, San Francisco, CA.

Perry, B.D. (1993) Dog ecology in eastern and southern Africa: implications for rabies control. *Onderstepoort Journal of Veterinary Research* **60**: 429–36.

Pierce, R.J. (1984) Plumage, morphology and hybridisation of New Zealand stilts *Himantopus* spp. *Notornis* **31**: 106–30.

Project Isabela (2004) Galápagos Conservation Trust, London. http://www.gct.org/projectisa.html.

Rhodes, C.J., Atkinson, R.P.D. Anderson, R.M. & Macdonald, D.W. (1998) Rabies in Zimbabwe: reservoir dogs and the implications for disease control. *Philosophical Transactions of the Royal Society of London, Series B* **353**: 999–1010.

Rhymer, J.M. & Simberloff, D. (1996) Extinction by hybridization and introgression. *Annual Review of Ecology and Systematics* **27**: 83–109.

Robertson, A.W., Kelly, D., Ladley, J.J. & Sparrow, A.D. (1999) Effects of pollinator loss on endemic New Zealand mistletoes (Loranthaceae). *Conservation Biology* **13**: 499–508.

Rocha-Camarero, G. & De Trucios, S.J.H. (2002) The spread of the Collared Dove *Streptopelia decaocto* in Europe: colonization patterns in the west of the Iberian Peninsula. *Bird Study* **49**: 11–16.

Roelke-Parker, M.E., Munson, L., Packer, C., et al. (1996) A canine distemper virus epidemic in Serengeti lions (*Panthera leo*). *Nature* **379**: 441–5.

Roemer, G.W. (2004) The evolution, behavioural ecology and conservation of Island foxes (*Urocyon littoralis*). In *Canid Biology and Conservation* (Eds D.W. Macdonald & C. Sillero-Zubiri), pp. 173–84. Oxford University Press, Oxford.

Rosenzweig, M.L. (2003) Reconciliation ecology and the future of species diversity. *Oryx* **37**: 194–205.

Rushton, S.P., Barreto, G.W., Cormack, R.M., Macdonald, D.W. & Fuller, R (2000) Modelling the effects of mink and habitat fragmentation on the water vole. *Journal of Applied Ecology* **37**: 475–90.

Sainsbury, A.W. & Ward, L. (1996) Parapoxvirus infection in red squirrels. *Veterinary Record* **138**: 400.

Sale, P.F. (1994) Overlap in resource use, and interspecific competition. *Oecologia* (Berlin) **17**: 245–56.

Sax, D.F., Gaines, S.D. & Brown, J.H. (2002) Species invasions exceed extinctions on islands worldwide: A comparative study of plants and birds. *American Naturalist* **106**(6): 766–83.

Sidorovich, V., Kruuk, H. & Macdonald, D.W. (1999) Body size, and interactions between European and American mink (*Mustela lutreola* and *M. vison*) in eastern Europe. *Journal of Zoology* **248**: 521–7.

Sillero-Zubiri, C. & Macdonald, D.W. (1997) *Ethiopian wolf: an Action Plan for its Conservation.* Canid

Specialist Group, International Union for the Conservation of Nature and Natural Resources, Gland, Switzerland.

Sillero-Zubiri, C., Gottelli, D. & Wayne, R. K. (1994) Hybridization of the Ethiopian wolf. *Canid News* **2**: 33–4.

Sillero-Zubiri, C., King, A.A. & Macdonald, D.W. (1996) Rabies and mortality in Ethiopian wolves (*Canis simensis*). *Journal of Wildlife Diseases* **32**: 80–6.

Simpson, G.G. (1980) *Splendid Isolation*. Yale University Press, New Haven and London.

Smirnov, V.V. & Tretyakov, K. (1998) Changes in aquatic plant communities on the island of Valaam due to invasion by the muskrat *Ondatra zibethicus* L. (Rodentia, Mammalia). *Biodiversity and Conservation* **7**: 673–90.

Smith, A.P. & Quin, D.G. (1996) Patterns and causes of extinctions and decline in Australian conilurine rodents. *Biological Conservation* **77**: 243–67.

Soulè, M.E. (1990) The onslaught of alien species, and other challenges in coming decades. *Conservation Biology* **4**: 233–39.

Taylor, R.H. (1979) How the Macquarie Island parakeet became extinct. *New Zealand Journal of Ecology* **2**: 42–5.

Thompson, H.V. & King, C.M. (Eds) (1994). *The European Rabbit: the History and Biology of a Successful Colonizer*. Oxford University Press, Oxford.

Tompkins, D.M., White, A.R. & Boots, M. (2003) Ecological replacement of native red squirrels by invasive greys driven by disease. *Ecology Letters* **6**: 189–96.

Towns, D.R. & Broome, K.G. (2003) From small Maria to massive Campbell: forty years of rat eradications from New Zealand islands. *New Zealand Journal of Zoology* **30**: 377–408.

Towns, D.R, Simberloff, D. & Atkinson, I.A.E. (1997) Restoration of New Zealand islands: redressing the effects of introduced species. *Pacific Conservation Biology* **3**: 99–124.

Tuyttens, F.A.M. & Macdonald, D.W. (2000) Consequences of social perturbation for wildlife management and conservation. In *Behaviour and Conservation*, Vol. 4 (Eds L.M. Gosling & J.W. Sutherland), *pp.* 315–29. Cambridge University Press, Cambridge.

Vahlenkamp, M., Muller, T., Tackmann, K., Loschner, U., Schmitz, H. & Schreiber, M. (1998) The muskrat (*Ondatra zibethicus*) as a new reservoir for puumala-like hantavirus strains in Europe. *Virus Research* **57**: 139–50.

Veitch, C.R. & Clout, M.N. (Eds) (2002) Turning the tide: the eradication of invasive species. *Proceedings of the International Conference on Eradication of Island Invasives*, p. 414. Occasional Paper 27: Species Survival Commission, International Union for the Conservation of Nature and Natural Resources, Gland, Switzerland.

Vitousek, P.M., D'Antonio, C.M., Loope, L.L., Rejmanek, M. & Westbrooks, R. (1997) Introduced species: a significant component of human caused global change. *New Zealand Journal of Ecology* **21**: 1–16.

Wangchuk, T. (2005) *The evolution, phylogeography, and conservation of the Golden Langur* (Trachypithecus geei) *in Bhutan*. PhD dissertation, Department of Biology, University of Maryland, College Park, MD.

Wauters, L.A., Gurnell, J., Martinoli, A. & Tosi, G. (2002a) Interspecific competition between native Eurasian red squirrels and alien grey squirrels: does resource partitioning occur? *Behavioral Ecology and Sociobiology* **52**: 332–41.

Wauters, L.A., Tosi, G. & Gurnell, J. (2002b) Interspecific competition in tree squirrels: do introduced grey squirrels (*Sciurus carolinensis*) deplete tree seeds hoarded by red squirrels (*S. vulgaris*)? *Behavioral Ecology and Sociobiology* **51**: 360–7.

West, E.W. & Rudd, R.L. (1983) Biological control of Aleutian Island arctic fox: a preliminary strategy. *International Journal for the Study of Animal Problems* **4**: 305–11.

Williamson, M. (1993) Invaders, weeds and the risk from genetically manipulated organisms. *Experientia* **49**: 219–24.

Williamson, M. (1999) Invasions (50th anniversary mini-review). *Ecography* **22**: 5–12.

Wilson, K.J. (2004) *Flight of the Huia*. Canterbury University Press, Christchurch.

Woodroffe, R, Ginsberg, J.R. & Macdonald, D.W. (1997) *The African wild dog – Status Survey and Conservation Action Plan*. International Union for the Conservation of Nature and Natural Resources, Gland, Switzerland.

Woods, M., McDonald, R.A. & Harris, S. (2003). Predation of wildlife by domestic cats *Felis catus* in Great Britain. *Mammal Review* **33**: 174–88.

Zavaleta, E.S., Hobbs, R.J. & Mooney, H.A. (2001) Viewing invasive species removal in a whole-ecosystem context. *Trends in Ecology and Evolution* **16**: 454–9.

14

Bushmeat: the challenge of balancing human and wildlife needs in African moist tropical forests

John E. Fa, Lise Albretchsen and David Brown

T.C. Brownell, a native catechist in south-east Liberia, wrote that, on 29 March 1857: 'At 8 o'clock, we stopped at a Nyambo town for refreshments. [A man] brought me something to eat which he called bush meat, but it had such a human aspect that I laid it aside, and awaited the repast which was preparing.'
(Quoted in A. M. Scott, *Day Dawn in Africa* 1858, p. 295.)

Introduction

Bushmeat hunting is the single most geographically widespread form of resource extraction in tropical forests and can affect the core of even the largest and least accessible nature reserves (Peres & Terborgh 1995). Exploitation of bushmeat by tropical forest dwellers has increased dramatically in recent years in many of the important source areas (Robinson & Bennett 2000). Game harvests in South America and Africa are believed to substantially exceed production (Robinson & Bodmer 1999), even in the case of traditional aborigine societies still using rudimentary hunting technology (Alvard et al. 1997). Such uncontrolled exploitation is likely to bring about marked population declines, even, eventually, the extinction of a number of game species. Coupled with threats from habitat loss, from historical deforestation (Cowlishaw 1999), global extinctions of the most sensitive species such as primates are likely to occur as an accumulation of local disappearances. This may result in long-term changes in tropical forest dynamics through the loss of seed dispersers, large granivores, frugivores and 'habitat landscapers' such as large forest mammals (Dirzo & Miranda 1991; Chapman & Onderdonk 1998; Wright et al. 2000).

At the same time, bushmeat has long been a critical component of the diet of forest dwellers in tropical forest regions. Some published estimates of bushmeat consumption suggest that rural people in Africa obtain at least 20% of their animal protein from wild animals (Chardonnet et al. 1995). There is evidence that the rural poor are particularly dependent on the income from bushmeat sales, which can make a substantial contribution to discretionary income in areas where there are few alternative income-generating opportunities.

The issue of overexploitation of wild species is complex and influenced by factors such as poverty, food insecurity, slow development, economic market failures, but also the lack of political and institutional understanding.

Thus, solutions to 'the bushmeat crisis', as it is sometimes known, cannot be tackled by single-discipline approaches alone but require multidisciplinary efforts. The purpose of this chapter is to highlight the nature and dimensions of the problem in tropical moist forests in Africa, and suggest possible integral solutions which involve melding economic and biological parameters. The consequences of uncontrolled overexploitation are extinction of animal species, but also – of equal importance – the loss of the mainstay of millions of people. It follows that the solutions that are offered must be ones that satisfy not only conservation criteria but also the development needs of the human populations involved.

Hunter offtakes – the view at ground level

There is strong evidence that there is a 'bushmeat crisis' of sizeable proportions, at least in the key source areas. Overhunting ranks as a major problem for one-third of the mammals and birds threatened with extinction, according to a recent analysis of the Red List, in which the International Union for Conservation of Nature and Natural Resources reports the degree of peril for plant and animal species worldwide. For 8% of mammals in greatest peril, overexploitation is the major threat. The bushmeat trade takes dozens of species, from elephants to birds. For example, Fa et al. (2005) lists 71 species of mammals that are traded in seven countries of west and central Africa: Cameroon, Equatorial Guinea, Gabon, Congo, Democratic Republic of Congo (DRC), Central African Republic (CAR) and Ghana. Hunters at the 36 villages and towns sampled were killing about 200 animals per hunter per year. Of that meat, nearly three-quarters, by weight, came from hoofed animals such as bay and blue duikers. However, the list also includes aardvarks, pangolins and 22 species of primates. A pattern emerges in that smaller carnivores and ungulates, the larger rodents and medium-sized primates are hunted most frequently (Figs 14.1 & 14.2). The primates

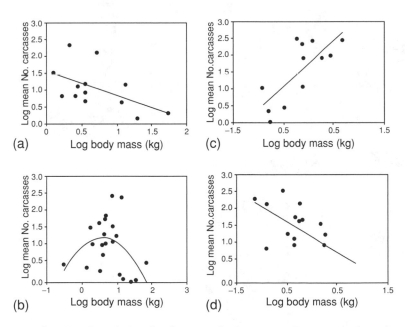

Fig. 14.1 Regressions showing the relationship between body mass and mean number of carcasses extracted per year of: (a) carnivores ($r^2 = 0.29$, d.f. 8, $P = 0.05$); (b) primates ($r^2 = 0.21$, d.f. = 20, $P = 0.000$); (c) rodents ($r^2 = 0.57$, d.f. 12, $P = 0.005$); (d) ungulates ($r^2 = 0.55$, d.f. 17, $P = 0.004$). (From Fa et al. 2005.)

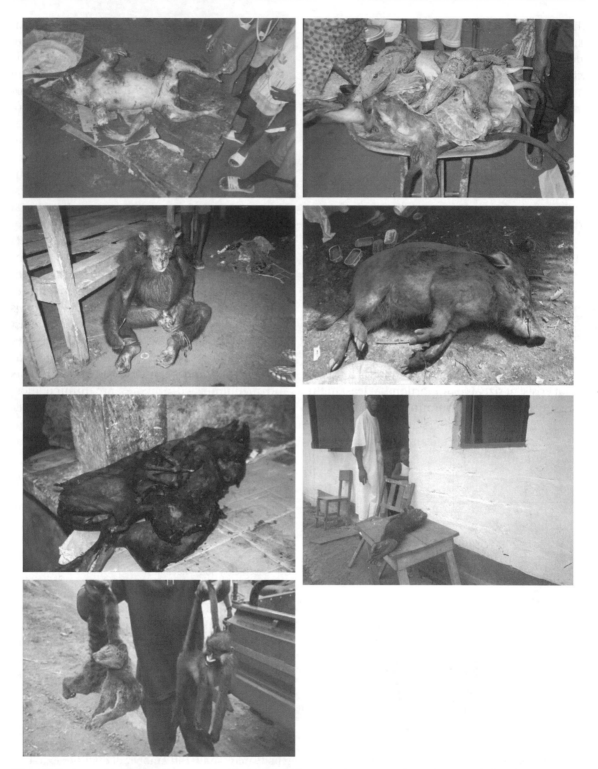

Fig. 14.2 Examples of bushmeat species hunted within moist forests in Africa, these include a large number of species ranging from reptiles to chimpanzees and gorillas.

have raised the most public alarm. The Ape Alliance, an England-based coalition of primate-related organizations, has reported that although primates do not represent a large proportion of the whole trade, hunting is a serious threat to chimpanzees, gorillas, bonobos, colobus monkeys and some other species.

The rapid recent acceleration in losses of tropical forest species owing to unsustainable hunting occurred in Asian forests first; for example, within the past 40 years, 12 large vertebrate species have been extirpated in Vietnam largely because of hunting. Africa is now experiencing species losses over wide areas and, in the next 10–20 years, losses are likely to be recorded in even the remotest parts of Latin America. This pattern follows the major impacts of development and forest loss on the three continents linked to dramatic human population growth: there are 522 people per square kilometre of remaining forest in South and South-East Asia, 99 in West/Central Africa and 46 in Latin America (Peres 2001; Milner-Gulland et al. 2003). However, the emphasis on hunting of forest species differs by continent. For example, Fa & Peres (2001) compared game harvest profiles from South America and Africa and showed that although 55% of African forest mammals were game species, significantly fewer numbers (28%) were hunted in Amazonian forests. These differences are due to the greater number of larger-bodied and substantially higher mean body mass of African game mammal species, compared with those in Amazonia. The prominent role of large-bodied mammals in African game harvests can also explain their greater vulnerability to indirect hunting techniques (e.g. traps, nets, snares), which opens the possibility for hunters to pursue more efficiently a greater range of animals (Bahuchet & de Garine 1990; Wilkie & Curran 1991; Noss 1998). The use of snares, especially cable snares, currently widespread in African forests, accounts for the extraction of more game species (and biomass) than firearms. Because cable snares are more affordable and accessible to local hunters than are firearms, extensive areas can be operated at very high snare densities (Colell et al. 1994; Noss 1995). Cable snares are known to capture virtually all species in African rain forests, except elephant (*Loxodonta cyclotis*) and hippopotamus (*Hippopotamus amphibius*), as well as several species of birds and reptiles.

Scaling up – regional extraction rates

One recent estimate puts the extraction of bushmeat in the Congo Basin at 1.1 million tonnes per annum (Wilkie & Carpenter 1999), although this could well be a major underestimate. Fa et al. (2002) in contrast calculated an annual bushmeat harvest within the main countries in Central Africa alone as 3.4 million tonnes of undressed meat (the subregional human population is 33 million – hence, this would imply between 30 and 150 kg per person per year, if all is consumed locally).

In West Africa, figures vary. Production for Liberia, immediately pre-civil war in 1989, gave an estimate of 165,000 t yr^{-1}, including both subsistence and commercial production (Anstey 1991). This was for a market value of $66 million (equivalent to per capita consumption of at least 55 kg yr^{-1} on a national scale). Production in Côte d'Ivoire is valued at almost $120 million per year (Bowen-Jones & Pendry 1999). Recent estimates put the figure for Ghana at 305,000 t of bushmeat sold annually at a net value in the region of $275 million. More than 3500 carcasses were recently counted at Kumasi Central Market in one month. In Takoradi, another Ghanaian market town, monthly bushmeat trade was estimated at almost 1.6 t, half of it rodents (Cowlishaw et al. 2005). Although these estimates offer some idea of the volume of bushmeat harvested in some specific but relatively small African moist forest areas, estimates based on extensive and simultaneous sampling, within larger geographical regions, are currently not available. Fa et al. (2006) present results of the first reported study of this kind (see Box 14.1).

Box 14.1 Seeking better estimates of the problem – the Cross–Sanaga rivers study

Multiple-site surveys are necessary to obtain accurate assessments of the magnitude of bushmeat extraction in large tracts of forests such as the Congo Basin. However, such methods are very expensive and labour-intensive at such large scales. An alternative method of estimating the state of hunted faunal assemblages is to conduct carcass counts of species at bushmeat markets, as these are found in almost every town and village and are important concentration points of wildlife harvests in surrounding catchment areas (Juste et al. 1995; Fa & Garcia Yuste 2001). Although accuracy may be compromised by variation in hunting effort, initial faunal assemblages and their densities, market data can be useful as a measure of hunting pressure within supply areas.

In a study funded by the UK Government's Darwin Initiative Fund, information about volume and identity of bushmeat taxa available at market sites in moist tropical forests between the Cross River in Nigeria and the Sanaga River in Cameroon were collected (Fa et al. 2006). These forests are rich in animal species, many of which are endemic. Some parts of the forest are protected although the area is heavily affected by human use, including logging and plantation agriculture.

During a period of 5 months, we counted bushmeat carcasses deposited in 89 urban and rural markets in a 35,000 km^2 area between the Cross River in Nigeria and the Sanaga River in Cameroon (Box Fig. 14.1).

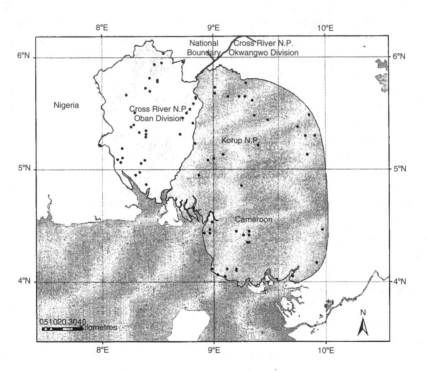

Box Fig. 14.1 Distribution of study points in the Cross–Sanaga rivers study.

We used these data to calculate annual bushmeat volume traded by site, species and overall study area.

Mammals represented > 90% of the bushmeat carcasses sold at all sites. Reptiles were also abundant, but birds and amphibians were relatively scarce. Estimates of carcasses extracted and crude biomass per site varied significantly between countries. In Nigeria, biomass (kg) extracted for sale per square kilometre per year was three times greater (600 kg km^{-2}) than in Cameroon. Conservative estimates for the entire study area indicate that > 900,000 reptiles, birds and mammals are sold each year by the rural and urban population, corresponding to around 12,000 t of terrestrial vertebrates. We also assessed the relationship between bushmeat harvested for sale and distance of the study settlements from the main protected areas (Cross River and Korup National Parks). The number of carcasses and biomass sold was negatively related to the proximity to the national parks in > 50% of species in Nigeria, and in 40% of species in Cameroon.

Box 14.1 Seeking better estimates of the problem – the Cross–Sanaga rivers study (Continued)

Our cross-site comparison documents the staggering volume of wild species affected by hunting in the region. We also conclude that species within the main protected areas in both countries are likely to be negatively affected by the current and future demand for bushmeat in the surrounding areas.

Commodity or necessity?

Wild meat appears in markets in almost every village as well as in large towns and cities. The contribution that such volume of bushmeat (see above) makes to overall protein supply and to food security of peoples living in the Congo Basin countries is disputed, although detailed field studies suggest that it can play a major role in livelihoods and livelihood security (see e.g. Fa et al. 2003). Estimates of wild animal protein versus non-bushmeat protein (from livestock and plant products) show that the situation is likely to be not only catastrophic for wildlife but also for the people who rely on it (Fa et al. 2003). If extraction continues at current levels, there will be a significant decline in available wild protein by 2050 (Fig. 14.3), and insufficient non-bushmeat protein produced to replace the amounts supplied by wild meats. The latter statement is derived from the grossly limited agricultural sector existing in all Congo Basin countries. Strong intercountry differences in future trajectories of protein supply are clear – the CAR, DRC and Congo show rapidly escalating extraction to production ratios in contrast to Cameroon and Gabon. This suggests that the most critical areas are in the central part of the Congo Basin. In terms of protein supply, only Gabon is able to depend on bushmeat. This is because Gabon has large tracts of forests still intact and low human population densities. All the other countries will have to find other sources of protein from the agricultural sector. The Fa et al. (2003) projections indicate that even if bushmeat protein supply were reduced to a sustainable level, non-bushmeat protein could not supply enough to cover the needs of the population in all countries except in Gabon.

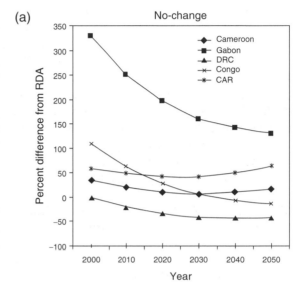

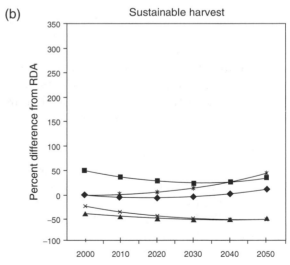

Fig. 14.3 Projected percentage deviation from g person^{-1}day^{-1} of the recommended daily allowance (RDA) of protein for (a) a no-change and (b) a sustainable bushmeat-harvest scenario. (From Fa et al. 2003.)

For how much longer?

Estimates of hunting sustainability are complicated by the efficacy of the theoretical models used to determine production of hunted species (see Milner-Gulland et al. 2003). Those studies that have measured extraction and production indicate that most bushmeat species are overexploited by hunting(Fa et al. 1995; Fitzgibbon et al. 1995; Noss 1998; Muchaal & Ngandjui 1999; Fa 2000). The efficacy of hunting and the rapid rate at which animal populations are depleted can be seen from Fa & Garcia Yuste (2001) in which 42 hunters were followed for more than a year. Within a period of less than four months, prey numbers and biomass fell by almost 80% due to the dramatic impact of such intensive 'strip-mining' of wildlife, as Robinson & Bennett (2000) have described bushmeat hunting. But, if hunting pressure is not too heavy, and large neighbouring tracts of undisturbed forest can buffer and replenish those hunted areas, wildlife populations can readily bounce back after exploitation.

Although a decline in hunted vertebrate densities is expected in overharvested areas, comparisons of quarry species in hunted and non-hunted Amazonian forest sites showed no significant differences (Peres 2000). In these forests, because hunting is highly selective towards larger-bodied species, abundance of small and mid-sized species remained largely unaffected or experienced an increase in numbers, whereas that of the largest size classes was significantly depressed at moderately and heavily hunted sites. For Amazonian primates, for instance, Peres & Dolman (1999) found reasonably good evidence of density compensation (or undercompensation) of the residual assemblage of non-hunted mid-sized species, where their large-bodied (ateline) counterparts had been severely reduced in numbers. The available data for Africa, although limited in comparison with the neotropics, points to a similar depletion pattern. From estimates of mammal abundance in non-hunted and hunted sites in Makokou, Gabon, Lahm (1993) showed that body mass and population density were negatively correlated with impact on the species. Whether or not density compensation occurs in African hunted sites, as observed in the neotropics, remains to be demonstrated. Shifts to smaller prey are the expected consequence in overhunted areas. Counts of the number of animal carcasses arriving at Malabo market, Bioko Island, Equatorial Guinea, were made during two 8-month study periods in 1991 and 1996. Between-year comparisons of these harvests showed that the number of species and carcasses in 1996 was greater than in 1991. In biomass terms, the increase was significantly less, only 12.5%, when compared with almost 60% more carcasses entering the market in 1996. A larger number of carcasses of the smaller-bodied species, such as rodents and the blue duiker, were recorded in the latter study period. Concurrently, there was a dramatic reduction in the larger-bodied species, the Ogilby's duiker and the seven diurnal primates (Fa & Garcia Yuste 2001).

Economic theory and bushmeat

Evidence has been presented above which suggests that:

1 consumption of bushmeat is an important aspect of household economies;
2 demand for this protein source results in unsustainable hunting levels.

Pressure on wildlife for meat can be reduced either by increasing supply or diminishing demand of wildlife. In principle, reducing demand can be achieved either by restricting supply or by providing consumers with other options or (to the extent that this is feasible) educating them about the options that already exist. Increasing supply of wildlife is possible but unlikely because of the low productivity of tropical moist forests. Also, any change or manipulation of forests to increase wildlife production (e.g. reduction of natural predators, or modification of habitats to favour certain bushmeat species) would be undesirable because it would adversely affect forest structure and composition. Increasing bushmeat supply by raising wildlife species in captivity makes little sense for low-productivity species such as most antelopes and primates (see below).

The interaction of prices and wealth on the consumption of domestic versus wild meats has been documented in a number of studies (de Merode et al. 2004; Wilkie et al. 2005; East et al. 2005). The general finding has been that increases in household wealth appear to drive a preference shift from bushmeat to the meat of domesticated animals or to narrow the range of bushmeat species. The suggestion that residents of the Congo Basin prefer the taste of bushmeat over the other meats is undocumented, with the only established fact being that consumers note 'meat hunger' when their diet is composed primarily of starches (de Garine & Pagezy 1990; de Garine 1993).

Economic theory proposes that by increasing the price of a good in a market, the suppliers are likely to supply more of this particular good, however, at the same time the amount of the good demanded by the consumers will drop. At a particular price, there is equilibrium between what is supplied and demanded. The hunters' harvest is what is supplied to the markets and is dependent upon several factors such as the profitability of hunting (subtracting all costs), the opportunity cost of the hunter (the availability of alternative employment), and the hunter's effort level (often measured in the amount of time and/or technology used during the hunt). For demand, the shape of its curve is dependent on its own price elasticity – that is, to what extent a change in the price will alter the current demand for that particular good. Understanding the price elasticity of demand indicates how policy changes to the supply and price of bushmeat can affect the amount and type of meat demanded. Important in this

evaluation are consumer preferences, social status of the good, and the availability of substitute products. With substitutes available, one is able to estimate the cross-price elasticity indicating how easily a consumer can change between goods. That is, how much of good X will be demanded when there is an increase/reduction in the price of good Y. On the other hand, consumption of a product is also directed by the income available. The income elasticity of a consumer will raise or lower their demand curve, indicating that at a higher income level, the consumer will demand more of the good for the same price (Milner-Gulland 2001; Wilkie & Godoy 2001).

Wilkie & Godoy (2001) examined the bushmeat demand of South and Central American Indians in Bolivia and Honduras. They found that a decrease in the price of alternative meat is likely to significantly decrease the consumption of fish, but not bushmeat (i.e. alternative meat and fish are substitutes, but not bushmeat). They also showed that the demand for bushmeat might follow an inverted U shape with income – as the consumer income increases, so will their bushmeat consumption, but only up to a certain point. At that point, the consumption will decrease as the income increases. This phenomenon is often referred to as a 'Kuznets' Curve' (Kuznets 1955). The use of these curves is highly debated within the whole field of environmental sciences (see Panayotou 1993). For the Central African/Congo Basin scenario, we are unsure whether such a curve exists. As the population is becoming richer there are yet to be found significant signs of slowing down or altering the bushmeat consumption habits. Rather, there is evidence that bushmeat functions in such areas as a 'luxury good', as illustrated in Box 14.2

Divining solutions

Because there are natural limits to the level of harvesting that wildlife populations can sustain,

---

Box 14.2 When bushmeat is less important as a food source

In a study undertaken in the Zande area of eastern Democratic Republic of Congo, de Merode et al. (2004) indicate that although significant variations in patterns of consumption and sale of bushmeat were found, correlated with relative wealth, all families in the study could be classed as 'living in extreme poverty' by the standard international test (income of less than US$1 per day). Thus the variations were only relative.

De Merode's study sought to address three questions:

1 Whether wild foods (including bushmeat) were valuable to households, in terms of both consumption and sales'
2 Whether the value varied according to the season;
3 Whether the value was greater to the poorer or less poor.

In summary, it was found that although wild foods in general formed a significant proportion of household production, most was sold on the market and not consumed. This was particularly the case with bushmeat and fish, where more than 90% of production was sold. Consumption levels varied by household, with both seasonality and wealth effects.

Consumption of wild foods increases significantly during the hungry season (particularly bushmeat, where consumption rose on average by 75%). Bushmeat and fish consumption were fairly even across all wealth ranks, except the poor (who consumed very little of their own production, although they made up for this through bushmeat gifts); bushmeat sales were exceptionally influenced by the wealth rank of the household, with the richer households more likely to be involved in market sales. This was unrelated to questions of land access and tenure (all families had equal theoretical access to the production zones, and – unlike with fishing – there were no restrictions on activity related to non-membership of a craft guild). However, it was strongly correlated with access to capital (shotguns, nets) and to the wealth required to generate a surplus over consumption needs. Interestingly, both fish and bushmeat exhibited the characteristics of 'superior goods' (i.e. luxury items which consumption increased exponentially with increasing wealth). By contrast, wild plants were 'inferior goods', in that increasing wealth implied decreasing household consumption.

A particularly interesting finding of the study was that, for families living in extreme poverty, market sales of bushmeat were more important than household consumption. This would appear to confound the frequent proposition that the welfare of the poor can be secured by conservation strategies that permit home consumption but prohibit market sales.

---

it is clearly possible that the bushmeat trade could result in the extinction of animal populations. Thus, if hunting wildlife for food is unsustainable and is jeopardizing the long-term survival of some species, understanding how economics and preferences influence consumer choice, and therefore demand for bushmeat, is fundamental. Multiple root causes drive the wildlife and bushmeat trades. The principal driver is a complex of consequences of what might loosely be termed 'development' and the need to meet modern consumption demands from a natural-based economy or from primary agricultural production, in areas where there are few alternative economic opportunities. At present, this is taking place against a background of subsistence needs, high human population growth and significant economic decline.

Given this, it may not be possible to stop biodiversity loss altogether, but management innovations and policy reform and implementation of legislative changes can hopefully slow the process.

Legal control and incentives to limit hunting might be achieved through improved training of control agents (e.g. eco-guards, customs agents, police, etc.), and this is often advocated. However, such an approach fails to recognize the low levels of ownership of the legislation in the producer states. This has often been the product of either the European colonial inheritance or post-colonial politics, and imposed with little regard to the welfare or interests of the populations involved. Indigenous tenure systems for the control of land and the resources on it (trees, wildlife, non-timber plant products) were often suppressed by the colonial regime, and ownership transferred to the state. As post-colonial governments have rarely sought to rectify this situation there are few incentives for resource users to invest in management. Further distortions in governance come from the extractive industries, which, through the distortions of governance structures and the concentration of wealth, marginalize the interests of local residents. There are also management challenges which derive from the very nature of wildlife. Wild animals are unusual in their 'fugitive' qualities, meaning (in the present context) lack of ownership until the point of being killed. This diminishes incentives to manage the resource; most notably, there is less incentive to invest in management where there is no certainty that the benefits will be felt by the individual concerned. This problem of 'non-ownership' tends to be phrased in the literature as favouring the 'free-rider syndrome' – i.e. behaviour of individuals or groups that benefit from the investments of others in environmental management, without themselves having to suffer the costs (Ostrom 1990).

In such circumstances, the management of wildlife in source areas has tended to involve conceding limited exploitation rights to categories of local users in return for respect of the boundaries of production forests and protected areas and, in some cases, absolute interdiction of 'commercial' sales outside of the locality. Variants of this approach include the World Wild Fund for Nature (WWF) programme in southern Cameroon and the involvement of the Wildlife Conservation Society (WCS) with the Congolaise Industrielle des Bois (CIB) timber concession in the northern parts of the Republic of Congo–Brazzaville. Such measures have considerable appeal to international timber companies with important 'green' markets, for whom a positive conservation image gives sound commercial advantage. But they may also involve denying access rights to local people which have been enjoyed for many generations. As these people rarely derive much benefit from the existence of the timber companies, it is unsurprising that such restrictive access policies often enjoy little if any local support.

A different administrative option is policy and legislative reform. Often seen as two facets of the same problem, legislative change is not the sole dimension of policy reform, but it is an important one, to the extent that policy ultimately must be expressed in appropriate legislation. Often proposed as a wildlife law is the combination of reasonable subsistence use of wildlife but banning all commercial trade to urban centres (Ly & Bello 2003; Bushmeat Crisis Task Force (BCTF) website). This has conservation logic, as we have shown above; it is the high commercial volumes traded to feed the fast growing urban centres that represent the main threat to sustainability.

Although attractive in theory, these proposals rarely stand up to rigorous analysis. At the moment, national legislation in most of the range states already permits hunting for subsistence use of non-listed species. But livelihood-oriented research tends to show that the poorer villagers are more likely to sell any game captured for the obvious reason that wild meat has a much higher price to weight ratio than most other forest products, and thus is a much more tradable commodity (de Merode et al. 2003). Although the total volume of sales by the poor may not be particularly high, the incidence of these sales may be crucial as they provide important social safety nets and help them cope with crises and large expenditures such as children's' school fees (Arnold & Ruiz Perez 2001). This indicates that attempting to ban commercial sales could have severe negative implications for local livelihoods (de Merode et al. 2003).

By contrast, attempts to link wildlife management to wider legislative reform – giving resident populations real influence over their natural resources – are more promising. Efforts in countries such as Cameroon to involve communities not only in wildlife management but also in all aspects of forest management, including exploitation of high valued timber, commend themselves in this regard. Their chances of success are enhanced by the economic and political benefits of joint enterprise. Although not without challenges, there are opportunities to build alliances between development assistance (poverty focus) and conservation (wildlife focus) around the joint enterprise management of the full range of forest resources.

A number of proposals have been made for the provision of alternative opportunities for the actors/agents involved in the bushmeat trade, as well as the production of alternative protein sources that include captive breeding schemes (Smythe 1991; Diamond 1999; Auzel & Wilkie 2000), increasing fish harvests (Redford & Robinson 1987), or more efficient rearing of domesticated species (Fa 2000). These opportunities often figure in aid-funded 'integrated conservation and development projects' (ICDPs), implemented as a way of 'selling' conservation goals to local forest dwellers (Brown 1998). With either of these initiatives, there are pros and cons attached.

Captive breeding schemes for the production of wild meat in large quantities have been largely unsuccessful. Even where technically feasible, the economics have often been wrong for peasant livelihoods. Whereas the typical peasant family has a preference for a range of activities that cut risk and reduce capital and labour requirements, the captive breeding schemes have often high-risks attached and are both capital and labour intensive. Fishing on the other hand is thought to be undeveloped and is often not a preferred economic activity. Fishing does tend to become more attractive with increasing human population densities (Boserup 1965), although this prospect is probably more likely in the urban and semi-urban areas rather than in the rural bushmeat source areas. The notion that local actors do not exploit the fisheries resources in such areas because of lack of knowledge or dietary preferences often does not stand up to economic analysis. Where fishing is economically feasible in forest areas, then this usually will be apparent already,

in terms of the large numbers of fishermen already exploiting the resource. The converse is equally true.

Although rearing domesticated livestock is a potential economic opportunity, there is rarely any direct link between this substitution activity and the hunting activity to be foregone, as the two target populations are likely to differ in their social characteristics. In addition, the potential for increasing domestic livestock production may be much less than the casual observer assumes. Free-range animals can usually survive quite well in what are essentially domestic foraging conditions around forest villages, but constraints such as sufficient food and animal enclosures inhibit the scaling-up of production. Confining animals in tropical conditions is also associated with disease, and is only feasible where veterinary services and medicines are easily accessible, and affordable – in other words, it demands a highly monetized economy. Additionally, an unfavourable political and economic environment constrains bushmeat policy development. From a tourism perspective, it is difficult to develop and maintain the standards that tourists require when the environments are not safe and the infrastructure is not developed. Without easy and sustained access, wildlife enterprises can be very vulnerable to the volatilities of the international tourism market. The lack of a tourist market coupled with the low productivity of the tropical forests makes the management options, such as Campfire in Zimbabwe, unlikely to generate high enough revenues to satisfy both entrepreneurs and the host communities.

An equally cautious assessment must be taken of other conservation options, such as classic protected area policies, of the exclusion type. Bushmeat source areas often have very low resident human populations and very poor communications. Thus, securing protected areas can be done only with the support of the populations involved. However, 'fines and fences' approaches tend to achieve precisely the reverse result. They risk alienating the local populations whose support is critical for success, while offering them no economic benefits or realistic alternatives (Cernea & Schmidt-Soltau 2003). Without local support, these areas are likely to be heavily encroached. Indeed, to the extent that any successes are achieved in concentrating animal populations, then this tends to be self-defeating, because of the way that it reduces the opportunity costs of the hunters involved. Thus, such protected areas tend to be feasible only as long as outside conservation agencies are funding their maintenance, including paramilitary protection costs, at a high level and at considerable expense.

An interesting alternative approach to protected area development is based on the spatial harvest theory (McCullough 1996). This management theory advocates a division of areas into hunted and non-hunted (protected) zones ('sinks' and 'sources'), with animals moving without restriction between the two (see also Novaro et al. 2000). A generous estimate of the source area relative to the sink areas allows wide margins for potential overharvest, and acts as a counterbalance to the lack of animal population estimates (Bodmer & Puertas 2000; Fimbel et al. 2000). Similar approaches are already in operation in marine fisheries where protected zones are defined in relation to estimated future harvest needs and not independently of them (Milner-Gulland 2001). Policy development using spatial harvest principles would increase the attractiveness of conservation to local populations, and make the notion of protection much more saleable to them. Economic rationality commends them in bushmeat source areas, provided the long-term trust commitment of the resident communities can be secured.

The route from development policy?

Until now, bushmeat policy development has been heavily conditioned by its very narrow association with conservation interests. There is a strong case for widening the institutional engagement in bushmeat policy by bringing in development aspects (Brown & Williams 2003). Donor interest in bushmeat as a policy theme is likely to be increased if the link between the trade and poverty eradication can be made apparent (Arnold & Ruiz Perez 2001). Paradoxically, it may be partly because of the strengths of bushmeat as a livelihoods asset (low thresholds of entry, leading to broad participation, but also tight margins) that it is unlikely to figure strongly in rural transformation. There seem to be few opportunities to add value in processing, through technical sophistication or increased investments of labour (in this respect, bushmeat may differ from, say, artesian woodworking), particularly when the trade is treated as de facto illegal, pushing it to be maintained largely underground. From the perspective of volume, bushmeat is a discouraging prospect; even if the projections of sustainable offtake are overcautious, they are often so far below the existing offtake levels, which makes it unlikely that sufficient capital could be generated from the sector to sustain long-term economic change.

In terms of improving livelihood security, most of the range states do not have any form of publicly funded social protection (i.e. dimensions of social security), thus the poor are heavily dependent on natural social safety nets. In policy development terms, this means that policy must be much more focused on securing the rights of the poor and marginal, not just to use wildlife in subsistence strategies but also to generate income and long-term security (linking conservation policy directly to livelihoods concerns).

Conclusions

Through the past decade, the bushmeat crisis has been brought to our attention through different groups of experts working within countries where bushmeat consumption flourishes. Whereas the anthropologists focused on the dependence of the local people on this resource, it was the conservation biologists who first recognized that the volume traded could not be sustainable. The different angles of these fields of research made the initial focus on wild animal use confrontational (be it human development and poverty eradication or animal conservation and ecological research).

A conclusion that can be drawn from the above discussion, however, is that a people **or** animal approach is not applicable for viable solutions to the bushmeat crisis – we have to start working from a people **and** animal agenda. As we have shown above, the bushmeat crisis contains a diverse set of facets needed to be taken into consideration in wildlife conservation policies. Until recently the human development aspects were often sidelined because the main focus was on the conservation of animals. Although the experts now agree that the local people must be part of the solution – the question remains as to the optimal way of including them effectively. Opening communication channels between the fields of conservation, economics, anthropology and development, might lead us closer to the answer. Experts within each of these groups need to come to the table with their concerns and through discussions we might be able to find win-win situations for the conservation of the animals and development and prosperity for the human populations. The dynamics of the bushmeat hunting and trade system is full of positive and negative feedback loops affecting variables ranging from ecosystem composition and health to education and health care of people. The complexity of the problem is what makes the viable solutions far-ranging.

It is important to emphasize that even with free dialogue and exchange of ideas between the natural and social scientists, each bushmeat market and crisis is different. The common theme is the unsustainable use of the resources, but the factors contributing to this unsustainability are diverse and change both within and between countries.

In this policy setting, it is most important to remember that the emergence of the bushmeat crisis is not an original condition, but part of a process of historical change that has led to failures of governance. Improved bushmeat management will need to address these issues and search for solutions that both preserve biodiversity and are socially just. Improved wildlife policy is a matter of concern for both conservation and development.

The world is big. Some people are unable to comprehend that simple fact. They want the world on their own terms, its peoples just like them and their friends, its places like the manicured little patch on which they live. But this is a foolish and blind wish. Diversity is not an abnormality but the very reality of our planet. The human world manifests the same reality and will not seek our permission to celebrate itself in the magnificence of its endless varieties. Civility is a sensible attribute in this kind of world we have; narrowness of heart and mind is not.
(Chinua Achebe, *Bates College Commencement Address*, 27 May 1996.)

References

Alvard, M.S., Robinson, J.G., Redford, K.H. & Kaplan H. (1997) The sustainability of subsistence hunting in the Neotropics. *Conservation Biology* **11**: 977–82.

Anstey, S. (1991) *Wildlife Utilisation in Liberia*. WWF/FDA Wildlife Survey Report, World Wide Fund for Nature, Gland, Switzerland.

Arnold, J.E.M. & Ruiz Peres, M. (2002) Can non-timber forest products match tropical forest conservation and development objectives? *Ecological Economics* **39**:437–47.

Auzel, P. & Wilkie, D.S. (2000) Wildlife use in Northern Congo: hunting in a commercial logging concession. In *Hunting for Sustainability in Tropical Forests* (Eds J.G. Robinson & E.L. Bennett), pp. 413–26. Columbia University Press, New York.

Bahuchet, S. & de Garine, I. (1990) The art of trapping in the rain forest. In *Food and Nutrition in the African Rain Forest* (Eds C.M. Hladik, S. Bahuchet & I. de Garine), pp. 24–35. UNESCO/MAB, Paris.

Bodmer, R. & Puertas, P.E. (2000) Community-based Comanagement of Wildlife in the Peruvian Amazon. In *Hunting for Sustainability in Tropical Forests* (Eds J.G. Robinson & E.L. Bennett), pp. 395–412. Columbia University Press, New York.

Boserup, E. (1965) *Conditions of Agricultural Growth.* George Allen & Unwin, London.

Bowen-Jones, E. & Pendry, S. (1999) The threat to primates and other mammals from the bushmeat trade in Africa, and how this threat could be diminished. *Oryx* **33**(3): 233–46.

Brown, D. (1998) *Participatory Biodiversity Conservation: Rethinking the Strategy in the Low Tourist Potential Areas of Tropical Africa.* Natural Resource Perspectives No. 33, Overseas Development Institute, London.

Brown, D. & Williams, A. (2003) The case for bushmeat as a component of development policy: issues and challenges. *The International Forestry Review* **5**(2): 148–55.

Cernea, M.M. & Schmidt-Soltau, K. (2003) National parks and poverty risks: is population resettlement the solution? *Paper presented at the World Park Congress*, Durban, September. An abbreviated version was published as: The end of forced resettlements for conservation: conservation must not impoverish people. *Policy Matters* **12**: 42–51.

Chapman, C.A., & Onderdonk, D.A. 1998. Forests without primates: primate/plant codependency. *American Journal of Primatology* **45**: 127–41.

Chardonnet, P., Fritz, H., Zorzi, N. & Feron, E. (1995) Current importance of traditional hunting and major constraints in wild meat consumption in sub-Saharan Africa. Integrating people and wildlife for a sustainable future – proceedings of the 1st International Wildlife Management Congress. J.A. Bissonette and P.R. Krausman. Bethesda, MD, USA, The Wildlife Society: 304–307.

Colell, M., Maté, C. & Fa, J.E. (1994) Hunting among Moka Bubis in Bioko: dynamics of faunal exploitation at the village level. *Biodiversity and Conservation* **3**: 939–50.

Cowlishaw, G. (1999) Predicting the pattern of decline of African primate diversity: an extinction debt from historical deforestation. *Conservation Biology* **13**: 1183–93.

Cowlishaw, G., Mendelson, S., & Rowcliffe, J.M. (2005) Evidence for post-depletion sustainability in a mature bushmeat market. *Journal of Applied Ecology* **42**: 460–68.

De Garine, I. (1993) Food resources and preferences in the Cameroonian forest. . In *Food and Nutrition in the African Rain Forest* (Eds C.M. Hladik, S. Bahuchet & I. de Garine), pp. 561–74. United Nations Educational, Scientific, and Cultural Organization-Man and the Biosphere Programs, Paris, France.

De Garine, I. & Pagezy, H. (1990) Seasonal hunger or 'craving for meat'. In *Food and Nutrition in the African Rain Forest* (Eds C.M. Hladik, S. Bahuchet & I. de Garine), pp. 43–4. United Nations Educational, Scientific, and Cultural Organization- Man and the Biosphere Programs, Paris, France.

De Merode, E., Homewood, K. & Cowlishaw, G. (2004) The value of bushmeat and other wild foods to rural household living in extreme poverty in Democratic Republic of Congo. *Biological Conservation* **118**: 573–81.

Diamond, J (1999) *Guns, Germs and Steel: The Fates of Human Societies.* Norton, WW & Company, New York.

Dirzo, R. & Miranda, A. (1990) Contemporary neotropical defaunation and forest structure, function and diversity a sequel to John Terborgh. *Conservation Biology* **4**: 444–7.

East, T., Kümpel, N.F., Milner-Gulland, E.J. & Rowcliffe, J.M. (2005) Determinants of urban bushmeat consumption in Río Muni, Equatorial Guinea. *Biological Conservation* **126**: 206–15.

Egbe, S. (2000) *Communities and Wildlife Management in Cameroon.* Consultancy Report presented to the DFID–Cameroon Community Forestry Development Project, Yaoundé, 20 pp.

Fa, J.E. (2000) Hunted animals in Bioko Island, West Africa: sustainability and future. In *Hunting for Sustainability in Tropical Forests* (Eds J.G. Robinson & E.L. Bennett), pp. 168–98. Columbia University Press, New York.

Fa, J.E. & Garcia Yuste, J.E.G. (2001) Commercial bushmeat hunting in the Monte Mitra forests, Equatorial Guinea: extent and impact. *Animal Biodiversity and Conservation* **24**: 1–22.

Fa, J.E. & Peres, C.A. (2001). Game vertebrate extraction in African and Neotropical forests: an

Intercontinental Comparison. In: *Conservation of Exploited Species* (Eds J.D. Reynolds, G.M. Mace, K.H. Redford & J.G. Robinson), pp. 203–41. Cambridge University Press, Cambridge.

Fa, J.E., Juste, J., Perez del Val, J. & Castroviejo, J. (1995) Impact of market hunting on mammal species in Equatorial Guinea. *Conservation Biology* **9**(5): 1107–15.

Fa, J.E., Peres, C.A. & Meeuwig, J. (2002) Bushmeat exploitation in tropical forests: an intercontinental comparison. *Conservation Biology* **16**(1): 232–7.

Fa, J.E., Currie, D. & Meeuwig, J. (2003) Bushmeat and food security in the Congo BasIn linkages between wildlife and people's future. *Environmental Conservation* **30**: 71–8.

Fa, J.E., Ryan, S. & Bell, D.J. (2005) Hunting vulnerability, ecological characteristics and harvest rates of bushmeat species in Afrotropical forests. *Biological Conservation* **121**: 167–76.

Fa, J.E., Seymour, S., Dupain, J., Amin R., Albrechtsen, L. & Macdonald, D. (2006). Getting to grips with the magnitude of exploitation: Bushmeat in the Cross-Sanaga rivers region, Nigeria and Cameroon. *Biological Conservation* **129**: 497–510.

Fimbel, C., Curran, B. & Usongo L. (2000) Enhancing the sustainability of duiker hunting through community participation and controlled access in the Lobéké region of southeastern Cameroon. In *Hunting for Sustainability in Tropical Forests* (Eds J.G. Robinson & E.L. Bennett), pp. 356–74. Columbia University Press, New York.

FitzGibbon, C.D., Mogaka, H. & Fanshawe, J.H. (1995) Subsistence hunting in Arabuko-Sokoke forest, Kenya, and its effects on mammal populations. *Conservation Biology* **9**: 1116–26.

Harden, G (1968) The tragedy of the commons. *Science* **162**: 1243–8.

Juste, J., Fa, J.E., Perez del Val, J. & Castroviejo, J. (1995) Market dynamics of bushmeat species in Equatorial Guinea. *Journal of Applied. Ecology* **32**: 454–67.

Kuznets, S. (1955) Economic growth and income inequality. *The American Economic Review* **45**: 1–28.

Lahm, S.A. (1993) Utilization of forest resources and local variation of wildlife populations in northeastern Gabon. In *Tropical Forests, People and Food: Biocultural Interactions and Applications to Development* (Eds C.M. Hladik, A. Hladik, O.F. Linares, et al.), pp. 213–226. UNESCO, Paris.

Ly, I. & Bello, Y (2003) *Etude sur les lois et politiques sur la faune dans les pays d'afrique centrale*. The CITES Central Africa Bushmeat Working Group, International Union for the Conservation of Nature and Natural Resources, Gland, Switzerland.

McCullough, D. (1996) Spatially structured populations and harvest theory. *Journal of Wildlife Management* **60**: 1–9.

Milner-Gulland, E.J. (2001) Assessing sustainability of hunting: insights from bioeconomic modeling. In *Hunting and Bushmeat Utilization in the African Rain Forest: Perspectives towards a Blueprint for Conservation Action* (Eds M.I. Bakarr, G.A.B. de Fonseca, R. Mittermeier, A.B. Rylands & K.W. Painemilla), pp. 113–51. Conservation International, Washington, DC.

Milner-Gulland, E.J., Bennett, E.L. and the SCB 2002 Annual Meeting Wild Meat Group (2003) Wild meat: the bigger picture. *Trends in Ecology and Evolution* **18**(7): 351–7.

Muchaal, P.K. & Ngandjui, G. (1999) Impact of village hunting on wildlife populations in the Western Dja reserve, Cameroon. *Conservation Biology* **13**: 385–96.

Noss, A.J. (1995) *Duikers, cables and nets: a cultural ecology of hunting in a Central African Forest*. Unpublished PhD thesis, University of Florida.

Noss, A.J. (1998) The impacts of cable snare hunting on wildlife populations in the forests of the Central African Republic. *Conservation Biology* **12**: 390–8.

Novaro, A.J., Redford, K.H. & Bodmer, R.E. (2000) Effect of hunting in source-sink systemd in the neotropics. *Conservation Biology* **14**: 713–21.

Ostrom, E. (1990) *Governing the Commons: the Evolution of Institutions for Collective Action*. Cambridge University Press, Cambridge.

Panayotou, T. (1993) *Empirical Tests and Policy Analysis of Environmental Degradation at Different Stages of Economic Development*. International Labor Office, Geneva.

Peres, C.A. (2000) Effects of subsistence hunting on vertebrate community structure in Amazonian forests. *Conservation Biology* **14**: 240–53.

Peres, C.A. (2001) Paving the way to the future of Amazonia. *Trends in Ecology and Evolution* **16**: 217–9.

Peres, C.A. & Dolman, P.M. (2000) Density compensation in neotropical primate communities: evidence from 56 hunted and unhunted Amazonian forests of varying productivity. *Oecologia* **122**: 175–89.

'Peres, C.A. & Terborgh, J.W. (1995) Amazonian nature-reserves – an analysis of the defensibility status of existing conservation units and design criteria for the future. *Conservation Biology* **9**: 34–46.

Redford, K.H. & J.G. Robinson (1987) The game of choice – patterns of Indian and colonist hunting in the Neotropics. *American Anthropologist* **89**: 650–67.

Robinson, J.G. & Bennett, E.L. (Eds) (2000) *Hunting for Sustainability in Tropical Forests*. Columbia University Press, New York.

Robinson, J.G. & Bodmer, R.E. (1999) Towards wildlife management in tropical forests. *Journal of Wildlife Management* **63**: 1–13.

Scott, A.M (1858) *Day Dawn in Africa: Progress of the Protestant Episcopal Mission at Cape Palmas, West Africa*. Protestant Episcopal Society, New York.

Smythe, N. (1991) Steps toward domesticating the Paca (Agouti = Cuniculus paca) and prospects for the future. In *Neotropical Wildlife Use and Conservation* (Eds J.G. Robinson & K.H. Redford), pp. 202–216. University of Chicago Press, Chicage, IL.

Wilkie, D.S. & Carpenter, J.F. (1999) Bushmeat hunting in the Congo BasIn an assessment of impacts and options for mitigation. *Biodiversity and Conservation* **8**: 927–55.

Wilkie, D.S. & Curran, B. (1991) Why do Mbuti hunters use nets – ungulate hunting efficiency of archers and net-hunters in the Ituri rain-forest. *American Anthropologist* **93**: 680–9.

Wilkie, D.S. & Godoy, R.A. (2001) Income and price elasticities of bushmeat demand in lowland Amerindian societies. *Conservation Biology* **15**: 761–9.

Wilkie, D.S., Starkey, M., Abernethy, K., Effa, E.N., Telfer, P. & Godoy, R.A. (2005) Role of price and wealth in consumer demand for bushmeat in Gabon, Central Africa. *Conservation Biology* **19**: 268–74.

Wright, S.J., Zeballos, H., Dominguez, I., Gallardo, M.M., Moreno, M.C. & Ibáñez, R. (2000). Poachers alter mammal abundance, seed dispersal and seed predation in a Neotropical forest. *Conservation Biology* **14**: 227–39.

15

Does sport hunting benefit conservation?

Andrew J. Loveridge, Jonathan C. Reynolds
and E. J. Milner-Gulland

There is a passion for hunting deeply implanted within the human breast ......
(Charles Dickens, *Oliver Twist*, 1837–8)

Introduction

When a wildlife population is threatened, deliberately killing individuals from it may seem perverse. Yet some argue that, paradoxically, well-regulated sport hunting benefits wildlife populations, and may sometimes be the only way to ensure their persistence. In this essay we consider whether this assertion is supported by experience.

When poorly regulated, hunting can be – and historically often has been – damaging to the target population, with dramatic examples of extinction and population decline (Roth & Merz 1996). Even in the 1980s, hunting contributed to drastic reductions in populations of the dorcas gazelle (*Gazella dorcas*) and to extermination of the Nubian bustard (*Neotis nuba*) from Sahelian Africa (Newby 1990). The tally of millions of migratory birds shot and trapped annually by Mediterranean hunters has alarmed observers and caused them to question its sustainability (Lindell & Wirdheim 2001). However, the impact of hunting on population dynamics can be complex and difficult to quantify. For example, although there has been much concern about the

impact of hunting on migratory turtle doves (*Streptopelia turtur*), Browne & Aebischer (2004) found that the observed decline in UK breeding turtle doves could be entirely explained by changed UK farming practices with no direct evidence for a damaging impact of hunting. There are many examples all over the world where hunting has been regulated successfully (Tapper & Reynolds 1996). The wild turkey (*Meleagris gallopavo*; Dickson 1992), white-tailed deer (*Odocoileus virginianus*; Woolf & Roseberry 1998) and beaver (*Castor canadensis*; Novak 1987) in North America are all species whose fortunes have been dramatically improved by a programme of conservation measures that includes substantial regulated harvests.

Sport hunting and nature conservation have both been part of human culture from the earliest times. For example, Ancient Egypt had a strong tradition of sport hunting (Osborn & Osbornova 1998). Bogdkhan Mountain Reserve, Mongolia, was formally protected in 1778, but informal prohibitions on hunting and logging on this sacred site date from the thirteenth century (United Nations Mongolia Office 2004). Other similar examples include the Royal Chitwan National Park in Nepal and the New Forest in the UK. However, the roots of contemporary conservation are usually traced back to nineteenth century colonial sport hunting (Adams 2004). Pressure from hunters, alarmed at unregulated destruction of game habitats and populations, led to the establishment of parks and reserves in the British colonies and the USA (Fitter & Scott 1978; Adams 2004). Many of these were initially set aside as hunting grounds, and many national parks services and government conservation departments have origins in agencies established to defend hunting reserves and suppress poaching (Adams & Hulme 2001). This purely preservationist view of conservation has given way to a perspective that is more inclusive of humans (Duffy 2000; Hutton & Leader-Williams 2003; Jones & Murphree 2004). Modern conservation is about reducing extinction risks, maintaining essential ecological processes, preserving genetic diversity and ensuring that the use of species and ecosystems is sustainable (Convention on Biological Diversity 2003). Although this definition does not positively promote use as a conservation tool, the explicit inclusion of sustainable use is a recognition that human use of wildlife happens, and that the appropriate role of conservation is to ensure that it is sustainable, rather than to prohibit it.

Sport hunting is a multifaceted activity, occurring in many ecological and socio-political landscapes, variously motivated and generating a range of revenues. A single characterization of its impact on conservation is necessarily simplistic. We begin by clarifying what we mean by sport hunting, and discussing the ethical issues that bedevil the debate on its role in conservation. We then address the key issues summarized in Fig. 15.1.

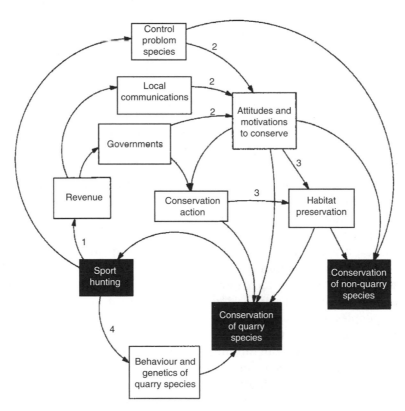

Fig. 15.1 Illustrating the relationship between sport-hunting and conservation discussed in this essay. The numbering corresponds to the key questions addressed in the main text.

1 We ask whether or not hunting raises revenue, and if so whether this revenue can be used to benefit conservation?

2 One important issue is whether or not revenue earned from hunting offsets the opportunity costs of not engaging in other activities, especially in the case of poor communities living in areas of rich biodiversity.

3 We explore whether hunting affects the conservation of habitat and biodiversity.

4 We explore whether there are any subtle side-effects affecting hunted populations.

What is sport hunting?

Hunting is often categorized into subsistence hunting, market (or commercial) hunting and sport (or leisure or recreational) hunting, the differences being primarily motivational. Whereas subsistence hunting provides food for hunters and their dependents, market hunting supplies food to a consumer community for cash. Sport hunting is undertaken primarily for leisure, motivated by 'the thrill of the chase'. However, complexities of motivation blur distinctions between these various kinds of hunting. For instance, there is a commercial element to sport hunting, because hunters are prepared to pay for it, and many components of the activity are saleable commodities. Neither is the distinction between subsistence and sport hunting clear-cut. Although some sport hunters do not link their hunting with personal consumption, others choose not to kill more than they can eat. It is also not just sport hunters who enjoy the experience of hunting.

Beyond subsistence, profit or recreation, hunting has profound cultural and spiritual importance for some peoples. As Canadian citizens with access to the welfare state, the Inuit no longer need to hunt for subsistence. But hunting remains intertwined with self-esteem, history and cultural identity (Canadian Arctic Profiles 2004). For this reason, hunting of threatened Arctic species (e.g. polar bears *Ursus maritimus*, walrus *Odobenus rosmarus*) is permitted, albeit under quotas. A seeming inconsistency is that traditional hunting methods have been largely replaced by exogenous technology such as snow-mobiles, motor boats and high-powered rifles (Stewart & Fay 2001; Stirling 2001). Even industrial societies have deeply rooted hunting traditions. European colonists in North America relied on hunting wildlife for subsistence up until the mid-nineteenth century, when the need to supplement protein requirements was reduced in an increasingly urban and wealthy society, and hunting became more recreational (Organ & Fritzell 2000). As a result, the right of citizens to hunt is still enshrined in law in many states within the USA (Muth & Jamison 2000; Grandy et al. 2003).

For some sport hunters, even killing the quarry may be unnecessary. For example, the hunting of foxes (*Vulpes vulpes*) with dogs in the UK, outlawed in 2004, evolved out of the wish to control foxes as a livestock/poultry pest, but became a ritualized socio-cultural activity in which the original aim of killing foxes was overshadowed or replaced by the thrill of the chase. Although hunting with dogs could – and in some situations did – control fox numbers (Heydon & Reynolds 2000), most hunts promoted a deliberately restrained strategy, resulting in a density of foxes that was acceptable to farmers yet ensured abundant hunting opportunities (Macdonald *et al.* 2000). At the extreme, 'green hunters' pay merely to immobilize wild animals, under the supervision of a veterinarian and usually as part of a management or research activity. Clients in South Africa pay up to US$25,000 to immobilize a trophy bull elephant, which is then fitted with a radio-collar as part of ongoing research. The revenue contributes to the research and management of the game reserve, and the hunter takes home moulded fibre-glass copies of the animal's tusks (Save the Elephants 2002).

Ethical considerations, inescapable elements of every conservation issue, are especially prominent in evaluations of sport hunting, polarizing debates on this topic. The acceptability of hunting as a component of conservation action is influenced by two central issues: the motivation of the hunter, and the extent to which the hunted animal suffers. Sport hunting is by definition pleasurable to the hunter. To others, killing living creatures is fundamentally wrong and therefore not to be enjoyed. Conservation is laden with non-utilitarian values (e.g. aesthetic, spiritual and bequest values), which are particularly strongly felt on both sides of the hunting debate. Our relationship with nature, simultaneously worshipped and exploited, has always been complex (Serpell 1986). Cultural and ethical issues do play a part in conservation success; for example, Oates (1999) argues that India has had more success than West Africa because its Government's conservation policy is based on recognizing the intrinsic value of nature rather than its utilitarian value. However, the ethical standpoint from which sport hunting is viewed is not easily subjected to rational analysis. We shall therefore lay aside motivational issues and take a utilitarian viewpoint in this essay, assuming that sport hunting is as valid as any other human relationship with nature.

The second issue is the distress and suffering caused to individual animals by hunting. Hunted animals may show measurable indications of stress (Macdonald et al. 2000), starting at first awareness of the natural (Chabot et al. 1996) or human (Jeppesen 1987) predator. At some point during a successful hunt, the hunted animal fails to cope with events, and stress becomes distress. In red deer (*Cervus elaphus*) hunted by hounds, Bateson & Harris (2000) found severe glycogen depletion, elevated cortisol levels and muscle damage. Although these symptoms are ambiguously associated with exercise whether voluntary or forced, it was clear that deer were attempting to cope with pursuit in unusual and costly ways. Equivalent indicators were not found in deer cleanly shot by deer stalkers (Bateson & Harris 2000), demonstrating that for such animals any period of distress was brief. So the probability, duration and magnitude of distress caused by

hunting depend on the method used. The welfare of animals at an individual level often conflicts with conservation of the species at a population level (Home Office 2000; Reynolds 2004). Furthermore, even non-hunting human activities can involve stress to wild animals. For example, translocation and release of animals for conservation purposes can be stressful for individuals (Mathews et al. 2005). So too can tourist viewing of endangered species (Sorice et al. 2003), and a supposedly neutral human activity such as orienteering also can cause significant stress (Jeppeson 1987). In this essay we try to deal dispassionately with the question of whether hunting can aid wildlife conservation at the population level, and not with the important, but different, question of whether hunting has welfare implications for individual animals.

Sport hunting as an economic driver

Sport hunting can generate substantial revenue. Some hunters spend extravagantly and travel extensively (PACEC 2000), and may pay high fees for syndicate membership, logistical support and guides. They may pay extremely high prices to shoot spectacular species (e.g. Table 15.1). Although the travel component of such hunting varies widely, even local hunters may bear large operating costs. Fox-hunters in the UK had operating costs (maintenance of hound packs, horses and associated equipment, hunt staff) of roughly £2600 per hunter per year (PACEC 2000).

Revenue from hunting affects economies at national and regional levels. Expenditure on field sports in the UK exceeded £1.4 billion in 1992 (Cobham Resource Consultants 1997). In 2000, red grouse (*Lagopus lagopus scoticus*) shooting in Scotland supported the equivalent of 940 full-time jobs, and £17 million worth of GDP in Scotland (FAIRSC, 2001). Employment related to fox-hunting in the UK is estimated at 6000 to 8000 full-time equivalent jobs (PACEC 2000).

Table 15.1 Trophy fees paid per animal shot in Botswana, illustrating the large fees that are charged to sport hunters for various species (Sources: http://www.gondala.co.za; Botswana Wildlife Management Association 2001; http://allafrica.com/stories/printable /200402050156.html)

Species	Trophy fee per animals ($US)
Warthog	200–300
Spotted hyaena	500–930
Zebra	900–1070
Giraffe	1800–3000
Crocodile	1850–2000
Buffalo	2500–10,800
Leopard	2800–6550
Cheetah	3000
Hippo	3000–6150
Lion	3000–30,000
Sable antelope	3100–10,000
Elephant	19,000–40,000
White rhinoceros	25,000–60,000

Jackson (1996) estimates that the monetary benefit of hunting to the USA economy in 1991 was US$35 billion, of which US$2 billion was spent annually on conservation and acquiring habitat. Hunters may also voluntarily contribute to conservation. For example Ducks Unlimited, a USA wildfowling charity, raised and spent $140 million in 2003, 79% of which was spent on conservation in the USA. Hunters pay up to US$160,000 for a single trophy bighorn sheep (*Ovis canadensis*, Marty 2002). Although trophy prices in North America are unequalled elsewhere, sport hunting in developing countries can be economically significant. South African game farms earned US$44 million in 2001 (Van der Merwe & Saayman 2003). In the year following the Botswana Government's moratorium on lion hunting in 2000, the hunting industry experienced US$1.26 million of lost revenue (BWMA 2001).

Hunting revenue can also accrue locally. The Botswana Wildlife Management Association estimates that 49.5% of hunting expenditure, totalling US$9.5 million per annum, remains in individual hunting districts; a further 25.7% remains in the country (BWMA 2001). Similarly, Humavindu & Barnes (2003) show that 24% of hunting revenue earned in Namibia (totalling US$19.6 million) accrues to the poorer segments of society in the form of wages, rentals and royalties. In Zambia, the ADMADE programme (Administrative Design for Game Management Areas) receives around 67% of all revenue generated by sport hunting activities in Zambia's Game Management Areas. Fifty-three per cent of ADMADE revenue is allocated directly to local wildlife management, the remainder to community development (Lewis & Alpert 1997). See Jones & Murphree (2004) for more examples.

Hunting revenue can in some instances be directly used by Government for conservation purposes. For instance, hunting revenue contributes to conservation throughout the USA via taxation: all sport hunting equipment sold in the USA is subject to an 11% tax under the Federal Aid in Wildlife Restoration Act (1937), which contributes directly to acquisition of habitat, research, conservation training and education and provision of access to hunting and recreational facilities (US Fish and Wildlife Service 1999). In the UK, angling licences generate substantial funds, which are spent on management of aquatic ecosystems by the relevant agency (Environment Agency). By contrast, the UK Game Licence, administered by central Government, is currently under review because it is widely evaded and does not generate significant funds. Virtually all game hunting in the UK takes place on land that is privately owned and managed. Although Government does influence conservation on privately owned land through agri-environment schemes, centrally raised taxes may not be 'ear-marked' for specific uses such as conservation. Nevertheless, game hunting in the UK is an economically significant land-use (Cobham Resource Consultants 1997), which influences both policy at a national scale and the implementation of conservation measures at a local scale.

Linking conservation benefits with attitudes to wildlife

There is both a practical and a moral imperative for conservationists to engage with local people, particularly in poor countries where people's livelihoods may be compromised by conservation actions (Adams & Hulme 2001; Hulme & Murphree 2001). Where people live alongside, and thus potentially in conflict with wildlife, their tolerance may be proportional to any financial benefits received from wildlife. Elephants (*Loxodonta africana*) raid crops, and lions (*Panthera leo*) and spotted hyaenas (*Crocuta crocuta*) kill domestic stock (e.g. Butler 2000). Jones (1999) and Duffy (2000) propose that these burdens might be compensated if problem animals can be sold to sport hunters. However, as Taylor (1994) notes, trophy hunting does not always target problem animals. For instance, crop raiding by elephants in Omay Communal land, Zimbabwe, occurs largely in the wet season (November to April), but the majority of sport hunting takes place during the dry season (May to October), so elephants shot as trophies are not necessarily the animals involved in crop raiding, nor does removal of these animals alleviate the problems of crop loss at other times of the year. However, hunt-

ing revenue and activity can contribute to local infrastructure (clinics, schools, roads), further enhancing its value. Some safari hunting companies have worked hard to ensure that local communities benefit from their activities, and are motivated to collaborate to protect hunted wildlife (e.g. Cullman and Hurt Community Wildlife Project 2004). The Bar Valley project in Pakistan, based on trophy hunting for ibex, is another example of successful community-based conservation using sport hunting as its main income generator (Garson et al. 2002).

Benefits derived from hunting affect attitudes, in ways that may further conservation goals. In 1982, Shangaan people in the Mahenye area, adjacent to Gonarezhou National Park, Zimbabwe, were allowed after protracted negotiation to sell two trophy elephant hunts to foreign hunters. They received both the financial profit and the meat from the elephants sport-hunted in their tribal area. In response to this relaxation of wildlife laws by the Department of National Parks the local community voluntarily relocated 100 people from Ngwachumene Island, an important wildlife habitat on the border of the National Park. The local community not only gained benefit from the hunting activity, but more importantly felt that they had reclaimed part of their ancestral ownership and rights to use local wildlife resources – an important factor in ameliorating ongoing enmity with wildlife authorities in the area (Murphree 2001). Based on experiences such as this, Jones & Murphree (2004) argue that revenue from hunting is not the only benefit to local communities: local institutions and management can be enhanced through participation in community-based natural resource management (CBNRM), strengthening traditional hierarchies and rights to common property and enabling interactions with external institutions.

The CAMPFIRE (Communal Areas Management Plan For Indigenous Resources) scheme in Mahenye, Zimbabwe is a widely-known CBNRM initiative. It attempts to alter people's perceptions of natural resources, from either a nuisance or a food source to a viable and sustainable revenue source (Murphree 2001). Trophy hunting is a key component of income generation for CAMPFIRE; Bond (2001) shows that in 1989–1996 Zimbabwean rural district councils earned US$ 8.5 million (93% of their income) from leasing out sport hunting concessions through CAMPFIRE. However, benefits were not always equally shared by all members of the community and individual households accrued significant benefits only in areas where there was low human population density and high abundance of trophy species. Similarly, Murombedzi (1999) suggests that corruption, poor representation and political marginalization prevented disbursement of revenue to individual households.

Sport hunting is just one way by which CBNRMs can obtain revenues from wildlife. It is a highly lucrative form of use, usually generating higher revenues per animal than, for instance, subsistence hunting. However, some question the value of CBNRM as a conservation strategy, suggesting that its underlying assumptions are flawed (Kiss 2004; du Toit et al. 2004). Local communities are not necessarily willing to bear the opportunity costs of conservation and may not be willing to reinvest gains derived from CBNRMs in conservation. In some cases earnings from CBNRMs have been reinvested in agricultural expansion, which is ultimately damaging to biodiversity (Murombedzi 1999). The CBNRMs require heavy investment with an uncertain outcome. Given this, there is a growing belief that the more cost-effective way to conserve biodiversity in poor countries is for wealthy states simply to compensate local people for not damaging sensitive sites or species (James et al. 1999; Nicholls 2004; Kiss 2004; du Toit et al. 2004). However, there is as yet no global commitment to financing the costs of conservation, nor the institutional capacity to distribute such payments. Until there is, encouraging local sustainable use of natural resources through high return, arguably low impact activities, such as sport hunting, may be preferable to more destructive alternatives, such as

agriculture, subsistence hunting or logging, which tend to extirpate wildlife populations and destroy habitat.

However, sport hunting faces all the difficulties that other forms of enterprise-based conservation have. These include getting the incentive structures and resource ownership correct, and ensuring effective and robust institutions for resource management and disbursement of benefits (James et al. 1999; Milner-Gulland & Mace 1998; Salafsky et al. 2001). Many countries also suffer from broader problems of corruption and poor governance, institutional failure, social and economic upheaval, which reduce the likelihood of long-term success for any conservation or development activity (Smith 2003).

Direct ecological effects of sport hunting

Sport hunting acts directly and indirectly on the ecology of the target species and other species. For example, where population management is thought desirable (e.g. red fox or roe deer *Capreolus capreolus*), hunters may take on the role of extirpated natural predators. Population management is often necessary when species have been reintroduced for conservation reasons (e.g. beaver (*Castor fiber*), lynx (*Felis lynx*) and wolf (*Canis lupus*) in several European countries), and revenue generated by sport hunting could potentially offset some of the losses caused by these species. This can in turn improve public acceptance of such reintroductions (Ericsson et al. 2004). One of the most important potential benefits is habitat preservation, which acts to conserve both the target species and associated species. Any kind of hunting affects the demographic structure of the target species, and this can have knock-on effects, including evolutionary change. Sport hunting may be particularly detrimental in this regard because it is often highly selective, targeting trophy individuals. We discuss these two issues in more detail below.

Habitat protection

Hunters are frequently instrumental in protecting habitats for hunted species. Oldfield *et al.* (2003) showed that landowners in the UK who allow shooting and fox (*Vulpes vulpes*) hunting on their property apportion a greater part of their farmland to woodland than average and take up government subsidies to plant woods or hedges as habitats for game species. In Norway, European beavers are hunted on centrally administered quotas. Revenues are distributed to landowners based on the amount of beaver habitat they maintain, encouraging habitat protection (Parker & Rosell 2003). It is difficult to distinguish causation from correlation in these examples; to what extent is hunting the cause of conservation behaviour by landowners? It could be argued that conservation-minded landowners would continue to protect habitats regardless of whether they are able to hunt on their land (Macdonald & Johnson 2000). However, experience within the UK game management sector (Game Conservancy Ltd advisory service, Reynolds personal observation) confirms that hunting is very often the motivation for investment in habitat management of wide conservation benefit, and for taking up agri-environment grants to assist with this.

In Africa, areas set aside for sport hunting and sustainable wildlife-use greatly increase the amount of habitat available to wild species. Without revenue from hunting, political pressure might be exerted to turn these areas over to domestic livestock production, which could irreversibly damage these ecosystems (e.g. Barnes 2001). Twenty per cent (140,000 km^2) of Zambia's land area is made up of Game Management Areas, whereas only half as much land is designated for National Parks (Lewis & Alpert 1997). The Zimbabwean rural district councils participating in the CAMPFIRE scheme set aside substantial areas for wildlife, estimated at 36,000 km^2 (Taylor 1998), whereas safari areas and private hunting land administered by National Parks make up another 50,000 km^2

(Cumming 2004). In South Africa from the late 1990s to 2002, land was converted from cattle ranching to extensive game ranching, largely for hunting, at a rate of 500,000 ha yr^{-1}. By 2002, 13% of the country's agricultural land was being used as game ranches (Van der Merwe & Saayman 2003).

In the USA, Ducks Unlimited conserves 10 million acres of waterfowl habitat across North America (Ducks Unlimited 2004). Proponents of hunting claim some remarkable recoveries of waterfowl species have occurred as a result (Jackson 1996). However, others claim that, despite reclamation of habitat, there is little evidence for population increases and some species (e.g. black duck, *Anas rubripes*) may have declined in number (Grandy 2003). In this case sport hunting may well contribute to habitat protection, but evidence is equivocal whether it has improved the fortunes of the species that it claims to protect. However, the net benefit of habitat protection and its associated biodiversity might outweigh this doubt.

Advocates of hunting often claim that hunters are effective custodians of wildlife habitat (Jackson 1996), providing support for anti-poaching teams and preventing poaching by operating in an area (Pasanisi 1996). Additionally although hunters are often instrumental in protecting species that they wish to hunt, this can sometimes be detrimental to habitats. In the USA some hunters put maintenance of artificially high quarry populations (e.g. white-tailed deer) ahead of ecosystem health and biodiversity and may be averse to or impede efforts to restore native wildlife and protect biodiversity (Holsman 2000; Peyton 2000). In another example, red deer (*Cervus elaphus*) are maintained in high numbers on some upland shooting estates in Scotland and also in New Zealand (where they are an alien species), with consequent damage to native habitat (Caughley 1983; Grandy et al. 2003). Sport hunting in these cases exacerbates the conservation problem. However, it could also be part of the solution, if it were structured to provide conservation revenues while keeping deer numbers low.

Direct impacts of hunting on the quarry population

Hunting is a selective force and must have consequences for demography and population genetics. Some hunters like to retain trophies, especially exceptionally fine, large or old specimens. Species with sexually selected features that are easy to preserve (such as antlers in deer, or tusks in elephants) may generate a particular demand among trophy hunters. This can lead to genetic change within populations due to highly selective removal or sometimes management of a population specifically to produce trophy animals. We discuss these issues in more detail below.

Species with harem breeding structures appear to be robust against quite intensive selective hunting, and trophy males have even been considered to be surplus to the population (Fairall 1985). Caro *et al.* (1998) found no consistent impact of sport hunting on ungulate population sizes in southern Tanzania. Nevertheless, intensive selective hunting pressure targeting adult males can cause sudden population collapse (Ginsburg & Milner-Gulland 1994). In saiga antelopes (*Saiga tatarica*), a harem-breeding species, conception rates remained normal (near 100%) when the sex ratio was highly distorted by commercial hunting (2.5% adult males in the population). However, when the sex ratio of adults fell below 1% males, only 20% of females conceived (Milner-Gulland et al. 2003). Fergusson (1990) describes intensive trophy hunting of male sable antelope (*Hippotragus niger*) in Zimbabwe, disturbing territorial and mating behaviour, leading to reduced calving rates and a protracted parturition period, which resulted in high calf mortality. However, there are many instances in which effects of trophy hunting are ambiguous. In Dall sheep (*Ovis dalli*), Heimer (1980) found reduced lamb production in a sport-hunted population, but Murphy et al. (1990) were unable to detect demographic differences between hunted and unhunted populations. In moose (*Alces alces*), Laurian et al. (2000) found no differences in mating behaviour or reproductive success between two populations, of which one had selective hunting of adult males and the other was unhunted; whereas Solberg et al. (2002) found that hunting-induced female-biassed sex ratios reduced fecundity in primiparous moose.

Social disruption has also been observed in highly skewed populations. If older bulls are removed from elephant populations, young bulls can show aberrant or delinquent behaviour. In Pilanesburg Game Reserve, South Africa, young bull elephants killed 40 white rhinoceros (*Ceratotherium simum*) over a period of 5 years, but this behaviour ceased when mature bulls were introduced to the population (Slotow et al. 2000). The disturbance caused by sport hunting can have an impact on movement behaviour. Ruth et al. (2003) showed that in hunting areas adjacent to Yellowstone National Park, cougars (*Puma concolor*) and elk (*Cervus elaphus*) avoided areas where hunting occurred, although grizzly bears (*Ursus arctos*) used hunting areas more frequently due to increased scavenging opportunities. In southern Quebec, snow goose (*Anser caerulescens atlanticus*) spring migrations were disturbed by introduction of a spring hunting season. Disturbance reduced feeding opportunities and prenuptial fattening prior to the 3000 km Arctic migration. Reduced body condition resulted in reduced breeding effort, lower clutch size and delayed laying. Hunting disturbance also caused geese to migrate westwards into sensitive agricultural land, causing a tenfold increase in damage compensation paid to farmers in these areas (Béchet at al 2003). Loveridge & Macdonald (2001) found that sport hunting in safari concessions surrounding Hwange National Park, Zimbabwe, removed around 67% of mature male lions from a study population covering 6000 km^2 of the National Park and reduced the proportion of males in the adult population from around 30% to 13%. Reduction in male lion density resulted in males expanding their ranges to include multiple prides of females. Furthermore, inflation of male home-range size increased the probability that males would leave the protection of the park and themselves become vulnerable to sport-hunting. In a similar situation in Savuti, Botswana, where male lions were also rare because of sport-hunting, female groups did not benefit from the protection of males and were exposed to high levels of kleptoparasitism by spotted hyaenas (Cooper 1991). However, notwithstanding high offtakes of male lions in some areas, well regulated sport-hunting does not appear to affect the viability of large healthy populations of this species (Whitman et al. 2004).

In social species, especially carnivores, killing one individual can have knock-on effects that result in unanticipated disturbance or death of other individuals in the population (Tuyttens & Macdonald 2000). Male felids and ursids enhance their reproductive success by killing a rival male's offspring. This brings newly acquired females into oestrus earlier than if they had successfully raised their offspring to maturation. In African lions and brown bears (*Ursus arctos*) removal of territorial males by sport hunters may result in the deaths of their offspring, killed by new males filling the newly opened space in the territorial hierarchy. If this occurs frequently it lowers population growth rates (Swenson et al. 1997; Greene et al. 1998; Whitman et al 2004). Similarly in a brown bear population in Alberta, removal of mature males by hunters and the resulting immigration of subadult males caused poor cub survival and population decline (Weilgus & Bunnell 1994). The mechanisms involved were infanticide and the use of suboptimal habitat by females with cubs to avoid the new males.

The highly selective nature of most sport hunting may also have genetic consequences. A priori, it is likely that hunting regimes that closely resemble mortality patterns in natural populations will have fewer long term evolutionary consequences than those with highly artificial or biased mortality patterns (Harris *et al.* 2002). This might suggest that alteration of mortality patterns by highly selective hunting could affect the gene pool of a hunted population, although it is not obvious that these alterations always have serious consequences for biodiversity. For example, hunting can cause increased gene flow and heterozygosity (e.g. for grey-winged francolin (*Francolinus africanus*); Little et al. 1993). Increased gene flow is problematic only if locally adapted gene complexes are threatened (Harris *et al.* 2002). It is even a moot point to what extent particular types of sport hunting actually lead to artificial patterns of mortality. Frati *et al.* (2000) showed that sport hunting of red foxes resembles predation by larger locally extinct predators such as wolves, leopards (*Panthera pardus*) and lynx.

Phenotypic changes caused by hunting have been linked to reductions in population performance. For example, Coltman *et al.* (2003) showed that body weight and horn size decreased significantly in a hunted population of big-horn sheep in response to hunter selection of large-horned rams (Fig. 15.2). Horn and body size are heritable traits closely linked to fitness, with larger-horned and -bodied rams able to defend and inseminate more ewes than smaller individuals. Likewise Shea & Vanderhoof (1999) found alteration of allele frequencies in white-tailed deer due to hunter selectivity. Larger-antlered bucks are those born earlier in the year and have fast growing antlers and high reproductive success. However, it is the large antlered animals that are chosen by hunters, leading to increased survival of late-born animals with slower growing antlers and lower reproductive potential.

In some instances, sport hunters select targets with the specific intention of "improving" the genetic stock of a population. There is a long-established tradition of attempting to improve the antler quality of red deer (*Cervus elaphus*) populations by selection as well as nurture (Thelen 1991). However, Kruuk *et al.* (2002) found no association between antler size and fitness, despite antler size being heritable, because of the large environmental influ-

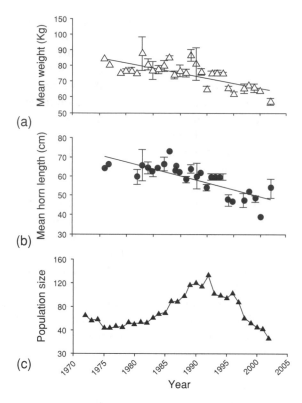

Fig. 15.2 Observed changes in mean weight and horn length and in the population size in big horn sheep from 1972 to 2002. (a) Relationship between weight (mean ± s.e.m.) of 4-year-old rams and year ($N = 133$ rams). (b) Relationship between horn length (mean ± s.e.m.) of 4-year-old rams and year ($N = 119$ rams). (c) Changes in population size (taken as the number of ewes aged at least 2 years plus yearlings) over time. (From Coltman et al. 2003; Copyright Nature Publishing Group.)

ence on each. This suggests that attempts to manipulate antler quality through selective shooting are unlikely to lead to an evolutionary response in antler size.

What role should sport hunting play in contemporary conservation?

In many parts of the world conservation had its roots in an era where wild species were preserved to be hunted by a small, wealthy elite. However, in contemporary conservation, there are many more interest groups with claims on wildlife and its habitats. Thus we need to evaluate sport hunting against viable alternative ac-

tivities. We take it as read that, from a conservation perspective, sustainable use of natural resources is preferable to extirpation of a population, either through overhunting or through conversion of habitat to alternative uses. Hence we compare sport hunting with other activities that contribute to conservation, rather than with economic activities that might be more lucrative, but which are ecologically damaging.

Sport hunting can clearly generate economic activity, but how does the magnitude and pattern of that activity compare with other conservation options? Sport hunting competes to varying extents with other land-uses (e.g. agriculture, subsistence hunting, logging, other tourism). Its significance as a conservation tool depends on each activity's relative contribution to human livelihoods and habitat degradation. Most commonly, sport hunting serves to preserve a mosaic of wildlife habitat within an agricultural landscape, and to mitigate the damaging effects of competing land-uses. Examples range from sport hunting in an agricultural landscape of communally owned land in Zimbabwe as part of CBNRM initiative (e.g. Bond 2001), to the conservation of grey partridges (*Perdix perdix*) within an agricultual landscape in the UK (Potts 1986).

Hunting is commonly compared with photographic tourism in its contrasting impact on wildlife populations and ecological systems (e.g. Leader-Williams 2002, Leader-Williams & Hutton 2005). An example from Zimbabwe indicates that photographic tourism can out-compete hunting as a revenue generator. In one CBNRM area a tourist lodge was established in 1994; by 1997, lodge revenue exceeded that from hunting by 100% (Murphree 2001). However, mass market tourism is also extremely volatile. When countries experience political instability, mass tourism rapidly disappears, but sport hunters often continue to travel (Bond *et al.* 2004; Muir & Bojo 1994). During the civil war in Zimbabwe in the late 1970s there was virtually no photographic tourism, but foreign hunters continued to come to the country (Martin 1996). In addition photographic tourism is not viable in some areas, even though wildlife is present (e.g. agricultural areas, areas where charismatic species are difficult to see or landscapes are unremarkable). Photographic tourists need to be accommodated in larger numbers than hunters for the same level of profit, and this causes greater incidental environmental damage (e.g. waste from tourist accommodation and collection of firewood can pollute or destroy wildlife habitat; Goodwin *et al.* 1998).

It is often assumed that demonstrating tangible economic benefits is the way to generate support for conservation (Goodwin *et al.* 1998). The degree to which any externally imposed revenue-generating activity genuinely influences public attitudes towards conservation depends on the incentive structure in place. African CBNRMs can contribute to livelihoods and, although small in global terms, incomes from CBNRM activities can make a significant difference to households. In some CAMPFIRE areas dividends from sport hunting exceeded income from agriculture (Bond 2001), but this does not necessarily mean that people are willing to abandon established activities. The fewer the competing activities, the lower human population levels and the more abundant the wildlife populations, the more likely it is that hunting can actually or potentially contribute to livelihoods and regional economic success. However, land use is dynamic. Currently protected areas have the potential to support other livelihood activities such as hunting or even agriculture and settlement (e.g. Murphree 2004). Likewise, agricultural land can be converted into wildlife sanctuaries (Goodwin *et al.* 1997), where sustainable sport hunting and wildlife conservation may become a primary economic activity. The actual or perceived value, at governmental or local levels, of conservation activities (including sport hunting) may be an important determinant of this change.

Conclusion

Sport hunting can benefit conservation in a number of ways, with acquisition and protection of habitat being a major benefit. Generation of substantial revenue is also possible, and there are cases where this revenue is used in conservation of biodiversity. There are also many cases where it is less clear that hunting revenue is reinvested in conservation. Genetic, behavioural and population impacts may need to be guarded against, although there are few clearcut cases where these have had significant impacts on the viability of populations. On balance it appears that the benefits of sport hunting can outweigh any disadvantage if responsibly managed and monitored. However, an institutional structure that is able to implement regulations effectively is a necessary precondition for success. All conservation and resource use requires that managers can monitor wildlife populations and ecosystems, set and enforce limits, and ensure that benefits are disbursed wisely so that conservation is competitive with alternative land-uses. Such a regulatory infrastructure can be expensive. Compared with other, non-consumptive uses of wildlife, sport hunting has the potential to generate large profits, and sport hunters often collectively demonstrate a responsibility and passionate concern to see their quarry species conserved. This makes sport hunting a potentially attractive option for conservationists.

Acknowledgements

The authors would like to thank David Macdonald, Patrick Doncaster, Jon Hutton and Nigel Leader-Williams for helpful comments.

If a bird's nest chance to be before thee in the way in any tree, or on the ground, whether they be young ones, or eggs, and the dam sitting upon the young, or upon the eggs, thou shalt not take the dam with the young: But thou shalt in any wise let the dam go, and take the young to thee; that it may be well with thee and that thou mayest prolong thy days.

(Deuteronomy 22: 6–7, written c.560 BC and attributed to Moses c.1300 BC on the fundamental tenets of moral law. King James translation.)

References

Adams, W. & Hulme, D. (2001) Conservation and community. Changing narratives, policies and practices in African conservation. In *African Wildlife and Livelihoods. The Promise and Performance of Community Conservation* (Eds D. Hulme & M. Murphree), pp. 9–21. James Currey, Oxford.

Adams, W.M. (2004) *Against Extinction: the Story of Conservation.* Earthscan, London, 311 pp.

Bateson, P. & Harris, R. (2000) The effects of hunting with dogs in England and Wales on the welfare of deer, foxes, mink and hare. *Report of the Committee of Inquiry into Hunting with dogs in England and Wales* (Burn's Report). United Kingdom Parliamentary Document CM4763. Available online at: http://www.huntinginquiry.gov.uk/mainsections/report.pdf. The Stationery Office, London.

Barnes, J.I. (2001) Economic returns and allocation of resources in the wildlife sector of Botswana. *South African Journal of Wildlife Research* **31**: 141–53.

Béchet, A., Giroux, J., Gauthier, G., Nichols, J.D. & Hines, J.E. (2003) Spring hunting changes the regional movements of migrating greater snow geese. *Journal of Applied Ecology* **40**: 553–64.

Bond, I. (2001) CAMPFIRE and the incentives for institutional change. In *African Wildlife and Livelihoods. The Promise and Performance of Community Conservation* (Eds D. Hulme & M. Murphree), pp. 227–43. James Currey, Oxford.

Bond, I., Child, B., de la Harpe, D., Jones, B., Barnes, J. & Anderson, H. (2004) Private Land Contribution to Conservation in South Africa. In *Parks in Transition: Biodiversity, Rural Development and the Bottom Line* (Ed. B. Child), pp. 29–61. Earthscan, London, 267 pp.

Browne, S.J. & Aebischer, N.J. (2004) Temporal changes in the breeding ecology of European turtle doves *Streptopelia turtur* in Britain, and implications for conservation. *Ibis* **146**: 125–37.

Butler, J.R.A. (2000) The economic costs of wildlife predation on livestock in Gokwe communal land, Zimbabwe. *African Journal of Wildlife* **38**: 23–30.

BWMA (Botswana Wildlife Management Association). (2001) *Economic Analysis of Commercial Consumptive Use of Wildlife in Botswana.* ULG Northumbrian Ltd, Leamington Spa, 48 pp.

Canadian Arctic Profiles. (2004) *Native Land Use and the Annual Harvesting Cycle.* http: //collections.ic.gc.ca/arctic/inu it/harvesti.htm. [Accessed on 20 May 2004]

Caro, T.M., Pelkey, N., Borner, M., et al. (1998) Consequences of different forms of conservation for large mammals in Tanzania: preliminary analyses. *African Journal of Ecology* **36**: 303–20.

Caughley, G. (1983) *The Deer Wars: the Story of Deer in New Zealand.* Heinemann, Auckland.

Chabot, D., Gagnon, P. & Dixon, E.A. (1996) Effect of predator odors on heart rate and metabolic rate of wapiti (*Cervus elaphus canadensis*). *Journal of Chemical Ecology* **22**: 839–68.

Cobham Resource Consultants (1997) *Countryside Sports: their Economic, Social and Conservation Significance.* Standing Conference on Countryside Sports, Reading.

Coltman, D.W., O'Donoghue, P., Jorgenson, J.T., Hogg, J.T., Strobeck, C. & Festa-Bianchet, M. (2003) Undesirable evolutionary consequences of trophy hunting. *Nature* **426**: 655–8.

Convention on Biological Diversity (2003) *Handbook of the Convention on Biological Diversity.* Secretariat of the Convention on Biological Diversity. http://www.biodiv.org/handbook/ [Accessed 4 November 2004]

Cooper, S.M. (1991) Optimal hunting group size – the need for lions to defend their kills against loss to spotted hyaenas. *African Journal of Ecology* **29**: 130–6

Cullman and Hurt Community Wildlife Project (2004) http://www.cullmanandhurt.org/april2000.html [Accessed 8 February 2005]

Cumming, D. (2004) Performance of parks in a century of change. In *Parks in Transition: Biodiversity, Rural Development and the Bottom Line* (Ed. B. Child00), pp. 105–24. Earthscan, London.

Dickson, J.G. (Ed.) (1992) *The Wild Turkey. Biology and Management.* Stackpole, Harrisburg, PA, 463 pp.

Ducks Unlimited (2004) *World leaders in Wetland Conservation.* http://www.ducks.org. [Accessed on 20 May 2004]

Duffy, R. (2000) *Killing for Conservation: Wildlife Policy in Zimbabwe.* African Issues, International African Institute. James Currey, Oxford, 209 pp.

Du Toit, J.T., Walker, B.H. & Campbell, B.M. (2004) Conserving tropical nature: current challenges for ecologists. *Trends in Ecology and Evolution* **19**(1): 12–17.

Ericsson, G., Heberlein, T.A., Karlsson, J., Bjarvall, A. & Lundvall, A (2004) Support for hunting as a means of wolf *Canis lupus* population control in Sweden *Wildlife Biology* **10**: 269–76.

FAIRSC (The Fraser of Allander Institute for Research on the Scottish Economy, University of Strathclyde) (2001) *An Economic Study of Scottish Grouse Moors: an Update.* Game Conservancy Limited, Fordingbridge, 26 pp.

Fairall, N. (1985) Manipulation of age and sex ratios to optimise production from impala (Aepyceros melampus). *South African Journal of Wildlife Research* **15**: 85–8.

Fergusson, R. (1990) *A preliminary investigation of the population dynamics of sable antelope in the Matetsi Safari Area, Zimbabwe.* MSc thesis, University of Zimbabwe, Harare.

Fitter, R.S.R. & Scott, P. (1978) *The Penitent Butchers: The Fauna Preservation Society 1903–1978.* Collins, London.

Frati, F., Lovari, S. & Hartl, G.B. (2000) Does protection from hunting favour genetic uniformity in the red fox? *Zeitschrift für Säugetierkunde* **65**: 76–83.

Garson, P.J., Arshad, M. & Ahmed, A. (2002) Trophy hunting and conservation: an audit of the Bar Valley Ibex project in Northern Pakistan. *Presented at the 16th Annual Meeting of the Society for Conservation Biology*, University of Kent, 14–19 July. http://www.kent.ac.uk/anthropology/dice/ scb2002/abstracts/WedPosters/posters01_11.html. [Accessed 9 February 2005]

Ginsburg, J. & Milner-Gulland, E.J. (1994) Sex-biased harvesting and population dynamics in ungulates: implications for conservation and sustainable use. *Conservation Biology* **8**: 157–66.

Goodwin, H.J., Kent, I.J., Parker, K.T. & Walpole, M.J. (1997) *Tourism, Conservation and Sustainable Development*, Vol. IV, *The South-east Low-veld, Zimbabwe.* Unpublished final report to the Department for International Development (DfID), London.

Goodwin, H.J., Kent, I.J., Parker, K.T. & Walpole, M.J. (1998) *Tourism, Conservation and Sustainable Development: Case Studies from Asia and Africa.* Wildlife and Development Series No 11, International Institute for Enviroment and Development, London, 88 pp.

Grandy, J.W. (2003) Data and observations on duck hunting in the United States. In *The State of the Animals*, Vol. II (Eds D.J. Salem & A.N. Rowan), pp 116–119. Humane Society Press, Washington, DC.

Grandy, J.W., Stallman, E. & Macdonald, D.W. (2003) The science and sociology of hunting: shifting practices and perceptions in the United States and Great Britain. In *The State of the Animals*, Vol. II (Eds D.J. Salem & A.N. Rowan), pp 107–130. Humane Society Press, Washington, DC.

Greene, C., Umbanhower, J., Mangel, M. & Caro, T. (1998) Animal breeding systems, hunter selectivity and consumptive use in wildlife conservation. In *Behaviour, Ecology and Conservation Biology* (Ed. T. Caro) pp. 271–305. Oxford University Press, Oxford.

Harris, R.B., Wall, W.A. & Allendorf, F.W. (2002) Genetic consequences of hunting: what do we know and what should we do? *Wildlife Society Bulletin* **30**: 634–43.

Heimer, W.E. (1980) A summary of Dall sheep management in Alaska during 1979 (or how to cope with a monumental disaster). *Symposium of Northern Wild Sheep and Goat Council* **2**: 355–80.

Heydon, M.J. & Reynolds, J.C. (2000) Demography of rural foxes (*Vulpes vulpes*) in relation to cull intensity in three contrasting regions of Britain. *Journal of Zoology, London* **251**: 265–76.

Home Office (2000) *Report of the Committee of Inquiry into Hunting with dogs in England and Wales*. United Kingdom Parliamentary Document CM4763. The Stationery Office, London, 223 pp. Also available online at http://www.huntinginquiry.gov.uk/mainsections/ report.pdf.

Holsman, R.H. (2000) Goodwill hunting? Exploring the role of hunters as ecosystem stewards. *Wildlife Society Bulletin* **28**: 808–16.

Hulme, D. & Murphree, M. (2001) Community conservation in Africa. An introduction. In *African Wildlife and Livelihoods. The Promise and Performance of Community Conservation* (Eds D. Hulme & M. Murphree), pp. 1–8. James Currey, Oxford.

Hutton, J.M. & Leader-Williams, N. (2003) Sustainable use and incentive driven conservation: realigning human and conservation interests. *Oryx* **37**: 215–26.

Humavindu, M.N. & Barnes, J.I. (2003) Trophy hunting in the Namibian economy: an assessment. *South African Journal of Wildlife Research* **33**: 65–70.

Jackson, J.J. (1996) An International Perspective on Hunting. In *Tourist Hunting in Tanzania* (Eds N. Leader-Williams, J.A. Kayera & G.L. Overton), pp. 7–11. Occasional Publication 14, International Union for the Conservation of Nature and Natural Resources, Cambridge.

James, A.N., Gaston, K.J. & Balmford, A. (1999) Balancing the Earth's accounts. *Nature* **401**: 323–24.

Jeppesen, J.L. (1987) The disturbing effects of orienteering and hunting on roe deer (*Capreolus capreolus*) at Kaloe. *Danish Review of Game Biology* **13**: 1–24.

Jones, B.T.B. (1999) Policy lessons from the evolution of a community-based approach to wildlife management, Kunene region, Namibia. *Journal of International Development* **11**: 295–304.

Jones, B.T.B. & Murphree, M.W. (2004) Community based natural resource management as a conservation mechanism: lessons and directions. In *Parks in Transition: Biodiversity, Rural Development and the Bottom Line* (Ed. B. Child), pp. 63–104. Earthscan, London.

Kiss, A. (2004) Is community based ecotourism a good use of biodiversity conservation funds? *Trends in Ecology and Evolution* **19**(5): 232–7.

Kruuk, L.E.B., Slate, J., Pemberton, J.M., Brotherstone, S., Guinness, F. & Clutton-Brock, T. (2002) Antler size in red deer: heritability and selection but no evolution. *Evolution* **56**: 1683–95.

Laurian, C., Ouellet, J., Courtois, R., Breton, L. & St-Onge, S. (2000) Effects of intensive harvesting on moose reproduction. *Journal of Applied Ecology* **37**: 515–31.

Leader-Williams, N. (2002) Animal conservation, carbon and sustainability. *Philosophical Transactions of the Royal Society of London, Series A* **360**: 1787–806.

Leader-Williams, N. & Hutton, J.M. (2005) Does extractive use provide opportunities to offset conflicts between people and wildlife? In *People and Wildlife: Conflict or Co-existence?* (Eds R. Woodroffe, S.J. Thirgood & A. Rabinowitz), pp. 140–61. Cambridge University Press, Cambridge.

Lewis, D.M. & Alpert, P. (1997) Trophy hunting and wildlife conservation in Zambia. *Conservation Biology* **11**: 59–68.

Lindell, L. & Wirdheim, A. (2001) Killing Fields. an increase in bird hunting in southern Europe. *Var Vagelwaerld* **5**. (Translated from Swedish.)

Little, R.M., Crowe, T.M. & Grant, W.S. (1993) Does hunting affect the demography and genetic structure of the grey–winged francolin (*Francolinus africanus*)? *Biodiversity and Conservation* **2**: 567–85.

Loveridge, A.J. & Macdonald, D.W. (2001) Impact of trophy hunting in Zimbabwe. In *Lion Conservation Research. Workshop 2: Modelling Conflict* (Eds A.J. Loveridge, T. Lynam & D.W. Macdonald), pp. 18–19. October, Okavango Delta, Botswana.

Macdonald, D.W. & Johnson, P.J. (2000) Farmers and the custody of the countryside: trends in loss and conservation of non-productive habitats 1981–1998. *Biological Conservation* **94**: 221–34.

Macdonald, D.W., Tattersall, F.H., Johnson, P.J., et al. (2000) *Management and Control of Populations of Foxes, Deer, Hares, and Mink in England and Wales, and the Impact of Hunting with Dogs*. Published with Home Office (2000). The Stationery Office, London,

206 pp. Also available online at: http://www.huntinginquiry.gov.uk/mainsections/research/macdonaldfinal.pdf

Martin, R.B. (1996) Sport hunting: The Zimbabwe Government viewpoint. In *Tourist Hunting in Tanzania* (Eds N. Leader-Williams, J.A. Kayera & G.L. Overton), pp. 43–9. Occasional Publication 14, International Union for the Conservation of Nature and Natural Resources, Cambridge.

Marty, S. (2002) Sacrificial ram. *Canadian Geographic* **November**: 37–50.

Mathews, F., Orros, M., McLaren, G., Gelling, M. & Foster, R (2005) Keeping fit on the ark: assessing the suitability of captive-bred animals for release. *Biological Conservation* **121**: 569–77.

Milner-Gulland, E.J. & Mace, R. (1998) *Conservation of Biological Resources*. Blackwell Science, Oxford.

Milner-Gulland, E.J., Bukreeva, O.M., Coulson, T.N., et al. (2003) Reproductive collapse in saiga antelope harems. *Nature* **422**: 135.

Muir, K. & Bojo, J. (1994) *Economic Policy, Wildlife and Landuse in Zimbabwe*. World Bank Enviroment Department, Working Paper Number 68, World Bank, Washington, DC.

Murombedzi, J.C. (1999) Devolution and Stewardship in Zimbabwe's campfire programme. *Journal of International Development* **11**: 287–93.

Murphree, M. (2001) Community, council and client. A case study in ecotourism development from Mahenye, Zimbabwe. In *African Wildlife and Livelihoods. The Promise and Performance of Community Conservation* (Eds D. Hulme & M. Murphree), pp. 177–94. James Currey, Oxford.

Murphree, M. (2004). Who and what are parks for in transitional societies? In *Parks in Transition: Biodiversity, Rural Development and the Bottom Line* (Ed. B. Child), pp. 217–32. Earthscan, London.

Murphy, E.C., Singer, F.J. & Nichols, L. (1990) Effects of hunting on survival and productivity of Dall sheep. *Journal of Wildlife Management* **54**: 284–290.

Muth, R.M. & Jamison, W.V. (2000) On the destiny of deer camps and duck blinds: the rise of the animal rights movement and the future of wildlife conservation. *Wildlife Society Bulletin* **28**(1): 841–51.

Newby, J.E. (1990) The slaughter of Sahelian wildlife by Arab royalty. *Oryx* **24**: 6–8.

Nicholls, H. (2004) The conservation business. *Public Library of Science Biology* **2**(9): 1256–9.

Novak, M. (1987) Beaver. In *Wild Furbearer Management and Conservation in North America* (Eds M. Novak, HJ.A. Baker, M.E. Obbard & B. Malloch), pp. 283–312. Ontario Trappers Association, North Bay, Ontario.

Oates, J.F. (1999) *Myth and Reality in the Rain Forest: how Conservation Strategies are Failing in West Africa*. California University Press, Berkeley, CA.

Oldfield, T.E.E., Smith, R.J., Harrop, S.R. & Leader-Williams, N. (2003) Field sports and conservation in the United Kingdom. *Nature* **423**: 531–3.

Organ, J.F. & Fritzell, E.K. (2000) Trends in consumptive recreation and the wildlife profession. *Wildlife Society Bulletin* **28**: 780–787.

Osborn, D.J. & Osbornova, J. (1998) *The Mammals of Ancient Egypt*. Aris and Philipps, Warminster.

PACEC (Public and Corporate Economic Consultants) (2000) The economic effects of hunting with dogs. *Report of the Committee of Inquiry into Hunting with Dogs in England and Wales* (Burn's Report). United Kingdom Parliamentary Document CM4763. Available online at: http://www.huntinginquiry.gov.uk/mainsections/report.pdf. The Stationery Office, London.

Parker, H. & Rosell, F. (2003) Beaver management in Norway: a model for continental Europe? *Lutra* **46**: 223–34.

Pasanisi, G. (1996) The outfitter's perspective of tourist hunting in Tanzania. In *Tourist Hunting in Tanzania* (Eds N. Leader-Williams, J.A. Kayera & G.L. Overton), pp. 17–18. Occasional Publication 14, International Union for the Conservation of Nature and Natural Resources, Cambridge, 138 pp.

Peyton, R.B. (2000) Wildlife management: cropping to manage or managing to crop? *Wildlife Society Bulletin* **28**: 774–9.

Potts, G.R. (1986) *The Partridge. Pesticides, Predation and Conservation*. Collins, London, 274 pp.

Reynolds, J.C. (2004) Trade-offs between welfare, conservation, utility and economics in wildlife management – a review of conflicts, compromises and regulation. *Animal Welfare* **13**: 133–8.

Roth, H.H. & Merz, G. (Eds) (1996) *Wildlife Resources. A Global Account of Economic Use*. Springer-Verlag, New York, 402 pp.

Ruth, T.K., Smith, D.W., Haroldson, M.A., et al. (2003) Large-carnivore response to recreational big-game hunting along the Yellowstone National Park and Absaroka–Beartooth Wilderness boundary. *Wildlife Society Bulletin* **31**: 1150–61.

Salafsky, N., Cauley, H., Balachander, G., et al. (2001) A systematic test of an enterprise strategy for community-based biodiversity conservation. *Conservation Biology* **15**: 1585–95.

Save the Elephants. (2002) *Save the Elephants Newsletter* September 2001 to September 2002, and www.eco-hunt.co.za [Accessed 11 August 2004]

Serpell, J. (1986) *In the Company of Animals*. Blackwell, Oxford.

Shea, S.M. & Vanderhoof, R.E. (1999) Evaluation of a five-inch regulation for increasing antler size of harvested deer in northwest Florida. *Proceedings of the Annual Southeastern Deer Study Group* **22**: 18–19.

Slotow, R., van Dyk, G., Poole, J., Page, B. & Klocke, A. (2000) Older bull elephants control young males. *Nature* **408**: 425–6.

Smith, R.J. (2003) Governanace and the loss of biodiversity. *Nature* **426**: 67–70.

Solberg, E.J., Loison, A., Ringsby, T.H., Saether, B-E. & Heim, M. (2002) Biased adult sex ratio can affect fecundity in primiparous moose *Alces alces*. *Wildlife Biology* **8**: 117–28.

Sorice, M.G., Shafer, C.S., Scott, D. (2003) Managing endangered species within the use/preservation paradox: understanding and defining harassment of the West Indian manatee (*Trichechus manatus*). *Coastal Management* **31**: 319–38.

Stewart, R.E.A. & Fay, F.H. (2001) Walrus. In *The New Encylopaedia of Mammals* (Ed. D.W. Macdonald), pp. 174–179. University of Oxford Press, Oxford.

Stirling, I. (2001) Polar Bear. In *The New Encylopaedia of Mammals* (Ed. D.W. Macdonald), pp. 76–7. University of Oxford Press, Oxford.

Swenson, J.E., Sandegren, F., Söderberg, A., Bjärvall J., Franzén, R. & Wabakken, P. (1997) Infanticide caused by hunting of male bears. *Nature* **386**: 450–1.

Tapper, S.C. & Reynolds, J.C. (1996) The wild fur trade: historical and ecological perspectives. In *The Exploitation of Mammal Populations* (Eds V.J. Taylor & N. Dunstone), pp. 28–43. Chapman & Hall, London.

Taylor, R. (1994) Elephant management in Nyaminyami District, Zimbabwe: turning a liability into an asset. *Pachyderm* **18**: 19–29.

Taylor, R. (1998) Wilderness and the CAMPFIRE Programme: the value of wildlands and wildlife to local communities in Zimbabwe. *Paper presented at the Wilderness Symposium, Waterberg Plateau Park, Namibia, 24 –27 June 1996*, International Union for the Conservation of Nature and Natural Resources, Gland, Switzerland.

Thelen, T.H. (1991) Effects of harvest on antlers of simulated populations of elk. *Journal of Wildlife Management* **55**: 243–9.

Tuyttens, F.A.M. & Macdonald, D.W. (2000) Consequences of social perturbation for wildlife management and conservation. In *Behaviour and Conservation*, Vol. 4 (Eds L.M. Gosling & S.W.J. Cambridge), pp. 315–329. Cambridge University Press, Cambridge.

United Nations Mongolia Office (2004) *Mongolia's Wild Heritage: Biological Diversity, Protected Areas and Conservation in the Land of Chingis Khan*. http://www-unmongolia.mn/archives/wild/bogdkhan.htm [Accessed on 20 May 2004]

US Fish and Wildlife Service (1999) *Conserving the Nature of America*. http://training.fws.gov/library/Pubs/ conserving.pdf [Accessed on 17 November 2004]

Van der Merwe, P. & Saayman, M. (2003) Determining the economic value of game farm tourism. *Koedoe* **46**: 103–12.

Whitman, K., Starfield, A.M., Quadling, H.S. & Packer, C. (2004) Sustainable trophy hunting of African lions. *Nature* **428**: 175–8.

Weilgus, R.B. & Bunnell, F.L. (1994) Dynamics of a small, hunted brown bear population in southwestern Alberta, Canada. *Biological Conservation* **67**: 161–6.

Woolf, A. & Roseberry, J.L. (1998) Deer management: our profession's symbol of success or failure? *Wildlife Society Bulletin* **26**(3): 515–21.

16

Can farming and wildlife coexist?

Ruth E. Feber, Elizabeth J. Asteraki and Les G. Firbank

What is a weed? A plant whose virtues have not yet been discovered.
(Ralph Waldo Emerson, *Fortune of the Republic*, 1878.)

Introduction

Energy for life is captured through photosynthesis. Agriculture involves capturing a proportion of this energy and diverting it to meet human needs, such as food or materials. This capture reduces the energy available to natural systems, creating an inherent competition for sunlight between agricultural production and the rest of biodiversity. In the long term, farming and wildlife can coexist only by recognizing and reconciling this inherent conflict of interest.

Biodiversity and the farmed landscape

The most obvious conflict is over the use of land. Agriculture has been expanding since the domestication of crop plants 10,000 years ago but, in the past three centuries, exponential human population growth has led to a 500% expansion in the extent of cropland and pasture world-wide. In Europe and North America, unchecked agricultural development has already transformed many natural habitats. The cultivation of agricultural export commodities has expanded rapidly within the developing world during the past half-century, notably for coffee, cocoa, sugar and (more recently) palm oil and soya, contributing to the current agricultural expansion in the tropics, resulting in unprecedented levels of habitat loss, particularly of tropical forests. Since this is where most of the world's biodiversity is found, there are huge implications for both wildlife populations and ecosystem functioning (Millenium Ecosystem Assessment 2005).

Forecasts by FAO indicate that the Earth could produce enough food and fibre to support the world's burgeoning population, but that this would come at considerable cost in terms of the environment. The Earth has about 4.2×10^9 ha (0.003% of total land area) of land that could, potentially, be used for arable cropping, but only around 1.5×10^9 ha are actually in use (FAO 2005). The proportions vary greatly across the globe, with north and east Asia, north Africa and North America already having over half their land in cultivation. Of the remaining 'potential' land, much is actually not available because it is already committed to other uses such as grassland, forestry or urban development. It has been estimated that at least an extra 120×10^6 ha will be needed by 2030 to meet the requirements of feeding the world's population and most of this land will have to be found in sub-Saharan Africa and central and southern Africa, where the use of 'potential' land is currently much lower. The greatest pressures are likely to be on forested areas, with their reduction leaving islands of remaining forested land on steep slopes and other inaccessible areas, within a matrix of agricultural land (Jenkins 2003).

While expansion of land, with the consequent loss of natural ecosystems to agriculture, is one option to meet the needs of a growing population, the other is to increase yields from land already under cultivation. This has already happened, to a dramatic extent, over the past four decades (Fig. 16.1a), accomplished largely through the use of chemical fertilizers and pesticides, irrigation, mechanization and crop breeding. Worldwide, agricultural policies have intensified farming in many countries,

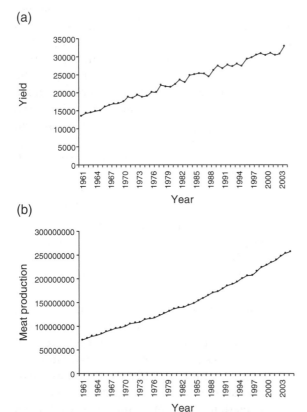

(a)

(b)

Fig. 16.1 Increases in (a) global cereal yields (hg ha^{-1}) and (b) global meat production (metric tons) 1961–2003. (Source: *FAO Statistical Databases*. Food and Agricultural Organization of the United Nations, Rome.)

with the result that much farmland has been environmentally degraded through, for example, soil erosion and lower soil fertility, pollution and eutrophication of rivers and lakes and pesticide bioaccumulation (Millenium Ecosystem Assessment 2005). Semi-natural habitat features have been lost from the farm landscape, particularly in arable areas, including strips of meadow, hedgerows, groves, small wetlands and tree stands along wetlands. One of the best studied taxonomic groups, birds, provides striking evidence of the toll that these changes have had across the temperate zone. Across Europe as a whole, almost 40% of bird species have small, declining or highly localized populations. Many of these species inhabit agricultural habitats and cannot be conserved solely within important sites such as nature reserves – indeed, around 90% of Europe lies outside such key sites. As well as habitat loss, agricultural intensification is also to blame in causing these declines among birds and other countryside biodiversity; 32% of Important Bird Areas (IBAs) in Europe are threatened by agricultural intensification (BirdLife International 2004). In the UK, research has shown that even relatively subtle changes, such as changes in cropping patterns (particularly a reduction in traditional rotations), spring rather than autumn sowing, hedgerow removal and pasture improvement are significant contributory factors in the huge recorded declines in farmland birds (Krebs et al. 1999, Donald et al. 2002).

Grassland as well as arable areas have undergone intensification; worldwide, meat production has more than trebled over the past 40 years (Fig. 16.1b). Many natural grasslands have been destroyed by cultivation or extensively modified by grazing from domesticated livestock or intensification. The range manager or pastoralist is typically keen to increase herbage production through, for example, the application of fertilizers, drainage or reseeding, but such intensification often leads to the loss of diversity of plants and invertebrates (e.g. Vickery et al. 2001). From a conservation perspective, grasslands are often important in terms of biodiversity, whether this is the large grazing mammals of the African savannas, or the plant diversity associated with traditional hay meadows in northern Europe, both of which depend critically upon the level of grazing or management. Direct conflicts with wildlife may also occur. India, for example, supports the world's largest livestock population (520 million; FAO 2002) and it is still increasing (6% increase between 1984 and 1994; WRI 1996). Wildlife reserves cover less than 5% of India's area, yet more than 3 million people are estimated to live in them. Livestock grazing is among the most widespread form of land use (an estimated three-quarters of Indian wildlife reserves support livestock populations); the evidence suggests high levels of resource competition between wildlife and livestock, with detrimental effects on wildlife populations (e.g. Charudutt et al. 2004).

The loss of land appears to be continuing unabated and, in many countries (particularly in the developing world), so does the intensification of existing farmland. Evidence for the devastating effects of these changes worldwide on other biota and ecosystems is now overwhelming (Matson et al. 1997; Millenium Ecosystem Assessment 2005). But creating a more sustainable future for agriculture raises unresolved questions about what we consider the relationship between farmed land and wildlife ought to be. In reality, perceptions of the importance of agriculture's role in delivering wildlife benefits vary greatly. In the USA and Canadian model, for example, large areas are dominated by intensive farming, both arable and livestock, with low associated biodiversity ('ecological sacrifice zones'; Jackson & Jackson 2002), while tracts of wilderness, particularly forested land, are preserved for biodiversity conservation.

In much of Europe, farming is more intimately linked with biodiversity, because of the long history of human modification of the entire landscape. The land has been farmed in ways that result in patchworks of habitats, each with their associated communities of plants and ani-

mals. In Britain, for example, hedgerows may retain the biodiversity of woodland edges, while cereal fields have integrated species of disturbed land with a flora and fauna introduced from steppe habitats further east. Suites of species that existed within the primary forest matrix have been inherited by grasslands and moulded into new assemblages according to details of soil, drainage and management. As Hails (2002) puts it, Europeans effectively live inside their national parks. Yet evidence from pollen records suggests that both the North American and the European landscapes were more similar than previously thought before human intervention and, as Sutherland (2002) suggests, lessons may be drawn from both models of nature conservation. Outside Europe, there are also examples of how tropical countryside of low to medium intensity of use may also function as important areas for biodiversity conservation; more than half Costa Rica's native bird species, for example, exist in largely deforested areas with remnants of native vegetation (Daily 2001), together with similar fractions of mammals and butterflies. Indeed, this argument is often used to justify a change of land use, despite total ignorance about the long-term viability of wildlife in such areas.

The costs of conservation

Increasingly, particularly in developed nations, agricultural policy is shifting away from maximizing production to a broader remit of sustainable rural development, which encompasses the protection of biodiversity. Sustainability involves preserving natural and social capital for future generations, rather than treating people or the environment as resources that can be used up. Although many (but not all) historic and traditional farming systems maintained levels of these types of capital effectively, current world food prices are simply too low, and trade too imbalanced, for this to continue. Instead, the economies of scale result in more food being produced more intensively by fewer producers, without bearing the environmental costs of loss of habitat and pollution (Pretty 1998; Pretty et al. 2000). However, focusing on the overall value of ecosystems will not only help preserve the integrity of ecosystem services, but may also be an effective way to preserve the diversity of species they contain (Edwards & Abivardi 1998).

If biodiversity and productive agriculture are to coexist, the costs of both need to be recognized. One option is to attempt to account for them through some form of payment system. The payment can be provided through a market mechanism or a subsidy, and may involve separating out biodiversity from agriculture or paying for some form of joint production. European agri-environment schemes are examples of subsidies intended to support both farm businesses and environmental objectives. Some, such as England's Countryside Stewardship (CS) scheme, have generated environmental gains (Carey et al. 2003), and options within the schemes undoubtedly have benefits for wildlife at the farm scale (see below). However, biodiversity benefits are often confounded with other environmental goods and services, especially landscape quality. In other European countries, evidence of the effectiveness of similar schemes for enhancing biodiversity is rather weak (Kleijn et al. 2001; Kliejn & Sutherland 2003) and scant attention has been paid to their cost-effectiveness (Watzold & Schwerdt 2005).

In the developing world, governments cannot afford to give agricultural subsidies and the domestic market is usually unable to support higher prices for local produce. The export market is the only opportunity for countries to recoup the real cost of production. In fact it is often argued that subsidies given to farmers in the developed world give them unfair market access, thus stopping developing countries from realizing much of their export potential. We need a better understanding of how much it really costs to enhance natural and cultural capital in rural systems, and mechanisms to make sure that these costs can be delivered to those people who have to pay. This is not easy in a marketplace that encourages payments to be driven down by externalizing costs and digging into natural capital.

Farming systems that, as well as providing environmental benefits, meet high social, animal welfare and food safety standards, are increasingly being marketed. Although organic production is the best known of these (see below), they also include Fair Trade, intended to give a fair price to local producers in the developing world, and a variety of much more local initiatives, such as the UK Countryside Agency's 'Eat the View'. The customer pays for a product that is perceived to be better, but also buys into a more sustainable vision of agriculture and rural development that encompasses healthy food, high landscape quality and animal welfare, as well as biodiversity. Such systems concentrate on reduced agrochemical inputs and greater emphasis on maintenance of diversity, and so it is reasonable to assume that they will be at least no worse than conventional systems. Some of the benefits that accrue from good practice can be taken up by other farming systems, whereas others might prove difficult to introduce to landscapes that have already been intensified, without expensive species reintroductions.

However, this vision hides a variety of tensions and conflicts. The outbreak of Foot and Mouth Disease among sheep in Britain in 2001 resulted in the closure of large areas of northern and South-western England to visitors, with catastrophic effects on the tourist economy. It clarified that the main economic role of agriculture in this area was to maintain the landscape quality needed for a successful tourism sector – an important lesson for many other parts of the world, where traditional landscapes are under threat from agricultural intensification or abandonment. However, there was anecdotal evidence that the reduction in tourists enabled a range of upland birds to breed more successfully than usual. The interactions between local business, agriculture, land management and biodiversity can be complex.

Moreover, more 'environmentally friendly' systems can result in the export of environmental damage in unexpected and perverse ways. For example, large areas of Brazilian rain forest were being cleared to supply the European market with GM-free soya, which is perceived by the consumer to be more environmentally friendly than its counterpart grown in the agricultural landscapes of US and Argentina (Vidal 2003).

Money alone is not always able to solve the problem of maintaining less intensive agriculture. Increasingly, young people are leaving the land by choice to pursue employment that provides them with a more appealing lifestyle so, as older farmers retire, the farm is no longer automatically handed down to the children. Consequently, it may be absorbed into larger units, or abandoned altogether. These changing social pressures are leading to systems such as mountain meadows, agroforestry and olive groves degrading rapidly across many parts of Europe, with loss of biodiversity and landscape quality (MacDonald et al. 2000). It is also true of the developing world, where the lure of the cities and modern life is resulting in an ever-aging population of farmers. It is unreasonable to deny the children of farmers the same opportunities that urban dwellers have, but much more can be done to encourage sustainable farming as a rewarding and attractive career option for the next generation and the situation provides a crucially important opportunity to rethink how farmed landscapes could be configured in a way that sustains both production and biodiversity.

Farmers as custodians of the countryside

The potential for agricultural landscapes to integrate food production and wildlife conservation within a single land management system is increasingly recognized as a viable option, but much depends on farmer attitudes. In the West, the farmer's role as producer alone is un-doubtedly changing; pressures from both within and outside the industry are placing much more emphasis on farmers to not only avoid damaging behaviour, but also adopt positive conservation management. Using questionnaire surveys of farmers in 1981 and 1998, Macdonald & Johnson (2000) found that the proportion of farmers who claimed to be 'very interested' in wildlife had increased from 40 to 62% over this period. These results illustrate the general view that, in Europe at least, conservation is becoming socially acceptable within farming communities. As a result, farmers' behaviour is becoming more integrated with the ecology of the land on which they live and work (Stoate 2002). The reasons for these changes are complex, but greater awareness through the media of the effects of farming on wildlife, high-profile meetings with high-profile outcomes such as the Earth Summit in Rio in 1992, and greater accountability with regard to the quality and origin of food have all probably played a part. Agri-environment schemes, too, may perform a role in influencing the attitudes of farmers towards environmental management. For example, Battershill & Gilg (1996) found that 'traditional' farmers were more reluctant than 'commercial' farmers to participate in agri-environment schemes, although, having done so, 56% of the traditional farmers said that the schemes had a positive effect on their attitude to conservation. The interaction of cultural values with financial considerations is of key importance in influencing the conservation behaviour of farmers and thus the future of the countryside.

In some instances it may take years of concerted effort by conservation organizations to change farmers' perceptions of the impacts of their industry on the environment. An example of this is presented by the oil palm industry. This, the world's second largest edible oil crop, next to soya oil, is grown mainly in Malaysia and Indonesia. Worldwide, 154 million metric tonnes of palm fruit producing 29 million metric tonnes of oil were produced in 2004. The WWF and other NGOs have, in recent years, been instrumental in rallying

European consumers to demand that palm oil be produced in an environmentally friendly manner. Concerns about the palm oil industry's impact on the environment intensified in 1997 and 1998 when South-east Asia was blanketed in thick smog because of massive land-clearing fires set by oil palm growers in Malaysia and Indonesia. Although NGOs have criticized the palm oil industry for damaging the environment, producers have insisted that they have adhered to sound practices. An attempt to break the deadlock has been made by bringing all interested parties together in a global initiative for sustainable palm oil. The Roundtable on Sustainable Palm Oil (RSPO) is an association created by organizations carrying out their activities in and around the entire supply chain for palm oil to promote the growth and use of sustainable palm oil through co-operation within the supply chain and open dialogue with its stakeholders. One of their key objectives is to research and develop definitions and criteria for the sustainable production and use of palm oil. However, the initiative is in danger of stalling before anything tangible is achieved. Although the industry feels the pressure from NGOs and consumers to become more sustainable, it has not yet suffered financially. If any real change is to take place, the sustainability criteria being developed need scientific verification and widespread adoption. In order to do this, the industry has to be willing to divert some of their profits to furthering the work of the RSPO.

Involving farmers at all levels of the debate will result in much more positive outcomes for sustainable methods of farming (the concept of 'social capital': Pretty & Smith 2004). Farmer education has a particularly important role to play. One example of this is provided by a major world commodity, cocoa, which is grown on over 7 million hectares. Cacao cultivation is concentrated in a band 15° north and south of the Equator and encompasses seven biodiversity hotspots (as defined by Conservation International). Almost 90% of world cocoa production is by smallholders (2.5 million).

However, farmers are moving away from traditional shade cacao to more intensive production. Also, farmers are expanding production into forest areas because of declining yields, due to environmental degradation of existing land. Shade cacao can provide a critical habitat for plants and animals. A successful project in Kakum National Park, Ghana showed that, with some basic training in low-input technologies and shade cacao best practices, farmers not only improved their incomes and the quality of their environment but also developed better environmental awareness and improved their understanding of ecological interactions at the farm field level. Another project in Sulawesi, Indonesia, involving industry and technical partners (SUCCESS Alliance of the ACDI/ VOCA) has trained more than 37,000 farmers in integrated pest management. Crop losses have dropped by nearly 30%, and incomes have increased by an average of $541 per year (ACDI/VOCA 2005).

Local solutions, particularly in areas of high biodiversity, can be extremely effective at mitigating conflict between farming and wildlife. For example, in Namibia, 95% of cheetahs live outside protected areas on commercial livestock farmland where they are perceived by farmers to be a significant cause of livestock losses (Marker et al. 2003). As removal of wildlife that predates on stock is one widely accepted practice within traditional methods of livestock husbandry, providing alternative management methods for farmers has become a major component of predator conservation strategies. One particularly interesting example is the use of livestock-guarding animals. Although livestock guarding dogs have been used in Eurasia for thousands of years, this practice had been all but forgotten until the 1970s. When Anatolian shepherd dogs were given to Namibian farmers, they proved to be very protective, with over 76% of farmers reporting an almost 90% decrease in livestock losses after placement of the dogs (Marker et al. 2005). The high level of satisfaction by the farmers towards the dogs, coupled with the dramatic decline in livestock losses, may result in Namibian farmers regarding predators as less of a threat on their land and make them less inclined to remove them. Other examples are given elsewhere in this volume (e.g. Chapter 17).

The role of research

Although economic and social factors have a crucial influence on patterns of agricultural production, intensification of agricultural methods across the globe have been made possible by advances in science and technology. In a change of emphasis since the Green Revolution, much research is now directed at reducing, or at least controlling, impacts on the environment, while continuing to achieve acceptable levels of production, rather than maximizing production at any cost (Tilman et al. 2001).

The research areas that seek to tackle these issues are diverse. For example, advances in plant breeding and biotechnology aim to improve the fundamental efficiency of crop nitrogen, phosphorous and water use, while methods of precision agriculture that will decrease inputs of nitrogen and phosphorous, and new methods for managing soil to reduce nutrient leaching and optimize soil fertility, are being developed. Ways to better control crop pathogens and pests, such as a greater use of natural enemies and the use of diverse cropping systems, are being investigated, as are methods to forecast quantitatively the impact on ecosystem functioning of loss of habitat, loss of biodiversity, changes in species composition and increased nutrient inputs.

Although the practical priorities on the agenda for research into conservation on farmland should not be tied to short-term issues related to the vagaries of policy, it is crucial that biodiversity research should be used to underpin policy decisions about changing farming practice. The issue of genetically modified (GM) crops in the UK provides evidence that

this approach is beginning to be adopted. Several crops (beet, maize and oilseed rape) have been modified to be resistant to a broad-spectrum herbicides, raising concerns that their management will accelerate declines of arable weeds that are important food items for birds. These concerns were tested in Government funded Farm Scale Evaluations, in which fields were split and half sown with GM crops, and half with conventional varieties, each managed according to commercial practice. Indeed, three of the GM crops (beet, winter and spring oilseed rape) were shown to reduce seed production by broad-leaved weeds, while the fourth crop, maize, resulted in an increase in seed production (Firbank et al. 2003a; Bohan et al. 2005). Such experiments allow new developments in agriculture to be tested before their large-scale introduction.

It seems unlikely that the potential of these scientific advances to benefit biodiversity can be fully exploited without detailed knowledge of the ways in which individual farming practices affect the flora and fauna of farmland. Such studies form the starting points from which to expand knowledge of the interrelationships between farming practice and the distribution and abundance of the animals and plants inhabiting the rural landscape. A key issue is that of scale. Research to date shows how biodiversity at the landscape scale is an emergent property of processes at smaller scales, and these processes are likely to be general (Macdonald et al. 2000). It is possible to create habitats for wildlife in farmland at a wide range of scales, from small patches of land within fields or at field margins, through whole-field management (e.g. set-aside or organic farming), to creating habitat networks across entire farmed landscapes, right through to partitioning blocks of land to wildlife and to intensive agriculture. For example, careful targeting of agri-environment scheme options across a landscape may increase their potential benefit for wildlife through increasing connectivity of habitats. In a partnership project across the Chichester flood plain in the UK, farmers have been encouraged to participate in agri-environment scheme agreements to benefit the rapidly declining water vole (Strachan & Holmes-Ling 2003). The water vole has declined in numbers by 95% in Sussex over the past 20 years, as a result of degradation of habitats and predation by introduced American mink. Instead of viewing farms as individual entities, the project considered the whole landscape of the coastal plain in terms of land-use and river catchment management. Through the delivery of whole-farm conservation plans that addressed the need for habitat creation and enhancement on individual farms, the partnership was able to create linked habitat corridors across and between neighbouring farms. In the first year of the project, water voles were found to be scarce within the coastal plain area, with the populations highly fragmented (Fig. 16.2a) but, 3 years later, the species had responded so well to the various habitat enhancements on farmland that populations had more than tripled (Fig. 16.2b).

Market forces are modified by government subsidies and tax incentives, which, together with social change, determine patterns of agricultural activity. However, conservation planning is often at inappropriate scales in relation to these patterns to protect biodiversity effectively. The focus for the future should be to identify and to integrate key observations over the range of relevant scales, and to move towards holistic and predictive models of farmland biodiversity. The adoption of a range of approaches, from large-scale manipulative experiments to detailed autecological studies, remains the strategy most likely to be successful in achieving this aim.

Changing farming systems

Since the late 1980s, environmental benefits from European agricultural policy have been delivered through agri-environment schemes. Agri-environment schemes have a number of objectives, such as enhancing landscapes, maintaining historical interest, and encouraging access to the countryside, but one key aim of the schemes is to encourage farmers to adopt more environmentally sensitive farming practices in order to promote farmland biodiversity. This is achieved through financial incentives to improve habitat management and create new habitats, with the aim of halting or reversing many of the changes that have been associated with declines in farmland wildlife.

One example is the widespread adoption of field margin restoration and management, which has arisen out of a combination of existing farm practice, research and policy development linked closely to these factors. Conservation Headlands were originally developed to increase the abundance of invertebrate food for gamebird chicks reared in arable field edges (Sotherton 1991), but other wildlife were found to be indirect beneficiaries of this commercial interest. At around the same time, burgeoning research into the management of uncropped arable field margins showed that relatively simple changes could result in radical improvements to the habitat for biodiversity (e.g. Feber et al. 1996; Fig. 16.3). As a result, field margin options are now included in the majority of agri-environment schemes in the UK and across Europe and North America. They have been adopted as one of the most widespread approaches for biodiversity enhancement on arable farmland in the developed world.

It does not always have to be the case that the preservation of biodiversity is the primary driver for policies that have successful conservation outcomes. One of the most important, long-term, large-scale environmental changes to occur in European arable landscapes in recent years has been the introduction of set-aside land. Set-aside was a major new policy, introduced in 1992 and aimed at manipulating (reducing) agricultural production within the European Union (EU); in 2003, 6.3 million hectares were in set-aside in the EU. Essentially, farmers are required to set aside a pro-

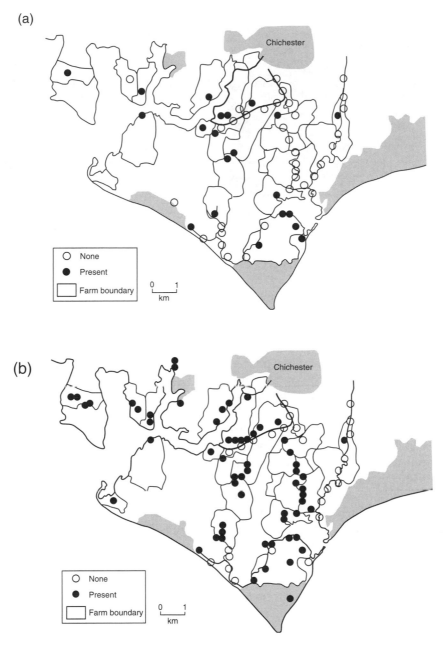

Fig. 16.2 Distribution of water vole colonies across the Chichester coastal plain farms, UK, in (a) 2000 and (b) following habitat restoration through agri-environment scheme agreements, in 2003. The most easterly watercourse (Pagham Rife) was a control area receiving little or no proactive management for water voles.

portion of their land from arable production each year by establishing a sown or naturally regenerated green cover over winter. It soon became clear that one consequence of set-aside was a benefit to wildlife (Firbank et al. 2003b), particularly farmland birds (Wil-

Fig. 16.3 Options for arable field margin management are now included in the majority of agri-environment schemes in Europe.

son et al. 1995). Through a process of research and monitoring, rules relating to set-aside management were modified to optimize set-aside as wildlife habitat, with substantial gains for biodiversity. Although set-aside may be phased out in the future, it shows that an agricultural policy primarily designed for another purpose can deliver environmental benefits.

One way in which biodiversity may be increased is through organic agriculture, which is supported by policies in many countries in the developed world. Organic farms demonstrate features that are now rare in northern European farming systems, such as crop rotations incorporating grass leys, exclusion of synthetic pesticides and fertilizers, and reliance on animal and green manures. Compared with non-organic farms they may also contain larger amounts of uncropped habitats such as hedgerows. All of these features may result in higher levels of habitat heterogeneity than non-organic farms, and the evidence suggests that there are indeed benefits for a wide range of taxa, including birds, bats, invertebrates and plants (Fuller et al. 2005; Hole et al. 2005). Demand for organic produce has increased, particularly in the UK, as a result of highly publicized food scares. Issues such as BSE (bovine spongiform encephalopathy) infected beef, and the foot-and-mouth outbreak,

which may or may not be direct consequences of intensive farming methods, have raised awareness about how our food is produced, the quality of produce we are willing to accept, and the price we are prepared to pay for it. For farmers outside the developed world, though, organic farming is still in its infancy. Often countries have not yet developed local certification schemes and the costs of certification by international bodies (e.g. IFOAM; International Federation of Organic Agriculture Movements) are often prohibitive. Usually the domestic market is unable to pay a premium for organic produce, so a farmer must find a buyer overseas. Much more could be done by certification bodies in the developed world to help developing countries not only to successfully grow organic produce but export it to lucrative markets.

Nonetheless, there are many other examples around the world where farmers have retreated from practices that destroy the capacity for food production towards systems that conserve soil, environmental and human resources, using strategies ranging from adding new elements to the farming system to introducing measures to conserve water and other natural resources (Pretty et al. 2003). In Thailand, for example, farmers have been developing farming systems that integrate annual rice crops with vegetable plots and tree crops alongside fish ponds. By emphasizing local species within carefully managed successions, the farmed landscape becomes much more complex, with elements of secondary forests. In China, traditional agroforestry systems are estimated to cover 45 million ha, with agrosilviculture being a dominant practice and aquasilvicultures, e.g. tree–fish–arable crop and tree–fish–livestock systems, as alternatives for land use in the wetlands (Huang et al. 1997). Compared with monocultures, well-managed agroforestry systems have many benefits, including the conservation of biodiversity. However, one must be careful not to think that a return to traditional farming is the 'magic bullet' in terms of conservation. Neither traditional nor wildlife-friendly farming can ever produce enough food to feed the global population. Although these approaches offer scope to increase the biodiversity value of farmed land on a per unit area basis, they may not result in a net benefit to biodiversity if they reduce crop yield. Recent modelling exercises using data from a range of taxa in developing countries suggest that high-yield farming may, overall, allow more species to persist, by sparing land which retains higher species biodiversity (Green et al. 2005).

Consumer choice

Consumer choice can have an important influence on the agricultural industry, particularly in the developing world. The coffee industry has huge significance both for livelihoods and biodiversity, covering just over 10 million ha worldwide (equivalent to the size of Iceland). The main production areas are in tropical rainforest zones, but the impact on biodiversity varies enormously depending on the type of production system used. Traditional low-intensity methods involve the planting of coffee bushes under a selectively thinned canopy of rainforest trees. Owing to their structural and floristic complexity, these shaded systems can support a high level of biodiversity (e.g. Moguel & Toledo 1999) and, of particular interest to Northern consumers, provide an important overwintering habitat for migrant bird populations. By contrast, sun coffee is labour and chemically intensive, is increasingly reliant on high yielding varieties, causes high rates of soil erosion, and reduces overall biological diversity (Fig. 16.4). Without the food and shelter that overstorey trees can provide, sun coffee supports up to 90% fewer bird species and 43% less mammalian diversity (Gallina et al. 1996). The conversion of shade to sun coffee occurred mainly as a short-term response to a fungal disease outbreak in the 1970s. Over the past few decades, in the drive to raise yields for export,

(a)

(b)

Fig. 16.4 The contrast between (a) shade and (b) sun coffee.

this conversion has continued, with over half the coffee plantations in Latin America using the sun method. For many farmers, the high productivity associated with sun coffee plantations outweighs the disadvantages of having to use large amounts of commercial fertilizers and chemical pesticides. Biodiversity is often sacrificed for the higher yields that allow producers to sell sun coffee at a cheaper price.

However, consumers can exert an influence on the type of production method used and hence the biodiversity it supports. For example, 'Bird Friendly' coffee is emerging as a niche market sector in America, Europe and Japan. 'Bird Friendly' has been trade-marked by the Smithsonian Migratory Bird Center to describe coffee grown in Latin America under defined environmental criteria, including a minimum percentage of shade cover and use of organic methods. Another scheme, Eco-OK, developed by the Rainforest Alliance and a network of Latin America environmental organizations, takes a more holistic approach, looking at minimum numbers of trees, the use of agro-chemicals, water resources, soil and waste management, hunting, working conditions and community relations. Although consumers pay what seem like premium prices for these products (shade coffee is usually more expensive to produce owing to lower yields and set up costs), many are willing to pay more to support a more environmentally and socially acceptable product. Connoisseurs say its taste is superior to sun coffee because the beans ripen more slowly. Nonetheless, although the rapid growth of these sectors in recent years is encouraging, they remain a tiny proportion of the wider coffee trade. Organic coffee, for example, accounts for just 0.9% of the USA coffee market (Gooding 2004). This may be because many farmers find the preparation and compliance for shade-grown certification just too difficult and costly and we, as consumers, are not prepared to pay the price.

Conclusions

As agricultural production requires more land and becomes more intensive to meet the needs of the world's increasing human population, space and solar energy is diverted away from the natural world towards farming. Finding strategies to reconcile the demands of food production, environmental services, rural development and the preservation of biodiversity will be helped by devising not only national, but international, links between the different components of these problems, blending the disciplines of economics, social and biological sciences, and bringing about the realization that environmental issues need to be addressed not only by land managers, policy makers and conservationists, but by everyone. It is all too easy to produce rural deserts that provide food at the lowest cash cost, ignoring the costs of damage to the environment, but this is not sustainable over the long term. By contrast, we could have farmed landscapes that, in varying forms, are rich in wildlife, beauty and heritage, but only if the right social and economic structures are in place.

Farming and biodiversity can coexist, and we have reviewed only a small number of the many examples worldwide of how such coexistence can be achieved at the levels of individual farms. However, the more serious challenge is to deliver increasing global food supplies while enhancing both rural livelihoods **and** the diversity and abundance of plants and animals. An integrated approach to planning, managing and valuing land resources is perhaps the only way to implement solutions that will allocate land to uses that provide the greatest sustainable benefit. This will involve challenging exercises in prioritization – as is characteristic of all issues in conservation. The real questions are much less about how to develop prescriptions for increasing numbers of skylarks on farmland, and much more about finding ways that farm businesses around the world can benefit appropriately from managing landscapes that provide a host of environmental services, including food, clean air and water, landscape quality as well as biodiversity. Both our fate, and the fate of the wildlife around us, depends on it.

Live as if you'll die tomorrow, but farm as if you'll live forever.

(Traditional farming proverb.)

References

ACDI/VOCA (2005) Agricultural Cooperative Development International and Volunteers in Overseas Cooperative Assistance, Washington, DC. http: //www.acdivoca.org/acdivoca/ Acdi-web2.nsf/1c326a779330eaac85256ac3005fc723/ f46482c96d1f658785256d3c006b397a?OpenDocument [Accessed 25 November]

Battershill, M.R.J. & Gilg, A.W. (1996) Traditional farming and agro-environment policy in southwest England: back to the future? *Geoforum* **27**: 133–47.

BirdLife International (2004) *State of the World's Birds 2004: Indicators for our Changing World.* BirdLife International, Cambridge.

Bohan, D.A., Boffey, C.W.H., Brooks, D.R., et al. (2005) Effects on weed and invertebrate abundance and diversity of herbicide management in genetically modified herbicide-tolerant winter-sown oilseed rape. *Proceedings of the Royal Society of London Series B – Biological Sciences* **272**: 463–74.

Carey, P.D., Short, C., Morris, C., et al. (2003) The multi-disciplinary evaluation of a national agri-environment scheme. *Journal of Environmental Management* **69**: 71–91.

Charudutt, M., van Wieren, S.E., Ketner, P., Heitkonig, I.M.A. & Prins, H.H.T. (2004) Competition between domestic livestock and wild bharal *Pseudois nayaur* in the Indian Trans-Himalaya. *Journal of Applied Ecology* **41**: 344–54.

Daily, G.C. (2001) Ecological forecasts. *Nature* **411**: 245.

Donald, P.F., Pisano, G., Rayment, M.D. & Pain, D.J. (2002) The common Agricultural Policy, EU enlargement and the conservation of Europe's farmland birds. *Agriculture, Ecosystems and Environment* **89**: 167–82.

Edwards, P.J. & Abivardi, C. (1998) The value of biodiversity: where ecology and economy blend. *Biological Conservation* **83**: 239–46.

FAO (2002) *FAO Statistical Databases.* Food and Agricultural Organization of the United Nations, Rome. http: //apps.fao.org [Accessed 12 May]

FAO (2005) *Terrastat Database.* Food and Agricultural Organization of the United Nations, Rome. http: //www.fao.org/ag/agl/agll/terrastat/ [Accessed 15 April]

Feber, R.E., Smith, H. & Macdonald, D.W. (1996) The effects on butterfly abundance of the management of uncropped edges of arable fields. *Journal of Applied Ecology* **33**: 1191–205.

Firbank, L.G., Perry, J.N., Squire, G.R., et al. (2003a) *The Implications of Spring-sown Genetically Modified Herbicide-tolerant Crops for Farmland Biodiversity: a Commentary on the Farm Scale Evaluations of Spring Sown Crops.* Department for Environment, Food and Rural Affairs, London.

Firbank, L.G., Smart, S.M., Crabb, J., et al. (2003b) Agronomic and ecological costs and benefits of set-aside in England. *Agriculture Ecosystems and Environment* **95**: 73–85.

Fuller R.J., Norton, L.R., Feber, R.E., et al. (2005) The scale of benefits to biodiversity from organic farming. *Biology Letters* **1**: 431–4.

Gallina, S., Mandujano, S. & Gonzalez-Romero, A. (1996) Conservation of mammalian biodiversity in coffee plantations of Central Veracruz, Mexico. *Agroforestry Systems* **33**: 13–27.

Gooding, K. (2004) *Sweet like Chocolate? Making the Coffee and Cocoa Trade Work for Biodiversity and Livelihoods.* Royal Society for the Protection of Birds, Sandy, and Birdlife International, Cambridge.

Green, R.E., Cornell, S.J., Scharlemann, J.P.W. & Balmford, A. (2005) Farming and the fate of wild nature. *Science* **307**: 550–5.

Hails, R.S. (2002) Assessing the risks associated with new agricultural practices. *Nature* **418**: 685–8.

Hole, D.G., Perkins, A.J., Wilson, J.D., Alexander, I.H., Grice, P.V. & Evans, A.D. (2005) Does organic farming benefit biodiversity? *Biological Conservation* **122**: 113–30.

Huang-W., Kanninen-M., Xu-Q. & Huang, B. (1997) Agroforestry in China: present state and future potential. *Ambio* **26**: 394–8.

Jackson, D.L & Jackson, L.L (2002) *The Farm as Natural Habitat: Reconnecting Food Systems with Ecosystems.* Island, Washington, DC.

Jenkins, M. (2003) Prospects for biodiversity. *Science* **302**: 1175–7.

Klein, D. & Sutherland, W.J. (2003) How effective are European agri-environment schemes in conserving and promoting biodiversity? *Journal of Applied Biology* **40**: 947–69.

Kleijn, D., Berendse, F., Smit, R. & Gilissen, N. (2001) Agri-environment schemes do not effectively protect biodiversity in Dutch agricultural land-scapes. *Nature* **413**: 723–5.

Krebs, J.R., Wilson, J.D., Bradbury, R.B. & Siriwardena, G.M. (1999) The second silent spring. *Nature* **400**: 611–2.

MacDonald, D., Crabtree, J.R., Weisenger, G., et al. (2000) Agricultural abandonment in mountain areas of Europe: environmental consequences and policy response. *Journal of Environmental Management* **59**: 47–69.

Macdonald, D.W. & Johnson, P.J. (2000) Farmers and the custody of the countryside: trends in loss and conservation of non-productive habitats 1981–1998. *Biological Conservation* **94**: 221–34.

Macdonald, D.W., Feber, R.E., Johnson, P.J. & Tattersall, F. (2000) Ecological experiments in farmland conservation. In *The Ecological Consequences of Environmental Heterogeneity* (Eds M.J. Hutchings, E.A. John, A.J.A Stewart). British Ecological Society Symposium Series, Blackwell Scientific Publications, Oxford.

Marker, L.L., Mills, M.G.L. & Macdonald, D.W. (2003) Factors influencing perceptions and tolerance toward cheetahs on Namibian farmlands. *Conservation Biology* **17**: 1290–8.

Marker, L. L., Dickman, A. J. & Macdonald, D. W. (2005) Perceived effectiveness of livestock guarding dogs placed on Namibian Farms. *Rangeland Ecology and Management* **58**(4): 329–36.

Matson, P.A., Parton, W.J., Power, A.G. & Swift, M.J. (1997) Agricultural intensification and ecosystem properties. *Science* **277**: 504–9.

Millennium Ecosystem Assessment (2005). *Synthesis Report*. Island Press, Washington.

Moguel, P. & Toledo, V.M (1999) Biodiversity conservation in traditional coffee systems of Mexico. *Conservation Biology* **13**: 11–21.

Pretty, J.N. (1998) *The Living Land*. Earthscan, London.

Pretty, J. & Smith, D. (2004) Social capital in biodiversity conservation and management. *Conservation Biology* **18**: 631–8.

Pretty, J.N., Morison, J.I.L. & Hine, R.E. (2003) Reducing food poverty by increasing agricultural sustainability in developing countries. *Agriculture, Ecosystems and Environment* **95**: 217–34.

Pretty, J.N., Brett, C., Gee, D., et al. (2000) An assessment of the total external costs of UK agriculture. *Agricultural Systems* **65**: 113–36.

Sotherton, N.W. (1991) Conservation headlands: a practical combination of intensive cereal farming and conservation. In *The Ecology of Temperate Cereal Fields* (Eds L.G. Firbank N. Carter, J.F. Darbyshire & G.R. Potts), pp. 373–97. Blackwell Scientific Publications, Oxford.

Stoate, C. (2002) Behavioural ecology of farmers: what does it mean for wildlife? *British Wildlife* **13**: 153–9.

Strachan, R. & Holmes-Ling, P. (2003) *Restoring Water Voles and other Biodiversity to the Wider Countryside: a Report on the Chichester Coastal Plain Sustainable Farming Partnership*. Wildlife Conservation Research Unit, University of Oxford.

Sutherland, W.J. (2002) Openness in management. *Nature* **418**: 834–5.

Tilman, D., Fargione, J., Wolff, B., et al. (2001) Forecasting agriculturally driven global environmental change. *Science* **292**: 281–4.

Vickery, J.A., Tallowin, J.T., Feber, R.E., et al. (2001) The management of lowland neutral grasslands in BritaIn effects of agricultural practices on birds and their food resources. *Journal of Applied Ecology* **38**: 647–64.

Vidal, J. (2003) Disappearance of Amazon rainforest brings pledge of emergency action. *The Guardian* **28 June**.

Watzold, F. & Schwerdt, K. (2005) Why be wasteful when preserving a valuable resource? A review article on the cost-effectiveness of European biodiversity conservation policy. *Biological Conservation* **123**: 327–38.

Wilson, J., Evans, A., Poulson, J.G. & Evans, J. (1995) Wasteland or oasis? The use of set-aside by breeding and wintering birds. *British Wildlife* **6**: 214–23.

WRI (1996) *World Resources Report 1996*. World Resources Institute, Washington, DC.

17

Living with wildlife: the roots of conflict and the solutions

Claudio Sillero-Zubiri, Raman Sukumar and Adrian Treves

Where there are sheep, the wolves are never very far away.

(Titus Plautus, 254–184 bc)

Introduction

Wildlife, particularly carnivores, ungulates, primates, rodents, raptors, granivores and piscivorous birds, come into conflict with people when they damage property or threaten human safety or recreation by feeding (killing, browsing, grazing), digging and burrowing. A further reason for conflict is that wildlife are carriers of diseases that can be harmful to people and their domestic animals (see Chapter 11). Because conflict applies, in one guise or another, to all sorts of organisms, including invertebrates and even plants, we opt for Conover's (2002) taxonomically mysterious, but nonetheless alluringly convenient, definition of wildlife as free-ranging vertebrates other than fish. That said the principles and dilemmas that we shall reveal here for creatures from sparrows to elephants can almost universally be transposed to humankind's dealings with organisms of every ilk.

In response to perceived wildlife damage or threat, people may retaliate in a manner that may be ineffective or biologically unsustainable, and political discord may ensue between those whose emphasis is conservation of biodiversity and/or the sustainable use of resources, and those defending the economic interests of affected people. In particular, people at the receiving end of wildlife damage tend to oppose conservation agendas, protected areas and conservation practitioners. Hence, the management of wildlife populations involved in conflict raises numerous issues relating to conservation, perceptions of nature, animal welfare, and the politics and economics of natural resources.

Conservationists face a critical challenge to develop workable measures for reconciling human activities and wildlife needs as a deliberate choice (as opposed to earlier views that were polarized between support for either wildlife or people and economic development), and thus minimize the severity or frequency of conflicts for both animals and people. There are strong economic and human health arguments for reducing the costs of plentiful species such as granivorous birds and rats threatening people's lives and livelihoods. Similarly, there are equally strong ethical arguments in favour of preserving species that are threatened as a consequence of human activity. Somewhere in between we may consider conflict between different sectors of society regarding a particular use of wildlife, such as town and country antagonizing over fox hunting in Britain or supporters of trophy hunting clashing horns with animal welfare flag-bearers.

We thus need general guidelines that can be tailored to each problem, gleaned from experience worldwide, and local strategies that integrate ecological, economic and social realities in the design and implementation of cost-effective interventions that can be monitored. In this essay we briefly review the patterns of human–wildlife conflict and the most commonly used approaches, or tentative solutions showing some promise, to its management. We then focus on gaps in our understanding that impede progress in mitigating human–wildlife conflicts as well as socio-political barriers to innovation that frustrate biodiversity conservation. For example, we still do not know if wild animals with a tendency to damage property or threaten human activities transmit these behaviours to their young, which hampers our analysis and use of negative conditioning from deterrents to lethal control. Affected people can also befuddle conservationists, as exemplified by the common claim that livestock loss to carnivores is more than economic because livestock producers love the animals they annually take to slaughter.

Characterizing conflict

Conflicts between wildlife and humans cost many lives, both human and wildlife, threaten the livelihoods of millions worldwide and jeopardize long-term conservation goals such as securing protected areas and building constituencies in support of wildlife conservation (Sukumar 1994; Treves & Newton-Treves 2005). Elephants, hippopotami, buffaloes, large carnivores (particularly bears and big cats) and crocodiles account for most human deaths or injury; the vast majority of attacks befall people harvesting resources from wildlife areas and those defending their farms from crop raiders (e.g. Treves & Naughton-Treves 1999; Rajpurohit & Krausman 2000). Wildlife dam-

age is widespread; in the USA, for instance, 80% of 2000 farmers surveyed suffered some damage, with 3% reporting losses in excess of $10,000 (Conover 2002). The federal agency charged with controlling agricultural damages caused by wildlife in the USA spent over $60 million in operations during 2000 and the agriculture industry estimated losses at nearly one billion dollars (National Agricultural Statistics Service 2002; see Breitenmoser & Angst (2001) for similar statistics for Europe). In communities with subsistence economies, even small losses can be economically important (e.g. Asian elephants – Sukumar 1989; African elephants – Naughton-Treves et al. 2000; snow leopards – Oli et al. 1994). Conflict sometimes may arise from unexpected quarters, such as tourists feeling threatened by begging macaques in China (and even one tourist dying as a result of a fall when fleeing a macaque – Zhao 1991), martens foraging under vehicle bonnets for plastic wiring in Germany, or fouling by pigeons in London almost bringing about the political downfall of the mayor.

Historically, and still largely today, solutions that are lethal to wildlife have been sought through bullets, poison or traps (Treves & Naughton-Treves 2005). This response is increasingly unpopular or illegal so interest has awakened in non-lethal techniques. In the past, one or two questions have not been answered about lethal versus non-lethal control: first, what is the magnitude of the problem relative to the proposed solution, second, how do lethal versus non-lethal alternatives measure up in cost-effectiveness, sustainability or socio-political acceptability? Furthermore, as values, especially of nature, are increasingly weighed with more than monetary dimensions, these questions, which were always technically difficult to answer, become intellectually and ethically hard too. For example, however threatening a predator or crop raider may be, and whether or not it costs you money, and irrespective of whether killing it diminishes your loss, in a world where biodiversity (and especially rarity) is valued, and suffering decried, is lethal control the best choice? (Fig. 17.1 also Chapter 18).

Predation on farm animals, game and fisheries

The most widespread source of human–carnivore conflict is competition for resources.

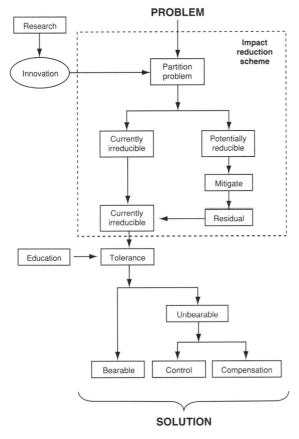

PROBLEM

Research

Innovation

Impact reduction scheme

Partition problem

Currently irreducible

Potentially reducible

Mitigate

Currently irreducible

Residual

Education → Tolerance

Unbearable

Bearable   Control   Compensation

SOLUTION

Fig. 17.1 Impact reduction scheme to mitigate conflict representing operational and iterative processes flowing from problem to solution. Problems can be partitioned notionally between reducible and irreducible elements, and the balance between these will shift as currently intractable elements are rendered reducible by new innovation. There could be overlap in the actions represented by the 'mitigation' and 'control' boxes, but these may loosely be partitioned as non-lethal and lethal interventions, respectively. (Redrawn from Macdonald & Sillero-Zubiri 2004.)

Wherever people exploit natural populations, rear livestock, game or fish outdoors, predation is a perennial and controversial complaint (reviewed in Gittleman et al. 2001; Sillero-Zubiri & Laurenson 2001; Conover 2002; Treves & Karanth 2003; Sillero-Zubiri et al. 2004; Woodroffe et al. 2005). The history of this conflict is the root of a deeply ingrained antipathy towards wild carnivores throughout the world that traces back to the development and spread of herding societies (Reynolds & Tapper 1996) and perhaps even further back in prehistory (Kruuk 2002). Domestication, via selection against 'wild' behaviours in stock, led to riches of clustered, accessible, unfit and generally dim-witted prey for opportunistic carnivores (Hemmer 1990).

The ecology of predation is an extremely complex issue. Recent analyses suggest predators can limit prey numbers or exert compensatory mortality depending on a complex array of environmental variables that defy global generalizations (Ray et al. 2005). Habitat loss and fragmentation, along with poaching and competition with domestic livestock can deplete the natural prey base (e.g. Saberwal et al. 1994; Mishra 1997; Jackson & Wangchuk 2001; Mishra et al. 2003), forcing predators to turn to domestic stock for food. The shifting balance of availability of livestock and natural prey can shift predator preferences and incidences of depredation (e.g. Meriggi & Lovari 1996; but see Treves & Naughton-Treves (2005) for some counterexamples).

Predation on domestic stock is affected by breed, stock management, the prey's previous enxperience of predators, predator density and individual predator behaviour (Jackson & Wangchuk 2001; Wydeven et al. 2004). Although larger carnivores are more conspicuous and attract particular wrath, the collective damage of smaller species such as jackals, foxes, coyotes, mustelids and small cats may be greater (e.g. Naughton-Treves 1998; Macdonald & Sillero-Zubiri 2002; Marker et al. 2003). Conflict with carnivores extends to other, 'non-traditional', stock such as cormorant and otters raiding carp pools and salmon fisheries (Kruuk et al. 1993; Cowx 2003; Britton et al. 2005), bears gorging on bee hives (Meadows et al. 1998) and wolverines and lynx killing semi-domestic reindeer (Pedersen et al. 1999).

Conflict has been exacerbated by changes in husbandry over the past 100 years. It is most acute where modern economic conditions preclude once-traditional livestock-guarding practices, which in many regions were relaxed, such as in the sheep milking regions of eastern Europe (Rigg 2001), or abandoned outright once large predators were removed, as in southern Europe (Boitani 1995; Breitenmoser 1998; Ciuci & Boitani 1998; Vos 2000). The true cost of livestock predation is higher where people's livelihoods depend entirely on livestock such as in many herding societies. Whereas a cow lost to a jaguar in a large South American ranch may be written off as part of this extensive husbandry practice, large carnivores can have disastrous consequences for the 400,000 people living in the Gir Forest Reserve, India, with the 250 odd remaining Asiatic lions (Divyabhanusinh 2005). Local enthusiasm for the lions is diminished by an average of nearly 15 attacks and over two human deaths annually; this may be as high as 40 attacks per year and seven deaths per year, as happened during 1989–91 (Saberwal et al. 1994). Livestock comprises about one-third of the lions' kills, and most villages report losses of about five cows annually to lions, with 61% of 73 villagers interviewed expressing hostility towards the lions, although one is in awe of the placid nature of the remaining villagers!

Humans are in competition with carnivores for prey, as exemplified by the estimated 3.4 million metric tonnes of bush meat extracted from Central Africa annually, which results in a diminished prey base to carnivores there (see Chapter 14), and piscivorous birds, sharks, seals and otters compete with humans for marine resources (Blackwell et al. 2000). Real or perceived competition has led moose and caribou hunters in Canada and Alaska to kill wolves in an attempt to increase the numbers of their quarry (Harbo & Dean 1983; Gasaway et al. 1992), and roe deer hunters in the Alps complain that their quarry populations have declined as a result of lynx reintroduction (Breitenmoser 1998). Raptors and small carnivores are persecuted in the developed world to protect game for humans (Reynolds & Tapper 1996; Thirgood et al. 2000). Interestingly, the nuisance value of these wild carnivores will vary markedly between arable farmers and those that grow livestock or game. Killing red foxes in parts of the UK may benefit the shepherd, but results in loss of income to cereal farmers per fox owing to the numbers of rabbits they thereby do not eat (Macdonald et al. 2003).

More recently, changing public opinion, legal protection, habitat recovery and conservation initiatives are allowing the return of predators such as grey wolves, bears and large cats in many areas, which tend to provoke furious public complaint and requests from farmers and hunters for compensation or carnivore population reduction (e.g. Mech 1995; Breitenmoser 1998; Treves et al. 2002). There is a widening urban–rural divide, with the lifestyles of minorities who live in contact with wildlife being increasingly influenced by city dwellers setting fashions (Naughton-Treves et al. 2003).

Crop damage by wild herbivores

Wild ungulates and primates tend to cause damage when agricultural crops are grown within or near their natural habitats. Crop damage is a major cause of conflict with wildlife, ranging in size from elephants to rodents, complicated by a mix of various ecological, social and political factors (Sukumar 1989; Naughton-Treves 1998; Nyhus et al. 2000; Conover 2002). Animals that damage crops may also injure or kill farm workers. Between 1980 and 2003, more than 1150 humans and 370 elephants died as a result of human–elephant conflicts in north-east India alone (Choudhury 2004), the majority of these incidents occurring within cultivation and settlement. Serious conflict may result in abandonment of otherwise profitable arable land (good for conservation, bad for the displaced farmers), or escalating costs of farming through investment in fencing and other non-lethal and lethal damage limitation measures (e.g. Studsrod & Wegge 1995; Naughton-Treves et al. 1998, 2000). Distance from the forest edge in Uganda explained the greatest amount of variation in crop damage by ungulates and primates (Naughton-Treves 1998). Farmers residing within < 500 m of protected areas experienced the majority of crop losses, losing 4–7% of their crops to wildlife per season on average. Ecological factors that correlate significantly with crop raiding by elephants include the degree of habitat fragmentation, the higher nutritive value of cultivated crops compared with analogous wild forage, and the higher risk-taking behaviour of individual bulls. Indeed, bulls that are normally solitary during the day often come together in the evening, and gang up before entering agricultural fields (Sukumar 1989, 1991; Hoare 1999). In that way they may be better able to tackle hostile farmers.

People's perception of wildlife damage

In addition to a scientific understanding of wildlife damage, people's perceptions of the conflicts are critical to managing the conflicts (Manfredo et al. 1998; Marker et al. 2003; Naughton-Treves et al. 2003; Naughton-Treves & Treves 2005). Indeed, the two are complementary because individual perceptions of conflict with wildlife are shaped more by catastrophic events than by regular, small-scale events (Naughton-Treves 1997, 1998; Naughton-Treves et al. 2000). Because we talk to neighbours and retain fact and fiction from past generations, alleged and real catastrophic events can shape perceptions for decades and spread across broad regions (Linnell & Bjerke 2002; Naughton-Treves & Treves 2005). Tolerance for losses is strongly influenced by socio-economic factors, such as the legality of retaliation, individual farmers' vulnerability and the availability of farming alternatives to palatable crops or susceptible stock. In the UK for instance red foxes are said to be tolerated by some farmers in a sense if they have any hunting interests. A fragment of evidence for the latter is that significantly more hunting farmers (28.9%) approve of the active conservation of foxes compared with non-hunters (14.7%) (Macdonald & Johnson 1996).

Conservationists should understand both scientific measures of damage and perceptions of the conflict because affected communities tend to value perceptions and anecdotes, whereas policy makers, scientists and outsiders tend to value scientific measures. In designing interventions we must carefully consider tolerance among affected communities for the proposed intervention and the affected wildlife (Manfredo & Dayer 2004). Often those complaining most loudly and bitterly are not the most sorely affected but those who have a voice (Naughton-Treves et al. 2000, 2003), hence one's response to conflicts must be tailored to the perceived losses as well as the actual losses in order to satisfy the politically influential and the politically marginal.

Mitigating human–wildlife conflict

Conflict can occur anywhere along a continuum of species abundance. For those species for which the problem is their abundance (e.g. livestock predation by ubiquitous carnivores), mitigation will seek to reduce contact or manage damage. In contrast, for rare or threatened species the emphasis will be protection, shifting towards sustainable management as a population recovers. A simple scheme (Fig. 17.1) presented by Macdonald & Sillero-Zubiri (2004) proposes a linked and iterative rational process to tackle conflict. A problem can be partitioned notionally between reducible and irreducible elements, and the balance between these will shift as currently intractable elements are rendered reducible by new innovation brought about by research and experimentation. Reducible problems can be mitigated (e.g. by non-lethal intervention), thereby minimizing the current level of conflict. The irreducible problem that poses the conflict can be partitioned into that which is bearable (more or less willingly) by the afflicted stakeholders, and that which is unbearable. The extent to which these stakeholders will bear a cost (such as predation or crop raiding) will depend on their tolerance which, in turn can be heavily affected by education and value. The latter, which is not merely financial, may be attributed to both a species or an ecological, cultural or political process of which it is a part.

Two interventions are relevant to the unbearable component of current conflict: either to control (most often lethal control) the problematic species, population or individual (Fig. 17.2), or to compensate in some way the aggrieved stake-holder. A third option is to protect the species/population and tell the aggrieved person they simply have to put up with it. Each option raises questions, which can be partly answered by research. In the proposed scheme every box interacts with every other, creating a web of links (e.g. access to compensation might be contingent on improved animal husbandry – a form of mitigation).

We use this scheme to visualize the integrated analysis of human–wildlife conflicts we advocate; combining efforts to reduce the damage caused by wildlife with attempts to increase people's tolerance for wildlife. One can:

1 prevent or, reduce the frequency or severity of encounters between humans and wildlife (e.g. barriers, guards, wild prey recovery, establishment of refuges for wildlife, Fig. 17.2);
2 deal with those individuals that cause conflict (e.g. lethal removal, deterrence, translocation);
3 raise tolerance for conflicts in the affected people through a variety of mechanisms (e.g. incentive schemes tied to conservation, compensation for losses, legal harvests).

The most successful projects to date combine at least two approaches. For example, Nagarahole National Park in India was the site of a voluntary resettlement project (Karanth 2002; Karanth & Madhusudan 2002). Hundreds of villagers residing within the park were beset by tigers, elephants and other smaller problem species, while at the same time lacking employment, schools, clinics and other services. Through a fully participatory and voluntary negotiated resettlement, the villagers were moved out of the park and closer to the infrastructure and employment opportunities they desired. As a by-product of resettlement conflict declined and fewer wild animals had to be destroyed or relocated by the authorities. The USA Government reintroduced 31 grey wolves from Canada into the Greater Yellowstone Area in 1995–1996 after years of public outreach and comment; the wolf population now numbers over 800 animals. The project is deemed a success not only for reaching the numerical target for wolves but for suffering fewer depredations than expected and bringing a net economic benefit to the area (Bangs & Fritts 1996; Duffield & Neher 1996). Although there was a cost to the local community through predation, the Government and non-government organizations (NGOs) partnered to mitigate it, including an NGO scheme that has paid out nearly half a million dollars to compensate ranchers for the loss of close to 2000 livestock.

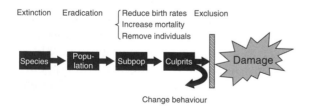

Fig. 17.2 Diagram representing the different levels at which the wildlife component of conflict may be managed by lethal and non-lethal approaches to mitigate damage. (Redrawn from Conover 2002.)

Increasing tolerance for damage by wildlife

The attitudes held by people towards wildlife in general and some species in particular, as well as their perceptions of management interventions, play an important role in conservation. For example, in Nepal, people living closer to the Royal Chitwan National Park were more negative towards it than those who visited the park less frequently and who lived further away in larger landholdings (Nepal & Weber 1995). Effective conservation requires government-backed institutions (e.g. legislation and protected area networks), but it also requires local cooperation (Jackson et al. 2001; Sillero-Zubiri & Laurenson 2001). Real local cooperation with government programmes is usually generated by human–wildlife conflicts that local groups see as requiring government intervention.

In the absence of institutions, the importance of individual attitudes is limited by the 'commons problem' dilemma (or 'collective action problem' see Macdonald et al. 2005). (This, by the way, is an understanding that cannot be claimed as new by biologists because it has been common knowledge for a long time, being well-articulated, for example, by Shylock in Shakespeare's *Merchant of Venice*). The commons problem can be exemplified by a forest hunter who, although believing that a primate population must be protected also suspects that if he refrains from shooting a monkey someone else will kill it, and the only practical outcome of his behaviour would be that his children get less than someone else's. The solution to such a social dilemma would lay with developing an incentive system – such as an agreement among hunters of a given village to hunt a certain quota or hunt only at certain times of the year, while preventing outside hunters from using their patch of forest, rather than a change in attitudes.

Having said this, we know little about how attitudes toward wildlife damage change.

These are deep-seated and reflect social settings more than individual experience (e.g. Bjerke et al. 1998; Vitterso et al. 1999; Naughton-Treves et al. 2003; Kaczensky et al. 2004). Attitudes may change from tolerance to hostility within one generation within the same community, as has happened in parts of India. For instance, in northern West Bengal an elder in a village community affected by elephant depredation had pleaded with a wildlife official to spare the life of the elephant because such depredation was only nature's way of extracting a tax from the people, and that this was no different from a tax extracted by the government. Two decades later in the same village the younger generation of farmers asserted that the offending elephant would have to be killed (V. Rishi, personal communication 2000). This is a reflection of rapidly changing socio-economic contours in the region. Conservationists often make the simplistic assumption that education and economic incentives can overcome upbringing and improve tolerance for wildlife, but social scientists are less sanguine about the plasticity of values associated with use of wildlife (Manfredo & Dayer 2004).

Increasing tolerance through education

Clearly, the value people place on wild animals will often depend heavily on their knowledge of them, and so education is a major tool in conservation (Sutherland 2000; Mishra et al. 2003). Indeed, Balmford (1999) argues that the most depressing conservation problem is not habitat loss or overexploitation, but human indifference to these problems. However, a dangerous fallacy is that opponents to wildlife conservation are merely ignorant. On the contrary, opponents to black-footed ferrets and prairie dogs in USA (Clark et al. 2001) were extremely knowledgeable, often with first-hand negative experience; changing the attitudes of well-informed individuals presumably requires very sophisticated education

(Reading & Kellert 1993; Kaczensky et al. 2004). We have almost no evidence for individual changes in wildlife valuation over time or following interventions (Manfredo & Dayer 2004).

Education and information in general can improve tolerance in another way if it reduces the perceived threat to more realistic levels. For example, many affected communities perceive the risk posed by wildlife out of proportion to its actual occurrence. Information on actual risk levels – if presented with due respect for the experiences of affected communities – can reassure affected communities and help reduce vulnerability by means of simple modifications to their behaviour or husbandry. In central Namibia, for example, farmers perceived cheetahs to be a major problem in livestock and game farms; farmers that considered them problematic killed an average of 29 cheetahs each year (whereas other farmers removed 14 cheetahs on average), but in a follow up survey after an education campaign had been established the number of annual removals had declined to 3.5 and 2, respectively (Marker et al. 2003).

Likewise, communities beset by wildlife damage problems may be empowered by accessing information on the steps they could take to reduce their own vulnerability. This would suggest that research undertaken by conservation biologists and the effective dissemination of their results to stakeholders is an intervention in itself, whereas research results communicated only to outsiders via the scientific literature would not be adequate solutions to human–wildlife conflict.

Conservation NGOs often are advocates of some issues – to that extent, like all advocates – they may be entirely happy to shift perceptions in the direction they wish irrespective of how that bears on reality? Conservation scientists, we would suggest, should be driven entirely by evidence and thus their current best description of reality.

Increasing tolerance through economic incentives

The prevailing view of nature conservation, at least in western societies, is to protect biodiversity for the benefit of the public as a whole and for future generations. It has been argued that the cost of conservation should be borne by many and not only by particular individuals that live, work or move in or near wildlife ranges (Sukumar 1994; Naughton-Treves 1999; Nyhus et al. 2003; Naughton-Treves et al. 2003; Naughton-Treves & Treves 2005). Mechanisms include direct cash compensation and indirect compensation through co-management, integrated conservation development programmes, or resource use such as ecotourism, game ranching and sport hunting. These measures are not necessarily so much interventions to resolve human–wildlife conflict, but also ways to address economic and social inequities that arise in conservation programmes. The same applies to damage-prevention approaches such as large-scale fencing (Thouless & Sakwa 1995), voluntary resettlement (Karanth & Madhusudan 2002), large-scale incentive schemes (Mishra et al. 2003) and community participation in conservation initiatives (Sillero-Zubiri & Laurenson 2001), which may result in direct economic and social benefits while addressing conflict.

When attempts to prevent wildlife attacks on people's property fail, or are half-hearted, many government wildlife protection programmes deal indirectly with damage by paying compensation for livestock and crop losses (Treves et al. 2002; Montag 2003; Naughton-Treves et al. 2003; Sukumar 1994), but these compensation schemes do not address the root causes of conflict: competition over resources. To be effective, compensation programmes require strong institutional support, clear guidelines, quick and accurate verification of damage, prompt and fair payment, sufficient and sustainable funds, and measures of success (Nyhus et al. 2003). The majority of compensation programmes fail to deliver one or more of these services (Montag 2003; Naughton-Treves et al. 2003). Moreover the long-term sustainability of compensation schemes is questionable (Hötte & Bereznuk 2001), especially where monetary values are relatively high, because people may eventually stop preventing conflict, make false compensation claims and increase the costs of administering such schemes. On the other hand, when compensation is inadequate or government response unsatisfying, producers take things into their own hands, as did an Israeli farmer who poisoned livestock carcasses in an effort to kill wolves but in the process killed a number of threatened, scavenging birds (Nemtzov 2003).

Managing wildlife to reduce damage

Reducing 'problem' populations

Intervention may take place at different levels in order to reduce the severity or frequency of encounters between humans and wildlife (Fig. 17.2). Some early attempts at reducing predation on livestock or crop raiding resulted in extinction of a species (i.e. Falklands wolves were clubbed and shot to death by early sheep farmers, and passenger pigeons were shot by the millions in the name of sport – Wilcove (1999) reviews such mismanagement), or eradication of whole populations (e.g. grey wolves in USA, Young & Goldman 1944; several carnivore species in Britain, Langley & Yalden 1977). Such mismanagement was accelerated when the wildlife had some value, as leopards and elephants did for colonial British in Uganda (Naughton-Treves 1999; Treves & Naughton-Treves 1999). Reducing predation or crop raiding losses through the systematic and widespread killing of native animals has become uncommon with rising concern over biodiversity loss (Treves & Naughton-Treves 2005).

Killing the competition has been humanity's way of coping for millennia. Lethal control is exerted in various ways, not all of which are a simple response to economic damage. For example, predator control is done to elevate next season's gamebird populations, and the killing of livestock predators is usually done proactively (Treves & Naughton-Treves 2005).

The decline in many wildlife populations along with changing perceptions of nature and a decrease in livestock and crop-based economies in many developed nations has prompted interest in non-lethal methods of preventing damage by wildlife. Non-lethal methods remain in one of two categories: novel and largely untested (e.g. Musiani et al. 2003; Shivik et al. 2003) or ancient and largely unstudied (e.g. Ogada et al. 2003). But lethal control predominates. For example, around Antesana, Ecuador, cattle producers killed nine spectacled bears – a globally threatened species – before they felt satisfied they had eliminated the one cattle-killing bear (Galasso 2002 – see Karanth & Madhusudan (2002) for a leopard example from India). Unnecessary destruction of wildlife occurs in the USA as well – in 2002–2003, USDA-Wildlife Services killed 235,000 wild carnivores to control agricultural damage.

Opponents of lethal control also criticize its indiscriminate use – killing target and non-target animals – and its use as a political palliative or hidden subsidy for economic activities that are inappropriately managed, situated or financed. On the other hand, proponents of lethal control maintain that even killing non-target individuals will reduce future problems. Conservation biologists do not have adequate data to address this debate currently, although evidence is mounting that livestock-killers and crop-raiders are a minority in their populations and removal operations eliminate non-targets in up to 81% of cases with prevention of subsequent conflicts lasting a mode of 1 year (Treves & Naughton-Treves 2005).

Nevertheless, non-lethal methods face an uphill battle against institutional inertia, affected individuals' desire for revenge or domination of offending wildlife, and the perception that lethal control is the easiest and cheapest method. An alternative to blanket lethal control is the reduction of animal populations by using fertility control methods, but these are still largely experimental (Tuyttens & Macdonald 1998; Bromley & Gese 2001; Chapter 12). Indeed, coyotes are probably the most studied conflict-causing species on the planet and decades of testing non-lethal methods emphasizes the short-lived nature of deterrence, the need for multiple simultaneous defences and the technical challenges of non-lethal controls (Knowlton et al. 1999).

Individual differences among predators are important to managing conflict because one widespread (and generally supported) belief has been that only a small proportion of individuals is responsible for most stock-damage (Knowlton et al. 1999; Linnell et al. 1999; Treves et al. 2002; Wydeven et al. 2004). It was once thought that inexperienced, juvenile, old, infirm and injured predators may be more prone to attack livestock but the vast majority of studies fail to support this conjecture (see Peterhans & Gnoske 2001). Young carnivores, especially males, are more likely to disperse from protected areas into habitats with no wild prey, and where interaction with humans and livestock is much higher (Saberwal et al. 1994). Body size may explain a greater role for male carnivores in killing large livestock, with male bears and large cat males shot or trapped more often following depredation (reviewed in Linnell et al. 1999). Gender-specific predatory behaviour such as the wider-ranging movements or higher risk-taking behaviour of adult males in polygynous mammalian carnivores might also play a part in disproportionate involvement of male cats and bears in livestock predation (e.g. Sukumar 1991; Peterhans & Gnoske 2001). Long-term studies of radio-collared carnivores suggest the majority can coexist with humans and domestic animals without being implicated in conflicts (Wydeven et al. 2004). Indeed, some avoid humans and domestic animals (e.g. Jorgensen 1979; Suminski 1982).

Translocation has often been used to manage problem wildlife despite serious reservations about its application and effectiveness (reviewed by Linnell et al. 1997). Most translocated animals end up causing problems again, fail to form social bonds or end up dead. Asian elephants translocated several tens or even over a hundred kilometres away from their capture locations in the Indian states of Karnataka and West Bengal have invariably gone back to their original homes within a few weeks (Sukumar 2003). Translocated grey wolves in north-western USA follow a similar pattern. Bradley et al. (2005) examined 63 individuals and nine cohesive groups of wolves (out of 105 translocated), mostly moved reactively in response to livestock conflicts. Nineteen wolves (27%) depredated after release, either creating new conflicts (18%) or returning home and resuming depredations in their original territory (9%). Wolves that were pre-emptively moved appeared no less likely to avoid conflicts; three of seven (43%) depredated after release. Most translocated wolves (67%) were never known to establish or join a pack.

Benefits of non-lethal control

Targeting problem animals with non-lethal methods (e.g. methods that alter individual behaviour include conditioned taste aversion, electric shock, sound, light and chemical repellents, diversionary feeding) could prove more effective than lethal control, because they tend to target problem animals and thus minimize population perturbation, for example, by retaining the predator in its original territory and social position (Jorgenson et al. 1978; Tuyttens & Macdonald 2000; Woodroffe & Frank 2005). For example, a traditional Polish hunting device, fladry, appears to deter grey wolves from entering fenced pastures (Musiani et al. 2003). Probably the single most effective non-lethal deterrent against crop-raiders and livestock-killers is human presence and supervision of property (Naughton-Treves 1997; Mertens & Promberger 2001; Knight 2003; Ogada et al. 2003; Osborn & Parker 2003) – with the possible and notable exceptions of incursion by elephants, lions and tigers. Surprisingly, the cost-effectiveness of guarding by humans has not been widely tested as a deterrent. This targeting may avoid the density dependent population responses and immigration that can result from culling (for a review see Treves & Naughton-Treves 2005), while allowing the animal to continue with whatever effect it has on limiting other prey numbers or excluding conspecifics (Baker & Macdonald 1999).

One of the simplest but most innovative examples of behavioural modification resulted from the observation that tigers tend to attack people from behind when they crouched to gather firewood in the jungle, possibly mistaking them for a natural prey species (Rishi 1988). A scheme in the Indian Sundarbans to persuade people to wear facial masks behind their heads when venturing into the jungle proved effective in reducing attacks by tigers until a superstitious belief led to people discarding these masks. Tigers are now conditioned through electrified dummies to avoid people in this region, perhaps one of the most effective means to reduce man-eating (Sanyal 1987). Better monitoring of these schemes could have provided objective measures of success.

Whether lethal or non-lethal, all control actions fail sometimes to prevent damage. Some habitual offenders go to great lengths to reach their target. Certain individuals will find ways to pass through electric fencing given enough time (e.g. coyotes, Thompson 1978; elephants, Thouless & Sakwa 1995) and certain individual predators with a taste for livestock or humans become vexingly hard to kill or capture (e.g. leopards, Corbett 1954; lions, Peterhans & Gnoste 2001).

Mitigating encounters

Where integration proves unworkable, limiting the intersection of wildlife and human activities remains one of the most effective ways to preempt conflict (Fig. 17.2). Barriers, guarding and managing livestock are some of the most ancient and still widespread techniques to mitigate conflict (e.g. Thouless & Sakwa 1995; Andelt 1999; Knight 2003; Ogada et al. 2003). Unfortunately fencing sufficiently robust, deep and high to prevent wildlife from digging under or climbing over can be very expensive (Thouless & Sakwa 1995; Angst 2001). However, in Tibet's Qomolangma National Nature Preserve, production doubled in 2 years following use of communal corrals, built cheaply by villagers. They used the time saved from guarding to improve their handicrafts and income generation, and attitudes towards conserving wildlife improved substantially (Jackson & Wangchuk 2001). At the other end of the management continuum some wildlife agencies or NGOs have provided support and capital for fences and deterrent devices (Coppinger et al.1988; Fox 2001; Nemtzov 2003).

Guarding is widely used in many parts of the world, and often does not require large investment of capital. Usually during pre-harvesting and harvesting time, farm family members would take turns guarding field crops using makeshift watchtowers (e.g. against elephants; Sukumar 1989). To avoid heavy losses or high guarding investment, highly palatable seasonal crops such as maize should not be planted on the forest edge (Naughton-Treves 1998). On the broader level, conserving large blocks of forests and reducing edge habitat should be a management priority. More often guarding is undertaken by guard animals (Andelt 1999; Meadows & Knowlton 2000; Rigg 2001), or more rarely electronic guards (Knight 2003; Shivik et al. 2003) and sound systems to scare away animals (Studsrod & Wegge 1995). Trials with potential chemical deterrents such as pepper spray have shown limited success against African elephants (Osborn & Rasmussen 1995).

In grazing systems where livestock are free-ranging and unattended, the presence of scattered livestock throughout a carnivore's home range may increase the likelihood of encountering, and consequently being killed by, the carnivore in question. This may explain why, even in areas with a good abundance of wild prey, livestock losses are high (Linnell et al. 1999). Small changes to husbandry practices, such as reducing herd size, keeping them in proximity to people and buildings and away from thick cover, not leaving carcasses out in the open and improving construction of holding pens, can improve livestock safety from wild predators (e.g. Naughton-Treves et al. 1998; Landa et al. 1999; Linnell et al. 1999; Naughton-Treves et al. 2000; Stahl & Vandel 2001; Ogada et al. 2003; Wydeven et al. 2004).

What else do we need to know?

With the exception of a handful of case studies, several reviewed above, we remain largely ignorant of the ecology and behaviour of problem wildlife, hence many management techniques often mistakenly encompass all wildlife as potential problems despite evidence to the contrary. There is a need to identify first whether problems are soluble or intractable. Second, how much more knowledge do we need in order to find solutions to many of the challenging cases of human–wildlife conflict that we have been occupied with? This lack of knowledge often promotes population reduction measures when we may really need problem animal identification and removal measures. The effectiveness of lethal control versus non-lethal control needs to be compared systematically and experimentally.

Too often researchers do not design studies in collaboration with managers who might be their immediate and critical audience. Likewise, managers often ignore good research

and stick to traditional methods of managing human–wildlife conflicts. For example, the incidence of a few cases of cervid chronic wasting disease was treated as an emergency by deer managers, who decided on widespread culling, ignoring the advice of veterinary epidemiologists about the speed of responses and human dimensions experts about the appropriate response (Heberlein 2004).

A similar gulf separates most social scientists from biological scientists (Manfredo & Dayer 2004). Human–wildlife conflict starkly illustrates how modern conservation problems are primarily people–people conflicts revolving around the use or protection of natural resources and biodiversity. Yet wildlife managers have been slow to appreciate or adopt methods from the social sciences such as participatory planning, co-management and economic analysis. Likewise social scientists have been slow to understand the need for applied research that addresses conservation dilemmas preferring instead to generate theoretical treatises. One interpretation of the generalized failure to deal with the underlying bases of human–wildlife conflict is the assumption that human behaviour and attitudes do not change. This conjecture demands some study and particularly experimental tests of different methods of changing human behaviour and attitudes.

Conclusions – a need to compromise

Most landscapes are now dominated by humans. Where wildlife and people coexist, particularly when large carnivores and ungulates are involved, their biology provokes conflict and the best we can hope for may be an uneasy tolerance (Sillero-Zubiri & Laurenson 2001). Conflict occurs between competing interests for environmental resources; and solutions need compromise and strategies that do not necessarily involve sealing people off from nature but, on the contrary involve a respectful engagement with wildlife (Macdonald 2001). Whereas this may once have typified the interaction of some knowledgeable country-people with wildlife – they killed wild animals when they had to, and tolerated them when they could, and could be at ease with both these outcomes – more recently, additional stakeholders have been added into the mix. These are bringing in a blend of conservation, perceptions of nature, animal welfare, politics and natural resource economics with them. There is an increasing urban-rural divide, brought about chiefly by the enormous political issues associated with city dwellers making decisions, and setting fashions, about the lifestyles of minorities who live in contact with wildlife. This is an extremely complex area requiring innovative, clear-thinking solutions. Thus dealing with conflict now often necessitates an orchestrated, multidisciplinary approach (Heberlein 2004).

Conflict between wildlife and people will continue to exist long into this century if not beyond, and necessitates management, for both imperilled and abundant species. The problems faced by these two categories clearly differ in detail, but both merit the attention of conservationists, and both may be susceptible to similar approaches using the same tools. Successful strategies will have to be based on the integration of many disciplines, including elements from the social and political sciences. Innovation and imagination are required to find solutions to conflict outside protected areas, and these most probably will require a mixture of strategies, including preservation, lethal and non-lethal control, changes in farming and animal husbandry, consumptive and non-consumptive uses, and complicated evaluations of costs and benefits (measured in such incommensurable currencies as biodiversity, money and ethics).

Conflict mitigation would be advanced by conservation initiatives that recognize the dual importance of large, linked areas of suitable habitat and of the protection of the economies and safety of human communities alongside wildlife. Crucially, an important requisite for success is often an involvement of the local community in the decision-making process and the sharing of any revenues accruing from wildlife. A traditional approach to conflict, now hopefully outmoded, characterized rural people as the problem; although this may be partly true it seems essential that they become part of the solution (Sillero-Zubiri & Laurenson 2001). In many cases, education must challenge deeply engrained cultural prejudices, whereas the sources of genuine conflict must be identified, understood and dealt with. Where conflict remains it will often be fitting for wider society to lift the burden, or risk, off individual producers in the interest of preserving species.

Acknowledgements

We are grateful to Jorgelina Marino, Paul Johnson, students and other anonymous reviewers for helpful contributions to this article.

We should find out as much as possible about someone before coming into conflict with him.
(Aesop's Fables, in *Collected Tales from Aesop's Fables*, Smithmark, 1988.)

References

Andelt, W.F. (1999) Relative effectiveness of guarding-dog breeds to deter predation on domestic sheep in Colorado. *Wildlife Society Bulletin* **27**: 706–14.

Angst, C. (2001) Procedure to selectively remove stock raiding lynx in Switzerland. *Carnivore Damage Prevention News* **4**: 8.

Baker, S.E. & Macdonald, D.W. (1999) Non-lethal predator control: exploring the options. *Advances in Vertebrate Pest Management* (Eds P.D. Cowan & C.J. Feare), pp. 251–66. Filander Verlag, Furth.

Balmford, A. (1999) (Less and less) great expectations. *Oryx* **33**: 87–8.

Bangs, E.E. & Fritts, S.H. (1996) Reintroducing the gray wolf to central Idaho and Yellowstone National Park. *Wildlife Society Bulletin* **24**: 402–13.

Bjerke, T., Odegardstuen, T.S. & Kaltenborn, B.P. (1998) Attitudes toward animals among Norwegian adolescents. *Antrozoos* **11**: 79–86.

Blackwell, B.F., Dolbeer, R.A. & Tyson, L.A. (2000) Lethal control of piscivorous birds at aquaculture facilities in the northeast United States: effects on populations. *North American Journal of Aquaculture* **6**(2): 300–7.

Boitani, L. (1995) Ecological and cultural diversities in the evolution of wolf-human relationships. In *Ecology and Conservation of Wolves in a Changing World* (Eds L.N. Carbyn, S.H. Fritts & D.R. Seip), pp. 3–11. Canadian Circumpolar Institute, Edmonton.

Bradley, E.H., Pletscher, D.H., Bangs, E.E., et al. (2005) Evaluating wolf translocation as a non-lethal method to reduce livestock conflicts in the northwestern United States. *Conservation Biology* **19**: 1498–508.

Breitenmoser, U. (1998) Large predators in the Alps: the fall and rise of man's competitors. *Biological Conservation* **83**: 279–89.

Breitenmoser, U. & Angst, C. (2001) Statistics of damage caused by large carnivores in Europe. *Carnivore Damage Prevention News* **4**: 11–13.

Britton, J.R., Shepherd, J.S., Toms, S. & Simpson, V. (2005) Presence of carp, *Cyprinus carpio*, in the diet of the otter, *Lutra lutra*. *Fisheries Management and Ecology* **12**: 221–3.

Bromley, C. & Gese, E.M. (2001) Effect of sterilization on territory fidelity and maintenance, pair bonds, and survival rates of free-ranging coyotes. *Canadian Journal of Zoology* **79**: 386–92.

Choudhury, A. (2004) Human–elephant conflicts in Northeast India. *Human Dimensions of Wildlife* **9**: 261–70.

Ciucci, P. & Boitani, L. (1998) Wolf and dog depredation on livestock in central Italy. *Wildlife Society Bulletin* **26**: 504–14.

Clark, T.W., Mattson, D.J., Reading, R.P. & Miller, B.J. (2001) Some approaches and solutions. In *Carnivore Conservation*, (Eds J.L. Gittleman, S.M. Funk, D.W. Macdonald & R.K. Wayne), pp. 223–240. Cambridge University Press, Cambridge.

Conover, M.R. (2002) *Resolving Human–Wildlife Conflicts*. Lewis Publishers, Boca Raton, Florida.

Coppinger, R., Coppinger, L., Langeloh, G., Gettler, L. & Lorenz, J. (1988) A decade of use of livestock guarding dogs. *Proceedings of the Vertebrate Pest Conference* **13**: 209–14.

Corbett, J. (1954) *The Man-eating Leopard of Rudraprayag*. Oxford University Press, London.

Cowx, I.G. (Ed.) (2003) *Interactions between Fish and Birds*. Fishing News Books, Blackwell Science, Oxford.

Divyabhanusinh, C. (2005) *The Story of Asia's Lions*. Marg Publications, Mumbai, India.

Duffield, J.W. & Neher, C.J. (1996) Economics of wolf recovery in Yellowstone National Park. *Transactions of the North American Wildlife and Natural Resources Conference* **61**: 285–292.

Fox, C.H. (2001) Taxpayers say no to killing predators. *Animal Issues* **32**: 1–2.

Galasso, L. (2002) The spectacled bear's impact on livestock and crops and use of remnant forest fruit trees in a human-altered landscape in Ecuador. Department of Zoology. University of Wisconsin-Madison, Madison.

Gasaway, W.C., Boertje, R.D., Grangaard, D.V., Kelleyhouse, D.G., Stephenson, R.O. & Lar-sen, D.G. (1992) The role of predation in limiting moose at low densities in Alaska and Yukon and implications for conservation. *Wildlife Monographs* **120**: 5–59.

Gittleman, J.L., Funk, S.M., Macdonald, D.W. & Wayne, R.K. (Eds) (2001) *Carnivore Conservation*. Cambridge University Press, Cambridge.

Harbo, S.J., Jr. & Dean, F.C. (1983) Historical and current perspectives on wolf management in Alaska. In *Wolves in Canada and Alaska: their Status, Biology and Management* (Ed. L.N. Carbyn), pp. 51–64. Canadian Wildlife Service, Edmonton.

Heberlein, T.A. (2004) 'Fire in the Sistine Chapel': how Wisconsin responded to chronic wasting disease. *Human Dimensions of Wildlife* **9**: 165–79.

Hötte, M. & Bereznuk, S. (2001) Compensation for livestock kills by tigers and leopards in Russia. *Carnivore Damage Prevention News* **3**: 6–7.

Hemmer, H. (1990) *Domestication: the Decline of Environmental Appreciation*. Cambridge University Press, Cambridge.

Hoare, R. (1999) Determinants of human–elephant conflict in a land-use mosaic. *Journal of Applied Ecology* **36**: 689–700.

Jackson, R.M. & Wangchuk, R. (2001) Linking snow leopard conservation and people–wildlife conflict resolution: grassroots measures to protect the endangered snow leopard from herder retribution. *Endangered Species Update* **18**: 138–141. [http: // www.snowleopardconservancy. org/publications. htm]

Jackson, R.L., Hillard, D. & Wangchuk, R. (2001) Encouraging local participation in efforts to reduce livestock depredation by snow leopard and wolf in Ladakh, India. *Carnivore Damage Prevention News* **4**: 2–6.

Jorgensen, C.J. (1979) Bear–sheep interactions, Targhee National Forest. *International Conference on Bear Research and Management* **5**: 191–200.

Jorgensen, C. J., Conley, R. H., Hamilton, R. J. & Sanders, O. T. (1978) Management of black bear depredation problems. *Proceedings of the Eastern Workshop on Black Bear Management and Research* **4**: 297–321.

Kaczensky, P., Blazic, M. & Gossow, H. (2004) Public attitudes towards brown bears (Ursus arctos) in Slovenia. *Biological Conservation* **118**: 661–74.

Karanth, K.U. (2002) Nagarahole: limits and opportunities in wildlife conservation. In *Making Parks Work: Identifying Key Factors to Implementing Parks in the Tropics* (Eds J. Terborgh, C.P. van Schaik, M. Rao & L.C. Davenport), pp. 189–202. Island Press, Covelo, CA.

Karanth, K.U. & Madhusudan, M.D. (2002) Mitigating human–wildlife conflicts in southern Asia. In *Making Parks Work: Identifying Key Factors to Implementing Parks in the Tropics* (Eds J. Terborgh, C.P. van Schaik, M. Rao & L.C. Davenport), pp. 250–64. Island Press, Covelo, CA.

Knight, J. (2003) *Waiting for Wolves in Japan.* Oxford University Press, Oxford.

Knowlton, F.F., Gese, E.M. & Jaeger, M.M. (1999) Coyote depredation control: an interface between biology and management. *Journal of Range Management* **52**: 398–412.

Kruuk, H. (2002) *Hunter and Hunted: Relationships between Carnivores and People.* Cambridge University Press, Cambridge.

Kruuk H., Carss D.N., Conroy J.W.H. & Durbin L. (1993) Otter *Lutra lutra* (L.) numbers and fish productivity in rivers in north-east Scotland. *Symposium of the Zoological Society of London* **65**: 171–91.

Landa, A., Gudvangen, K., Swenson, J.E. & Roskaft, E. (1999) Factors associated with wolverine *Gulo gulo* predation on domestic sheep. *Journal of Applied Ecology* **36**: 963–73.

Langley, P.J.W. & Yalden, D.W. (1977) The decline of the rare carnivores in Great Britain during the nineteeth century. *Mammal Review* **7**: 95–116.

Linnell, J.D.C. & T. Bjerke (2002) Fear of wolves: an interdisciplinary study. *NINA Oppdragsmelding* **722**: 1–110.

Linnell, J.D.C., Aanes, R., Swenson, J.E., Odden, J. & Smith, M.E. (1997) Translocation of carnivores as a method for managing problem animals: a review. *Biodiversity and Conservation* **6**: 1245–57.

Linnell, J.D.C., Odden, J., Smith, M.E., Aanes, R. & Swenson, J.E. (1999) Large carnivores that kill livestock: do 'problem individuals' really exist? *Wildlife Society Bulletin* **27**: 698–705.

Macdonald, D.W. (2001) Postscript: science, compromise and tough choices. In *Carnivore Conservation* (Eds J.L. Gittleman, S.M. Funk, D.W. Macdonald & R.K. Wayne), pp. 524–38. Cambridge University Press, Cambridge.

Macdonald, D.W. & Johnson, P.J. (1996) The impact of fox hunting: a case study. In *The Exploitation of Mammal Populations* (Eds N. Dunstonee & V. Taylor), pp. 160–207. Chapman & Hall, London.

Macdonald, D.W. & Sillero-Zubiri, C. (2002) Large carnivores and conflict: lion conservation in context. In *Lion Conservation Research. Workshop 2: Modelling Conflict* (Eds A.J. Loveridge, T. Lynam & D.W. Macdonald). pp. 1–8. Wildlife Conservation Research Unit, Oxford.

Macdonald, D.W. & Sillero-Zubiri, C. (2004) Conservation: from theory to practice, without bluster. In *The Biology and Conservation of Wild Canids* (Eds D.W. Macdonald & C. Sillero-Zubiri), pp. 353–72. Oxford University Press, Oxford.

Macdonald, D.W., Reynolds, J.C., Carbone, C., Mathews, F., Johnson, P.J. (2003) The bioeconomics of fox control. In *Conservation and Conflict: Mammals and Farming in Britain* (Eds F.H. Tattersall & W.J. Manley), pp. 220–36. Linnean Society Occasional Publication, Westbury.

Macdonald, D.W., Sillero-Zubiri, C., Wang, S.W. & Wilson, D. (2005) Lions and conflict – lessons from a wider context. In *Lion Conservation Research: Workshops 3&4: From Conflict to Socioecology* (Eds A.J. Loveridge, T. Lynam & D.W. Macdonald), pp. 1–16. Wildlife Conservation Research Unit, Oxford.

Manfredo, J. & Dayer, A.A. (2004) Concepts for exploring the social aspects of human–wildlife conflict in a global context. *Human Dimensions of Wildlife* **9**: 317–28.

Manfredo, M.J., Zinn, H.C., Sikorowski, L. & Jones, J. (1998) Public acceptance of mountain lion management: a case study of Denver, Colorado, and nearby foothill areas. *Wildlife Society Bulletin* **26**: 964–70.

Marker L.L. Mills M.G.L. & Macdonald D.W. (2003) Factors influencing perceptions of conflict and tolerance toward cheetahs on Namibian farmlands. *Conservation Biology* **17**: 1290–8.

Meadows, L.E. & Knowlton, F.F. (2000) Efficacy of guard llamas to reduce canine predation on domestic sheep. *Wildlife Society Bulletin* **28**: 614–22.

Meadows, L.E., Andelt, W.F. & T.I. Beck. (1998) *Managing Bear Damage to Beehives*. Colorado State University Cooperative Extension Report No. 6.519. http: //www.ext.colostate.edu/ pubs/ natres/06519.html.

Mech, L.D. (1995) The challenge and opportunity of recovering wolf populations. *Conservation Biology* **9**: 270–8.

Meriggi, A. & Lovari, S. (1996) A review of wolf predation in southern Europe: does the wolf prefer wild prey to livestock? *Journal of Applied Ecology* **33**: 1561–71.

Mertens, A., Gheorghe, P. & Promberger, C. (2001) Carnivore damage to Livestock in Romania. *Carnivore Damage Prevention News* **4**: 9.

Mishra, C. (1997) Livestock depredation by large carnivores in the Indian trans-Himalaya: conflict perceptions and conservation prospects. *Environmental Conservation* **24**: 338–43.

Mishra, C., Allen, P., Mccarthy, T., Madhusudan, M.D., Bayarjargal, A.& Prins, H.H.T. (2003) The role of incentive schemes in conserving the snow leopard, *Uncia uncia*. *Conservation Biology* **17**: 1512–20.

Montag, J. (2003) Compensation and predator conservation: limitations of compensation. *Carnivore Damage Prevention News* **6**: 2–6.

Musiani, M., Mamo, C., Boitani, L., et al. (2003) Using fladry barriers to protect livestock from wolves in western Canada and the western United States. *Conservation Biology* **17**: 1538–47.

National Agricultural Statistics Service (2002) US Department of Agriculture. http: //www.nass.usda.gov/census/

Naughton-Treves, L. (1997) Farming the forest edge: vulnerable places and people around Kibale National Park. *The Geographical Review* **87**: 27–46.

Naughton-Treves, L. (1998) Predicting patterns of crop damage by wildlife around Kibale National Park, Uganda. *Conservation Biology* **12**: 156–68.

Naughton-Treves, L. (1999) Whose animals? A history of property rights to wildlife in Toro, western Uganda. *Land Degradation and Development* **10**: 311–28.

Naughton-Treves, L. & Treves, A. (2005) Socioecological factors shaping local tolerance of crop loss to wildlife in Africa. In *People and Wildlife, Conflict or Coexistence?* (Eds R. Woodroffe, S. Thirgood & A. Rabinowitz), pp. 253–77. Cambridge University Press, Cambridge.

Naughton-Treves, L., Treves, A., Chapman, C.A. & Wrangham, R.W. (1998) Temporal patterns of crop raiding by primates: Linking food availability in croplands and adjacent forest. *Journal of Applied Ecology* **35**: 596–606.

Naughton-Treves, L., Rose, R.A. & Treves, A. (2000) *Social and Spatial Dimensions of Human–Elephant Conflict in Africa: a Literature Review and Two Case Studies from Uganda and Cameroon*. International Union for the Conservation of Nature and Natural Resources, Gland, Switzerland.

Naughton-Treves, L., Grossberg, R. & Treves, A. (2003) Paying for tolerance: the impact of livestock depredation and compensation payments on rural citizens' attitudes toward wolves. *Conservation Biology* **17**: 1500–11.

Nemtzov, S.C. (2003) A short-lived wolf depredation compensation program in Israel. *Carnivore Damage Prevention News* **6**: 16–17.

Nepal, S.K. & Webber, K.E. (1995) Prospects for coexistence: wildlife and local people. *Ambio* **24**: 238–45.

Nyhus, P.J., Fischer, H., Madden, F. & Osofsky, S. (2003) Taking the bite out of wildlife damage: the challenges of wildlife compensation schemes. *Conservation in Practice* **4**: 37–40.

Nyhus, P.J., Sumianto & Tilson, R. (2000) Crop-raiding elephants and conservation implications at Way Kambas National Park, Sumatra, Indonesia. *Oryx* **34**: 262–74.

Ogada, M.O., Woodroffe, R., Oguge, N.O. & Frank, L.G. (2003) Limiting depredation by African carnivores: the role of livestock husbandry. *Conservation Biology* **17**: 1521–30.

Oli, M.K., Taylor, I.R. & Rogers, M.E. (1994) Snow leopard *Panthera unica* predation of livestock: an assessment of local perceptions in the Annapurna Conservation Area, Nepal. *Biological Conservation* **68**: 63–8.

Osborn, F.V. & Parker, G.E. (2003) Towards an integrated approach for reducing the conflict between elephants and people: a review of current research. *Oryx* **37**: 80–4.

Osborn, H.F. & Rasmussen, L.E.L. (1995) Evidence for the effectiveness of an oleo-resin capsicum aerosol as a repellent against wild elephants in Zimbabwe. *Pachyderm* **20**: 55–64.

Pedersen, V.A., Linnell, J.D.C., Andersen, R., Andrén, H., Lindén, M. & Segerström, P. (1999) Winter lynx *Lynx lynx* predation on semi-domestic reindeer *Rangifer tarandus* in northern Sweden. *Wildlife Biology* **5**: 203–11.

Peterhans, J.C.K. & Gnoske, T.P. (2001) The science of 'man-eating' among lions Panthera leo with a reconstruction of the natural history of the 'Man-eaters of Tsavo'. *Journal of East African Natural History* **90**: 1–40.

Ray, J.C., Redford, K.H., Steneck, R.S. & J. Berger (Eds) (2005) *Large Carnivores and the Conservation of Biodiversity*. Island Press, Covelo, CA.

Rajpurohit, R.S. & Krausman, P.R. (2000) Human–sloth–bear conflicts in Madhya Pradesh, India. *Wildlife Society Bulletin* **28**: 393–9.

Reading, R.P. & Kellert, S.R. (1993) Attitudes toward a proposed black-footed ferret (*Mustela nigripes*) reintroduction. *Conservation Biology* **7**: 569–80.

Reynolds, J.C. & Tapper, S.C. (1996) Control of mammalian predators in game management and conservation. *Mammal Review* **26**: 127–56.

Rigg, R. (2001) *Livestock Guarding Dogs: their Current Use World Wide*. Occasional Paper 1, Species Survival Commission, International Union for the Conservation of Nature and Natural Resources, Gland, Switzerland. http://www.canids.org/occasionalpapers/

Rishi, V. (1988) Man, mask and man-eater. *Tigerpaper* **July–September**: 9–14.

Saberwal, V.K., Gibbs, J.P., Chellam, R. & Johnsingh, A.J.T. (1994) Lion-human conflict in the Gir forest, India. *Conservation Biology* **8**: 501–7.

Sanyal, P. (1987) Managing the man-eaters in the Sundarbans Tiger Reserve of India – a case study. In *Tigers of the World: the Biology, Bioploitics, Management and Conservation of an Endangered Species* (Eds R. L. Tilson & U. S. Seal), pp. 427–34. Noyes Publications, Park Ridge, New Jersey.

Shivik, J.A., Treves, A. & Callahan, M. (2003) Non-lethal techniques for managing predation: primary and secondary repellents. *Conservation Biology* **17**: 1531–1537.

Sillero-Zubiri, C. & Laurenson, M.K. (2001) Interactions between carnivores and local communities: conflict or co-existence? In *Carnivore Conservation* (Eds J.L. Gittleman, S.M. Funk, D.W. Macdonald & R.K. Wayne), pp. 282–312. Cambridge University Press, Cambridge.

Sillero-Zubiri, C., Reynolds, J. & Novaro, A. (2004) Management and control of canids near people. In *Biology and Conservation of Wild Canids* (Eds D.W. Macdonald & C. Sillero-Zubiri), pp. 17–40. Oxford University Press, Oxford.

Stahl, P. & Vandel, J.M. (2001) Factors influencing lynx depredation on sheep in France: problem individuals and habitat. *Carnivore Damage Prevention News* **4**: 6–8.

Studsrod, J.E. & Wegge, P. (1995) Park–people relationships: the case of damage caused by park animals around the Royal Bardia National Park, Nepal. *Environmental Conservation* **22**: 133–42.

Sukumar, R. (1989) *The Asian Elephant: Ecology and Management*. Cambridge University Press, Cambridge.

Sukumar, R. (1991) The management of large mammals in relation to male strategies and conflicts with people. *Biological Conservation* **55**: 93–102.

Sukumar, R. (1994) Wildlife-human conflict in India: an ecological and social perspective. In *Social Ecology* (Ed. R. Guha), pp. 303–316. Oxford University Press, New Delhi.

Sukumar, R. (2003) *The Living Elephants: Evolutionary Ecology, Behavior and Conservation*. Oxford University Press, New York.

Suminski, H.R. (1982) Mountain lion predation on domestic livestock in Nevada. *Vertebrate Pest Conference* **10**: 62–6.

Sutherland, W.J. (2000) *The Conservation Handbook: Research, Management and Policy*. Blackwell Science, Oxford.

Thompson, B.C. (1978) Fence-crossing behavior exhibited by coyotes. *Wildlife Society Bulletin* **6**: 14–17.

Thouless, C.R. & Sakwa, J. (1995) Shocking elephants – fences and crop raiders in Laikipia District, Kenya. *Biological Conservation* **72**: 99–107.

Thirgood, S., Redpath, S., Newton, I. & Hudson, P. (2000) Raptors and red grouse: Conservation conflicts and management solutions. *Conservation Biology* **14**: 95–104.

Treves, A. & Karanth, K.U. (2003) Human-carnivore conflict and perspectives on carnivore management worldwide. *Conservation Biology* **17**: 1491–9.

Treves, A. & Naughton-Treves, L. (1999) Risk and opportunity for humans coexisting with large carnivores. *Journal of Human Evolution* **36**: 275–82.

Treves, A. & Naughton-Treves, L. (2005) Evaluating lethal control in the management of human–wildlife conflict. In *People and Wildlife, Conflict or Coexistence?* (Eds R.W. Woodroffe, S. Thirgood & A. Rabinowitz), pp. 86–106. Cambridge University Press, Cambridge.

Treves, A., Jurewicz, R.R., Naughton-Treves, L., Rose, R.A. Willging, R.C. & Wydeven, A.P. (2002) Wolf depredation on domestic animals: control and compensation in Wisconsin, 1976–2000. *Wildlife Society Bulletin* **30**: 231–41.

Tuyttens, F.A.M. & Macdonald, D.W. (2000) Consequences of social perturbation for wildlife management and conservation. In *Behaviour and Conservation* (Eds L.M. Gosling & W.J. Sutherland), pp. 315–329. Cambridge University Press, Cambridge.

Tuyttens, F.A.M. & Macdonald, D.W. (1998) Fertility control: an option for non-lethal control of wild carnivores? *Animal Welfare* **7**: 339–64.

Vitterso, J., Bierke, T. & Kaltenhorn, B.P. (1999) Attitudes toward large carnivores among sheep farmers experiencing different degrees of depredation. *Human Dimensions of Wildlife* **4**: 20–35.

Vos, J. (2000) Food habits and livestock depredation of two Iberian wolf packs (*Canis lupus signatus*) in the north of Portugal. *Journal of Zoology* **251**: 457–62.

Wilcove, D. (1999) *The Condor's Shadow*. H. Freeman, New York.

Woodroffe, R. & Frank, L.G. (2005) Lethal control of African lions (*Panthera leo*): local and regional population impacts. *Animal Conservation* **8**: 91–8.

Woodroffe, R., Thirgood, S. & Rabinowitz, A. (Eds) (2005) *People and Wildlife. Conflict or Coexistence?* Cambridge University Press, Cambridge.

Wydeven, A.P., Treves, A., Brost, B. & Wiedenhoeft. J. (2004) Characteristics of wolf packs that prey on domestic animals. In *Predators and People: from Conflict to Conservation* (Eds A.D.N. Fascione, A. Delach & M.E. Smith), pp. 28–50. Island Press, Washington, DC.

Young, S.P. & Goldman, E.A. (1944) *The Wolves of North America*. Dover, New York.

Zhao, Q.K. (1991) Macaques and tourists at Mt. Emei, China. *National Geographic Research* **7**: 15–16.

18

Principles, practice and priorities: the quest for 'alignment'

David W. Macdonald, N.Mark Collins and Richard Wrangham

We shall never achieve harmony with land, any more than we shall achieve absolute justice or liberty for people. In these higher aspirations the important thing is not to achieve, but to strive.
(Aldo Leopold, *Round River*, Oxford University Press, New York, 1993.)

Introduction

Things have changed. Whether you consider the birth of modern conservation to be 1890 when Scottish naturalist John Muir persuaded the US Congress to create Yosemite National Park, or 1903 when the Society for the Preservation of the Wild Fauna of the Empire (latterly Fauna and Flora International) became the world's first international conservation organization, or even 1948 when the International Union for the Protection of Nature (IUCN – now The World Conservation Union) became the first international association designed to protect habitats and species – the fact is that the whole business of conservation is very different now from when it began. Once, the emphasis was protection. Now, it is integration.

The most obvious reason for this change is that the world's human population has more than trebled since Muir's day, from less than two billion to 6.5 billion. Furthermore, world economic output has meanwhile increased

twenty-fold. From the dawn of history to 1950 the world economy grew to six trillion dollars. It now grows by this amount every 5–10 years! The result, in short, is that humanity has stamped down so hard on the accelerator of change that our impact has outstripped the capacity of natural systems to adapt successfully.

However, the problems we face are not just about how many of us there are, nor how much we produce; they are also about how we are sharing wealth, utilizing our environmental and technological resources, expanding agriculture and fisheries, and building infrastructure. Inequity creates a political obstacle to the cooperative behaviour that will be essential to a satisfactory future for humanity. A more equitable world is a necessary, but not a sufficient, condition for sustainable development.

At one end of global society the rich consume without sufficient thought to future generations, whereas at the other the poor live unsustainably in environments that often fail to meet human needs. This disparity is starkly revealed by comparative income data. While 295 million Americans on average earn $40,100 per annum

(GDP per capita) and 457 million Europeans earn $26,900, the 4.94 billion inhabitants of the world's developing countries on average each earn only $4054 per annum (the latter figure is adjusted for purchasing power parity per US$ – without that adjustment the developing world's income per capita is $1264 per annum). These general figures paint a broad picture, but should not disguise the fact that alongside overconsumption in the developed world, deeply entrenched and huge inequities within developing countries are as great an obstacle to sustainable development as are those between rich and poor countries.

Unsustainable consumption and poverty are two great challenges for governments around the world. They are also the most important topics for wildlife conservation because their impact is immense not only for the future of humanity, but also for the future of other species. Although there is an emerging global determination to address the poverty issue by achieving more equitable international development, the environmental and nature conservation movements have previously tended to perpetuate a different approach. Whereas the structures of poverty alleviation have generally involved spending their way out of trouble, the first weapon of environmental protection has been to regulate. While the former involves either giving people money or helping them to make money for themselves, the traditional strategy of building regulatory fences to constrain their behaviour has the drawback that as inequities grow, so does the pressure to break the rules. In the future, as has increasingly been emphasized in the past decade (e.g. Balmford et al. 2002), maintaining the environmental conditions for development will require expenditure, both public and private.

If the problems facing conservation have changed, so have the solutions. Indeed, during our professional lifetimes the conservation philosophy that we three were learning as graduate students in 1972 (when the Stockholm Declaration on the Human Environment was agreed) was very different to that we were practicing in 1992 (when the Convention on Biodiversity was formulated). Today it needs to be different again. For example, conservationists recognize that their strategies must be integrated with poverty alleviation (see discussion of the Millennium Ecosystem Assessment, below). But questions such as how much to spend, how to spend it, and how to avoid conservation receiving no more than lip-service as development advances, remain unanswered.

So in much of what follows we identify tensions between the priorities of conservation and other elements of the human enterprise. Our intention is to avoid setting up false choices; rather we advocate the principle of 'alignment' – a concept used in the business community to describe the process of harnessing the disparate individual and institutional drivers within an enterprise to a common purpose. For us, this essay – and the future of wildlife conservation – is about the quest to align human and conservation imperatives – a challenge to human creativity and ingenuity.

In the Preface we explained that this book grew from a desire to identify what really matters to conservation – the key topics – and to do so bluntly. Now, in this postscript, our purpose is to identify broader issues emerging from the 17 more-focused essays that precede this one. Because our task is to take a view from a mountaintop that overlooks the panorama of those 17 hills, the issues we address are large-scale and coarse-grained. But if they lack in detail, they share a theme: whatever the rapidly changing problems faced by conservation, we see repeatedly that the solutions must involve 'alignment', as defined above. Doubtless there are book fulls of themes that should be aligned within the future conservation enterprise, but in the remainder of this essay we will explore 11 that are high on our list.

1 The aspirations of biodiversity conservation and development – particularly the alleviation of poverty – cannot be solved separately;
2 Global conservation institutions must punch their weight in the development ring, and to

do so they need to be better integrated with each other.

3 Although the bedrock of conservation is evidence, judgments based on that evidence inevitably draw the conservationist into the realm of politics. Indeed, although there is much common ground to be striven for, there remain genuine conflicts between conservation and other sectors of society, and even between different conservation interests.

4 Conservation needs to connect better with a wider public, and that involves demonstrating its structural importance within a framework for living that embraces not only utilitarian but also cultural and spiritual elements of human well-being.

5 As conservation becomes a mainstream political issue, conservationists must face up to a demanding need for transparency, and weed out inconsistencies in their arguments.

6 The burgeoning complexity of conservation issues is potentially paralysing but, although solutions must be customized to each case, time can often be saved by recognizing that many conservation projects follow a predictable trajectory.

7 Species with frail populations that straddle several countries – and these are mainly large species – are generally faring poorly, and need new international instruments to protect them.

8 The Shifting Baseline Syndrome is a major threat to conservation, and fighting this erosion is helped by a wider perspective of what has been lost, as well as what is at risk.

9 The goals of conservation are often arbitrary, and we should recognize the different arguments for conserving process and for preserving the products of that process;

10 Prioritization is difficult, compromise is necessary, and, whatever their market value, it is prudent to behave as if species have inherent value.

11 Although populations are often sensible units for conservation planning, conservationists should remember that populations are emergent properties of individuals and should not be dismissive of individual welfare.

Conservation, development and alignment

Economic development and poverty alleviation have traditionally been priorities for governments, often regardless of the consequences for conservation. Our question here is how to make room for conservation in a world focused on reducing poverty. By conservation we mean sustaining a desirable amount of biodiversity (sufficient not to deprive successor generations of ecosystems capable of maintaining or rebuilding their natural species richness).

The problem is most severe in the tropics. Developed countries still engage, rightly, in evermore heated debates about their natural areas (despite the irony that what stability they have achieved in their conservation strategies stems partly from the fact that they have long since lost many of the species that cause the greatest conflicts with humans). But it is with respect to the developing countries that such debates loom largest. Economies range from the large and rapidly developing (such as China, India or Brazil) to the small and relatively stable or declining (such as Haiti, Malawi or Mauritius) and, although their circumstances are very different, in all cases the challenges for biodiversity conservation are immense. The common theme among the tropical countries is that the poorest people live alongside the greatest biodiversity (Chapter 10). The overwhelming importance of poverty alleviation creates the risk that conservation priorities are demoted.

The primacy of development among developing countries is seen in such initiatives as the UN's Monterey Consensus of 2002, seeking a commitment of developed nations to give 0.7% of GDP in overseas aid, the 'Make Poverty History' campaign (2005) and the Report of the Commission for Africa 'Our Common Interest'. In these cases more aid, fair trade and debt-forgiveness are the three prongs of an emerging global consensus concerned primarily with promoting development. Conservation has mostly been set aside. Similarly the first objective for 2015 of the 2000 Millennium Development Goals (MDGs) of the UN was to reduce by 50% the number of people living on less than a dollar a day. This admirable aim offered no advice on how to achieve the transfer from poverty not only to a wealthier lifestyle, but also to a more sustainable one consistent with biodiversity conservation. 'Environmental sustainability' was mentioned only in Objective 7 of the MDGs, and then in a relatively disconnected way.

The political and humanitarian reasons why poverty alleviation strategies have primacy are obvious. But the routine failure to attend to environmental concerns appears short-sighted for a simple reason: without healthy ecosystems economic development fails. The destruction of the Aral Sea through unsustainable irrigation developments as recently as the 1960s illustrates this all too clearly, and the same lesson echoes through history and beyond. Humans living in subsistence or near-subsistence conditions have inadvertently transformed the natural history of whole continents and island systems long before modern tools, transportation and communication systems were developed. For example, the megafauna of the Australian continent was most likely hunted to extinction by aboriginal peoples, and the civilization of Easter Island destroyed itself through overconsumption of timber and other resources. These lessons are likely to be repeated with a vengeance until regulations for sustainability are adopted and enforced.

Admittedly, in recent decades conservationists have begun to cooperate with development planners but, at least outside Europe, the results have too often been unfortunate for conservation, because instead of 'aligning' they often 'conceded' (Oates 1999; Terborgh 1999). So for all the optimistic goals – a cynic might say platitudes – the political reality is that instead of simultaneously aiding humans and the environment, there is a risk of conservationists making the best of a bad job and

accepting strategies based on the perceived necessity to develop first and then to clean up afterwards. Such strategies gain momentum because there is little recognition in economic theory of the environment as a 'system condition' or prerequisite for the economy – environmental considerations are not merely 'externalities' to development, but are essential to it. An important alignment in conservation matters, therefore, is between those who suffer the negative 'externalities' and those who enjoy the positive ones. The principle of 'the polluter pays' (generalizable to the damager of biodiversity and Nature pays) attempts to internalize the negatives. A matching principle – 'the provider gets' – attempts to internalize the positives by the idea of payment for environmental goods and services (Dobbs & Pretty 2004). The traditional assumption that with a little help from NGOs or UN agencies, tropical countries can be relied on to set aside 5% of more of their land as protected areas, even at a financial loss, will not be the way of the future. If the global community, and particularly the developed nations, want to see biodiversity maintained or increased in the tropics, the likelihood is that they will have to pay for it, and find ways of doing so that align the interests of development and conservation.

Some forlornly conclude that the next age of conservation may therefore involve a 'holding pattern' where the priority for conservationists will be to keep options open by conserving and protecting key environments while engaging fully in the plans to defeat poverty, achieve social justice and contain overconsumption. One might draw an unhappy parallel with the constructing of monasteries in the Dark Ages – refuges built in the hope and expectation of a better future.

We prefer, instead, to emphasize the urgency of radically increasing public and political appreciation of the inseparability of biodiversity conservation and human well-being, and thus the need for alignment in seeking solutions that foster both. Such 'alignment' is certainly sometimes recognized as the best option. Thus in 2005 the Millennium Ecosystem Assessment – a UN-sponsored global audit of nature – concluded that ecological processes are breaking down and at risk of failing to deliver basic human needs. Poverty was seen as a major cause of environmentally unsustainable conditions, such as slums, agriculture on marginal lands, water and air pollution, and disease. So conservation initiatives, such as the 2004 Malahide Message, from more than 200 stakeholders in 25 European countries aspiring to halt biodiversity loss by 2010, should in theory pull closely together with development strategies.

It is not to diminish the urgency of plans for bringing people out of poverty to look forward, as we do, for a change in global attitudes and institutions that will routinely integrate conservation aspects into the top level of such plans. On the contrary, such 'alignment' is the best hope for biodiversity and for humanity too, because without healthy ecosystems economic development becomes impossible. Although the model suffers from a problem associated with political boundaries, the environmental Kuznets curves (e.g. Hoffmann 2004) may be useful to highlight the disparities amongst developing countries. Environmental Kuznets curves show an inversed 'U' shaped relationship between nature destruction and the wealth within one country – the worst would come, and last for a while, when the country would reach the middle stage of economic development.

Fragmented institutions: united we stand, divided we fall.

Conservation does not happen without planning. The market alone is not enough to sustain populations of fish, lobsters, elephants or wild timber, even when consumers and industries are aware of the problems of unsustainable use. Conservation, alas, necessitates regulation. At the global level, this means it needs global institutions. The question, then, is whether existing institutions are fit for this purpose. The answer, we believe, is that the global institutions currently battling for conservation are inadequate because the problems have changed faster than the structures and ways of thinking that have been erected to solve them.

The difficulty can be seen from the major conservation initiatives of the past three decades. In 1972 the United Nations Environment Programme (UNEP) was created at the Stockholm Conference on the Human Environment, and over the next two decades it became the prime engine of international conservation agreements. Then at the Earth Summit at Rio de Janeiro in 1992, the follow-up to the political outcome embodied in Agenda 21 was given to a new institution: the Commission on Sustainable Development (CSD), independent of UNEP. The all-important UN Framework Convention on Climate Change and the UN Forum on Forests are also independent of UNEP, while the UN Convention on Biological Diversity links to UNEP only through its basic administration. This fragmentation of the institutions of environmental sustainability prevailed through to the Johannesburg summit in 2002, and continues.

Today, international convention secretariats are scattered across the globe, each charged with addressing one or other fragment of the conservation agenda. There are five main biodiversity-related treaties. The Convention on Biological Diversity is in Montreal, Canada, the Convention on International Trade in Endangered Species is headquartered in Geneva, the Convention on Wetlands of International Importance is at IUCN in Switzerland, the Convention on Migratory Species is in Bonn, Germany, and the World Heritage Convention and Man and Biosphere Programme, dealing *inter alia* with protected areas, are at UNESCO in Paris. The Commission on Sustainable Development is in New York and UNEP, which is expected somehow to pull these organizations together but has no power or money to do so, is in Nairobi, Kenya. Once the convenor of environmental initiatives, UNEP, is no longer holding the ring. Its authority and budgets are overshadowed by newer institutions with their own budgets and headquarters, and its location in Kenya presents serious operational challenges. All these institutions, created and developed over several decades, indisputably encompass enormous talent, wisdom and commitment, but inevitably their efforts are at best fragmented, and at worst in competition or conflict with each other. It is not surprising if the institutions that today are tasked to oversee the environment occasionally creak – most were conceived at a time when the current reality of globalized markets and development plans would have been unimaginable. The childhood of conservation was naturalistic, unilateral and flamboyant; its adolescence was turbulent and utilitarian. We suggest that its maturity depends on alignment and that to fulfil the potential of this third age of conservation there is an urgent need to unify institutional and regulatory structures in line with today's global reality.

For the moment, the unfortunate result of institutional fragmentation is political weakness. For example, no environmental organization has the weight to counterbalance the arguments for unfettered trade coming from the World Trade Organization (WTO), or from the development programmes of the US-based UN Development Programme (UNDP) and World Bank (which have their own very large environmental divisions). Moreover, there is no environmental institution powerful enough to tackle some key global issues (deforestation and deep-sea natural resource exploitation being good examples). The creation of a World Environment Organization (WEO) to overcome these problems has been proposed and promoted, principally by the Government of France. This new structure would be designed to balance the power of the WTO, and could arise from strengthening, rather than replacing, UNEP. But most countries have been unsupportive; some because they host existing institutions that they do not want to lose, others perhaps because they see advantage in

a fragmented and less powerful environmental lobby. Without new energy the idea seems unlikely to mature in the near future, and risks being forgotten once its main advocate, President Chirac, is replaced.

Even if a World Environment Organization is unachievable, perhaps the five biodiversity-related organizations listed above could be reformed. Does the world still need a separate convention on migratory species when we have a framework convention covering all of biodiversity? Why are natural World Heritage Sites managed under the aegis of a UNESCO convention when protected areas are arguably the most important instruments available for the implementation of the conservation component of the Convention on Biodiversity (CBD)? Should the Ramsar Convention on Wetlands of International Importance be transformed into a protocol of the CBD? The CBD itself is suffering from this fragmentation.

The idea of rationalizing the many international conservation agreements may look like common sense, but to date such proposals have ended up in the 'too difficult' tray of international diplomacy, partly because attempts to improve multilateral agreements mean that the entire original agreement is available to be modified, risking a backslide, and partly because not all countries are signed up to all the agreements. But conservation is the loser. Energy, resources and public relations are scattered too thinly and the development-first, pro-free trade lobbyists divide and rule. A new level of integration – ideally embodied by a new integrated institution – is essential if conservation is to take a seat at the table alongside development and poverty alleviation.

The difficulty of establishing a truly global response to the problem of global warming reminds us again of how hard it is to create effective coalitions when the goals are longer term than this year's GDP or next year's oil supply. Different countries, and within them, different cultures, wealth classes, political parties and so on, have different interests. These lead to mistrust, and jeopardize united action.

Nevertheless, at least the headlines, if not the tides, are turning on the prospects for international action towards climate change, so they can turn, too, with respect to conservation. Turning headlines into substance will require inspired political leadership – and for that a prerequisite is that decision-makers, along with their citizens, are better educated in environmental (and especially ecological) sciences.

A generation ago an education in the Classics was considered a fine foundation for citizenship and leadership; how wonderful if a new Enlightenment made it *de rigueur* for the future renaissance politician (and other leaders of society) to have a thorough grounding in conservation sciences.

Conservation, wider environmentalism and politics

Conservation decisions must be based on the best possible empirical science, and sometimes the scientific questions are very difficult. Nonetheless, the hardest questions faced by conservation practitioners are not of the ilk that has a single right answer – rather the task is to decide upon a desired outcome. Naturally the best option will vary between circumstances, and the decision may be different between urban, rural and marine environments, between ecologically rich or impoverished systems, and between rare and more widespread environments. Modern biodiversity conservation involves the reconciliation of many conflicting goals and perspectives, from natural history to politics nested within an intricate web of wider environmental and societal issues. Thus the practice of conservation extends beyond the foundation of evidence (vitally important, and indeed complicated, though that is) to the even trickier ground of judgment. For an evidence-based profession, it is an awkward reality that conservationists cannot avoid their evidence pushing them towards a view, and

thus edging perilously towards subjective ground. Western medicine, for example, has long-since had to combine scrupulous scrutiny of evidence with the obligation to cure the sick patient, much as the conservationist may feel a duty to cure the sick environment. Just as doctors find it is easier to say what illness is than what health is, ecologists start to know what is unhealthy in ecosystems, although understanding ecosystem health will involve a far higher level of complexity than does public health (ecosystems, afterall, contain publics).

When we were students inspired by Paul Ehrlich and the Blueprint for Survival emanating from the 1972 Stockholm Conference on the Human Environment, the mushrooming population explosion seemed to herald a catastrophe for biodiversity, and for humanity itself. Demand for space and resources would obviously, increasingly, and necessarily, outstrip supply. Malthus, of course, had foreseen the same thing. Vast, illuminating and complicating detail has since been built on this simple starting point, and sensitivity to political correctness has burgeoned. While we argue that conservation and development must be aligned, we also acknowledge the tendency of some conservationists – perhaps fearful of being thought to value Nature too much and people too little – to downplay the costs of living alongside biodiversity. In our sphere, as vertebrate ecologists, it is not uncommon to meet conservationists who seem to believe that all conflict in Nature's garden would vanish if only their misinformed opponents 'understood' (see Chapter 16). This hopeful vision imagines that the lions that eat cattle (and a number of cattle-herders that is not inconsequential to those involved) would become welcome denizens of everybody's backyard if only people understood them better (often in the cornucopian embrace of eco-tourism). In reality, there are plenty of things about Nature that are not at all good for some people living alongside it (zoonoses perhaps being paramount amongst these, Chapter 8). The costs of tolerance cannot be dismissed as blinkered misunderstandings;

rather, tensions must be acknowledged, understood and solved through alignment.

Conflict is not confined to stand-offs between the enlightened cognoscenti and the red-necked (or dark-suited) ignorant, as it is so often, and so tawdrily, portrayed. Conflict is real, and it occurs amongst conservation goals as awkwardly as it does between conservation and different sectors. For example, the intriguing phenomenon of intraguild aggression reveals some nasty trades-off – Macdonald & Sillero (2004b) discuss how conservationists might have to make a consumer choice between, for example, spotted hyaenas and wild dogs, lions and cheetah, coyotes and swift foxes – in all of which cases the larger carnivore is inimical to the smaller. Similarly, chimpanzees in parts of Uganda are depressing numbers of endangered red colobus monkeys, while in Hawaii it is argued that conserving the quetzel might necessitate killing toucans. Endangered predators can threaten endangered prey (on Java the dhole (*Cuon alpinus*), an endangered canid, threatened the banteng (*Bos javanicus*), an endangered wild cow). Again, having grasped the awkward reality that much of conservation planning is about consumer choice (i.e. implicit value systems, see below), the crucial point is that choice should be informed by evidence and arrived at by 'alignment'. A benefit of this transparent process will be the weeding out of the outrageous double standards that still nestle in the supposedly evidence-based judgments of conservation. A not-in-my-back-yard attitude (NIMBY-ism) is rife, as is the use of value-laden vocabulary (eating bushmeat implies something bad, bagging rabbit for the pot sounds quaintly rustic – see Chapter 14).

Nature, interdisciplinarity and a philosophy for the twenty-first century

To develop the theme of alignment, it is obvious that biodiversity conservation is inescapably interdisciplinary (amongst which disciplines

biology is a necessary but not sufficient component, along with economics, development, governance and health, *inter alia*). It is less obvious, but perhaps even more important, that biodiversity conservation is essential to human well-being in more far-reaching ways than the fundamental provision of ecosystem services (crucial as these are), and furthermore that fundamentally ecological ideas are relevant to the whole human enterprise. Thus, the fact that biodiversity has existence value at least to those educated in, and sensitized to, its marvels (Chapter 4) can be a foundation for understanding how participation with nature is important to both physical and mental well-being.

An important sphere for the future, amenable to scientific study, is the value of experiencing, and understanding, Nature for individual and thus societal well-being (the natty phrase of 'conservation therapy' was recently coined). The proposition that Nature brings health benefits, and thus saves money to public health systems, resonates in the UN Commission on Human Rights (2003) statement that 'protection of the environment and sustainable development can also contribute to human well-being'. Furthermore, it has down-to-earth expression in the news that some doctors in southern England are prescribing heathland walks. The proposition that contact with Nature contributes to human well-being can be extended beyond health to spiriutuality. With the widespread drift away from religion in western Europe, an understanding of Nature offers a promising intellectual, even spiritual, framework for a twenty-first century philosophy for personal and political life (Berry 1999). For members of strongly theistic cultures, conservation philosophy will have to find its ground within existing belief systems, which to some extent it already does (Waldau 2002).

Furthermore, the science of stability and diversity in ecosystems (Macarthur & Wilson 1967) can be generalized to linking cultural, spiritual and biological diversity, and all of these linking to stability of human societies and of the planet itself (a new generation of biological conservationists might, for example, increasingly celebrate linguistic and spiritual diversity and probe these wider linkages (e.g. Pimm 2000)). This is a line of thinking encouraged by UNESCO as part of the follow-up to the 2001 Universal Declaration on Cultural Diversity, and would be a worthy focus for a new generation of interdisciplinary conservation thinkers. The cultural understanding of nature needs far more systematic attention.

We write in the year 2006, which is, perilously in this context, within a year or two of the first time in the history of humanity in which more than half the world's population is classed as 'urban'. Contemporary urban disconnection with Nature's processes strikes us as deeply precarious, and how vastly more valuable could reconnection be when it is based on a rich understanding of life science (and a parallel understanding of the dangers of an unplanned increase in population growth and resource-mining). As Stephen Jay Gould observed 'we cannot win this battle to save species and environments without forging an emotional bond between ourselves and nature as well, for we will not fight to save what we do not love'. These are dauntingly big (but we think not pretentiously big) topics – one might say they are for the future, but the imperative of the extinction crisis means we had better make it the immediate future. Even for the present, Macdonald & Tattersall (2001, pp. 264–6) argue for solutions that do not involve sealing people off from Nature (or imagining it as a false Utopia) but, instead, involve interacting with wildlife through what they call a 'respectful engagement'. This engagement seeks alignment between appreciating that sometimes it is necessary to tolerate individuals of a problematic species, and sometimes it is necessary not to do so. Furthermore, the optimal solution to the problem of alignment will vary with local circumstances, and so as evidence-based ecological thinking replaces faith-based thinking, it is important not to replace one dogma with another, but rather to customize solutions and compromises flexibly to local circumstances. The art of conservation should be the science of fruitful compromise.

Two initiatives could help in joining up all these aspects of biodiversity conservation. First, at the practical level, in each community a liaison officer could be tasked to link the activities of the constellation of separate organizations relevant to biodiversity (spanning conservation bodies, education authorities, health trusts, farmers, etc). Second, academic structures could usefully change to recognize that tackling cross-cutting questions needs interdisciplinary minds to understand the problems, and philosophical sophistication to make judgments about the solutions. Today, especially in academia, while training for interdisciplinary individuals is increasing, there is scant career structure to foster, nor funding to create, interdisciplinary teams, nor the right time-scale for enabling them to operate. There might be a model found in Departments of Public Health, which bring together physicians, epidemiologists, social services, etc. We look forward to the creation of a network of centres (even one would be a start) assembling top figures and their emerging disciples from diverse fields relevant to conservation science.

The mainstreaming of conservation

Conservation is becoming a mainstream subject, but as it takes its seat at the table of wider environmental and societal debate the new, broader perspective brings two related risks: one is that biodiversity conservation finds itself more transparently compared with enormous and sometimes competing development priorities, and the second is that mainstreaming leads to paralysis owing to the number, and diversity, of stake-holders to be accommodated in every decision. Indeed, the weaning of conservationists from their single-issue proclivities has presented them with difficult choices (Macdonald 2001). It may be uncomfortable to be against nuclear power and at the same time determined to do something about climate change (Chapter 6). It may be uncomfortable to be pro-conservation and anti-hunting or anti-sustainable use (Chapter 15). It may be uncomfortable to be for trade liberalization, which will be tough on developing countries competing with those with a mechanized edge, but against the greater blight of wide-spread subsidies (Chapter 16). The extraordinary truth is that within one professional lifetime biodiversity conservation has won its place at the international table and climbed high on the political menu: no longer seen as the preoccupation of quirky, if generally harmless, muddy booted, binocular toting, often bearded naturalists – the realization is taking hold that at all scales (from parochial to global) biodiversity is connected to everything that matters. The first survey of public attitudes to the environment across the enlarged European Union of 25 countries reveals that 9 out of 10 Europeans believe that policy makers should pay as much attention to environmental issues as to economic and social factors.

But with this opportunity for conservationists to escape from the green ghetto has come perverse outcomes – by growing up, the great majority of the burgeoning legions of conservation's foot soldiers have swapped their muddy boots for polished shoes, and now lurk not in the bush (of which many have little experience) but in the corridors of governments and institutions. Furthermore, with acceptance that biodiversity issues are inextricably linked to wider environmental issues, to societal issues, to development and indeed to just about everything else in the human enterprise, there will be no dodging of a new transparency that will ferret out the inconsistencies that were rife in a less sophisticated age. The hard-won, indeed justly-won, seat at the table makes transparent how the biodiversity advocate ranks alongside the other guests (the advocates of development, trade, health, sustainability, governance and the rest), and the outcome of this mainstreaming can be both awkward (e.g. a rare species may carry little weight when traded

against human livelihoods) and paralysing (compromises are disproportionately harder to strike the more participants are involved and the more the stakes are transparent to all).

As Tom Burke (2005) has pointed out, the easy politics of the environment (pollution, toxic chemicals) had easily identified villains and, more importantly, action led to more winners than losers; the new agenda (deforestation, ocean degradation, water shortage, climate change) with biodiversity loss at its heart, requires action likely to create more immediate losers than winners. To quote Burke, 'just to confuse matters more, the victims and villains are often simply ourselves oscillating haphazardly between our needs as citizens and our desires as consumers'. The daunting reality is that an initial concern for a dazzling butterfly on the forest floor, or a be-whiskered snout whiffling from its burrow, is soon catapulted into questions of human economic sustainability. For example, the most exciting development in British conservation in several decades is the launching of new agri-environment schemes that shift payments to farmers from productivity to custody of a biodiverse countryside. The prospect is for biodiversity pay-offs at the national level (see Chapter 16). However, the impacts of such schemes need to be calculated at several scales (e.g. farm, landscape, region, country, continent, globe, etc). For example, cutting production in the UK may have a biodiversity pay-off in that nation's flower meadows, but might create economic conditions that favour the destruction of the (more pristine) farmland of eastern Europe, or lead to a huge cost of food-air-miles importing food from the other side of the world. The new, educated, joined-up reality prompts questions that are awkward for almost everybody – where once the UK taxpayer seemingly swallowed without question the idea of giving, for example, the Hill Farm Allowance to a farmer simply because he happened to farm on a hill, the 360 degree vision of participants at the new environmental table may scrutinize the reforming Common Agricultural Policy and ask what benefits (e.g. in terms of biodiversity) that farmer gives to the public in return (in a world where freedom of information is a statutory right, perhaps the sign advertising cream teas at the farm gate will be joined by one recording how much the farmer has received from society's kitty, and what he has delivered in return).

The conversion of conservation from maverick to mainstream is indicated not only by the fact that world leaders are talking about it, but by the terms in which they speak. They show that its place is predicated almost entirely on biodiversity's value to people, as opposed to its inherent value. Thus, Tony Blair, Britain's Prime Minister, said 'Make the wrong choices now and future generations will live with a changed climate, depleted resources and without green space and biodiversity that contribute to our standard of living and our quality of life'(March 2005). What this means is that we need to find a way of providing economic opportunities for the eight billion people soon expected on Earth, while maintaining the ecological foundations without which significant economic development cannot take place at all.

Complexity, the primacy of the human dimension and Goedel's proof

Moore's Law proposes that the power of information technology doubles every 18 months. It is just as well that tools to help humanity solve its problems are so fecund, because the complexity of issues facing conservationists seems to double just as fast. Although the conservation practitioner may feel overwhelmed by complexity, there may be comfort in considering it at just two levels – very general or very particular – with rather little benefit gained from intermediate levels. At one extreme, local circumstances differ so greatly that the most effective projects are those where the research to crack the problem is undertaken at the same site where the solution must be

implemented. At another extreme, most projects distil down to just a few generalizations.

Macdonald & Sillero-Zubiri (2004a) recently analysed 120 projects submitted as potential priorities by advisors to the IUCN/SSC Canid Specialist Group. The diversity of problems these addressed at first seemed overwhelming, but on further scrutiny all boiled down to variations on just three themes (problems concerning habitat loss, predation and infectious disease), with emphases on one or more elements of what Macdonald (2000) has called the Conservation Quartet (research, community involvement, education and awareness and implementation). Furthermore, most proposals could be placed somewhere along what seemed to be an almost invariable trajectory (from starry-eyed ecology, through interdisciplinarity, then weary pragmatism to optimistic compromise). Macdonald & Sillero-Zubiri (2004a) concluded that a lot of effort could be saved if those planning projects appreciated that most fit into one of relatively few genres, and go through a similar ontogeny.

The burgeoning complexity of conservation problems arises due not just to the exponential increase in the numbers of people and their demands, but also to the still unfolding awareness of the connectedness of all environmental issues and, most especially, the emerging primacy of the 'human dimension' in conservation thinking. Environmental problems can no longer be solved by the traditional algebra that isolates issues one at a time – but must be treated as an ensemble that is addressed as a whole. Getting conservation right requires developing practical solutions to the most complicated simultaneous equation ever written!

Consider the history of the Government's conservation agency in England. In 1947, Command 7122 of the then Ministry of Town and Country Planning led to the creation of the Nature Conservancy – England's first statutory body for conservation. Although this document (HMSO 1947) is humbling in the far-sightedness of many of its considerations, it can be compared revealingly with the 2005 equivalent, which has led to the creation of a new agency, Natural England. In 1947 five functions were ascribed to protected areas, of which the last, and most briefly, mentioned was 'amenity'. In 2005 the strap line for Natural England is 'for people, places and nature' – people conspicuously listed first of these beneficiaries. Arguing from the position that the natural environment underpins economic and social well-being and is essential for wealth creation, Natural England's mission will be to secure a better natural environment for the benefit of all; it will strive for landscapes that 'will be rich in wildlife and contribute to our wealth and well-being' and the first bullet point in the route map to achieving this is to 'put people at the heart of our work'. Indeed, the sophisticated statement of strategic direction (richly peppered with words such as enable, empower, share, deliver and engage) is very much about people (mentioning specially those most disadvantaged in society). Furthermore, Natural England is to be formed by combining English Nature (the previous champion of biodiversity conservation), with the Landscape, Access and Recreation remit of the Countryside Agency (the previous champion of rural communities, landscapes and wider access to the countryside) and the environmental activities of the Rural Development Service. In short, the shift in emphasis from 1947 to 2005 reflects exactly the new perception of biodiversity conservation as just one player within the all-embracing political bandwagon of sustainable development.

'Sustainability' is the property of being continuable, so that in terms of intergenerational equity the needs of the present must not be allowed to out-compete the needs of future generations. Sustainable development is development that maintains the environmental conditions for its continuation. Economic development is economic growth that maintains the social conditions for its continuation. Ergo, sustainable development is economic growth which maintains both the social and environmental conditions for its continuation. Thus, sustainable development is unavoidably

about biodiversity conservation, and it is not merely a lazy shorthand for any sort of development one might want it to be. Judgements about what constitutes sustainability are also precariously at risk from the Shifting Benchmark Syndrome (discussed below).

So, if biodiversity is to be conserved primarily for the good of people, how much of it do we need? Of the global biodiversity cake, it is clear that a substantial slice is essential to provide the ecosystem services on which humanity depends. It is also clear that a further hefty slice of global biodiversity is necessary for the continuation of social and environmental conditions allowing sustainable development. Indeed, the more that is understood about sustainability and the intricate involvement of nature in, for example, human health, mental well-being and other societal assets, the thicker will become that indispensable slice of biodiversity. Of course, nobody yet knows exactly how much biodiversity is needed to safeguard all these contributions to the well-being of humans, but we do know that the costs of wrongly underestimating it are so awful that prudence pushes for allowing a wide margin for error. Nonetheless, there is a logical possibility that there remains a portion of biodiversity that is surplus to human requirements. Some small butterfly, scuttling mouse or obscure lichen might, in the face of commercial gain or human development, have no case to make. Yet some people, like ourselves, attach great value to species that are quite likely to be in this 'surplus' portion, and often we value them for a mysterious jumble of aesthetic, intellectual and emotional reasons akin to those that make us value, for example, a particular piece of art. This truth forces us to confess that when drafting this essay we strove to find compelling, logical reasons to conserve biodiversity, and because of our reductionist scientific perspective, we would have been happiest if a single, consistent rationale could have embraced every case. However, the reality is that no such single rationale exists: completeness and consistency are unattainable.

Nonetheless we have identified a single primary principle, and it is that no biodiversity should be avoidably lost. This is because nobody can be sure exactly which bits of biodiversity might be sacrificed without peril. Historically, bounded communities that have got this wrong have faced ruin. The consequences of shifting from fewer than 2 billion to more than 6 billion humans on the planet (and the enhanced technologies and affluence probably make it sensible to add a fivefold multiplier to their impacts) means that we all now live in a bounded community. Therefore the margins for error are much reduced so there should be extreme caution in any activity that might damage ecosystems (and there should be a premium on restoring those that have been most degraded).

If no biodiversity should be avoidably lost, what constitutes 'avoidability'? The answer depends on the circumstances and requires the exercise of judgement, from which flows the question of whose judgement. Ultimately, the answer is that it will be society's judgment and this explains why stake-holder and wider public enthusiasm for conservation is essential for conservation. This is a yet further reminder that conservation spans science and politics. When it comes to conserving species which currently seem so far outside the safety net offered by the prudent margin for error that no utilitarian case can be made for them, those who like these species can do no more than use whatever means are available to persuade others to agree with them. Attempting such persuasion is no less legitimate than any other piece of advocacy within society, and the fact that no single scientific argument does the job should not embarrass the advocates. Take heart from Goedel's mathematical proof that even the basic axioms of arithmetic may give rise to contradictions – in short, no theory can be both complete and consistent. Conservationists will thus find themselves striving to protect species for motives that may differ between two ends of a spectrum: at one end prudence, to protect the biodiversity we need, and at the other end preference, to protect the biodiversity they like. Along this spectrum conservationists should use scientific evidence wherever it is available, but there will be slices of the biodiversity cake where they may be driven from arguments based on knowledge to advocacy based on wisdom.

If conservationists wish to convince a wide public that it is important to protect even those species which seem furthest from the protection of the utilitarian umbrella then their best hope is surely to reveal their fascination and beauty. Paradoxically, perhaps the most powerful weapon for the applied conservation of such species is thus the fundamental research that exposes how enthrallingly interesting they are – a weapon whose effectiveness relies upon excellent communication.

Rare species across borders

In the context of persuasion, it may seem perverse that conservationists often appear embarrassed that the public likes some species much more than others. Of course, biodiversity has value as a commodity, for its functions and for the inherent value some people attach to its existence. Concerning the latter – which is a non-market value – historically the predominant quest in environmental ethics has been for a non-anthropocentric strategy that acts as if all species have equal inherent value (Hargrove 1989; Naess 1986). So, in general, the theoretical priorities for conservationists are to direct resources to those species that are most threatened, or to areas of highest biodiversity (Chapter 2), and are not influenced by a species' inherent appeal (Mittermeier et al. 1998; Harcourt 2000; Olson & Dinerstein 2000). However, the public, from whom comes much of the funding for conservation, tend to be more concerned with the future of particular species, particularly large and charismatic taxa as exemplified by tigers, rhinos, apes, whales, raptors, turtles and so on. Different species have different existence values. This is instantly obvious when the stage is broadened from the narrow set of organisms (often vertebrates) that spring to the public mind as conservation priorities – putting aside their utilitarian or monetized values, should the quest for consistency cause us to strive as passionately to conserve an anonymous species of microbe as we do for gorillas? The political reality is that as we bargain for nature's future, society does not give equal value to all species.

Many of the most charismatic species are large, and their large size causes them to have geographical ranges that span several countries. There is a general need for increased intergovernmental cooperation in conservation strategies, and this applies in particular for these large (and therefore low-density and sometimes dangerous) species. Their habitats and numbers have shown consistent decline. One practical argument for giving such species special treatment is that without deliberate global cooperation they will soon become extinct – specifically, failure of collective action will lead to a steady erosion of populations within each range state. Another practical reason for giving them special treatment is that the public consider them as special.

The particular ways in which global cooperation is achieved in the conservation of particular species will necessarily vary. It may depend on who the relevant stake-holders are. For example, the International Whaling Commission represents a system of global integration that has operated since 1948 to save several species of whale. Because of the pressures for and against whaling, the stake-holders include both commercial and non-commercial interests. For species in which public interest is high, but commercial interests absent, the inadequacy of current global institutions is leading to new ideas. A proposal that recognizes the differential appeal of species is to nominate certain taxa as World Heritage Species, i.e. species that (in parallel with World Heritage Sites) can be designated as being 'of outstanding universal value to all mankind'.

The value of a species might be assessed by an international panel, much as the value of a cultural or natural site is assessed by the World Heritage Convention. Such a designation would increase public recognition, often a critical factor for effective conservation within range states. Many such species have large home ranges that encompass many other species, and so can serve as umbrellas to protect those with less leverage. These ideas are being applied to the great apes. Globally, conservation of the great apes is the concern of UNESCO and UNEP's Great Ape Survival Project (GRASP), a partnership of range countries, donor countries, NGOs, the various biological Conventions and other organizations. In September 2005 GRASP established a Scientific Commission charged with evaluating priorities for great ape conservation. This commission aims essentially to conduct a population/habitat viability assessment (PHVA) for each of the great ape taxa, and thereby identify key populations and sites as priorities for conservation based on a clear biological rationale (Chapter 9). For this kind of cooperation to work, of course, both range and donor countries must want to participate, ideally both motivated by real rewards. If the great apes are designated as World Heritage Species, their key populations would be designated World Heritage Species Sites or some legal equivalent. Unlike the pristine World Heritage Sites, well-managed and monitored resource extraction may in some cases be compatible with maintaining a viable population. To date, unfortunately, World Heritage Sites have benefited from their global status more by renown than by direct support: UNESCO grants the average site about $3000 per year. Pride will not pay the bills for conserving World Heritage Species, so donor countries must be inventive in creating real rewards to induce range countries to participate.

The general point is that collective action problems need to be solved collectively. The greatest conservation successes tend to come from populations confined to a single country, where it is obvious that if no one takes responsibility, disaster looms. For example, the first time that a primate's threat status has been lowered was in the 2003 Red List, when Brazil's golden lion tamarins were downgraded from 'critically endangered' to 'endangered'. This happened because a wild population of 200 had risen to 1200 as a result of reintroduction and conservation. If golden lion tamarins had lived in more than one country, the pressure to do something about them would have been so much the less.

Conservation goals and a natural bench-mark for success

Just as the way conservationists are trained, assembled and assessed needs to be up-dated to match the new understanding of their function, so too do the criteria for success of conservation need overhauling. Many of the milestones on the road to conservation success are intangible, and the road itself may take many turns, but nonetheless many NGOs currently present success too much in terms of raising and spending money and too little in terms of habitat protected, populations enlarged or policies changed. There is a risk of confusing being business-like about conservation with making a business out of conservation.

The proliferation of professional conservationists raises a fascinating and so far insoluble paradox that has emerged in our professional lifetimes – one that parallels a shift, by ecological analogy, from r-selected to K-selected characteristics. Conservation biologists of our now aging generation were few, often idiosyncratic, opportunists that had little choice but to run before they walked, and could truly enter the fray with the battle cry that 'if I don't do this, it won't get done' – most of their energy went on playing the course, not the other players. The heightened awareness of a new generation – much of it the triumph of television documentaries popularizing science that was

(interestingly) more fundamental than applied – now populates western countries with hundreds of thousands of graduates motivated to undertake applied rather than fundamental work, and thousands of organizations, dedicated to conservation – much of their energy is consumed by competing for the privilege of contributing. And for those in a position to contribute, there is the problem of a lack of professional skills, especially in terms of management, that have not grown to keep pace with organizational capacity. A further danger is that the competitive scrum forces organizations to trumpet their successes, distracting energy from the less marketable reality that while some truly memorable skirmishes are being bravely won at the margins, overall the war to save biodiversity is going catastrophically.

Various measures of conservation success may be fit for the purpose. Yamaguchi et al. (in preparation) propose a rather sobering one: the 'natural benchmark'. They point out that while conservation planning and action is switching from preservation of the current (and thus both arbitrary and highly 'unnatural') pattern of biodiversity to preservation of the processes behind it, the ultimate non-arbitrary standard for failure remains extinction. Indeed, preoccupation with extinction risk has generated important conservation principles especially concerning rarity, distinctiveness and endemism. On the other hand, although focusing attention on the risk of losing what is left, it pays less attention to what has already been lost and is thus susceptible to what has been called the Shifting Baseline Syndrome (SBS) – we become satisfied with holding a line at a point that would, earlier, have seemed catastrophic (Pauly 1995). Yamaguchi et al. (in preparation) argue for keeping in mind the situation that would have occurred 'naturally', where 'natural' is defined with reference to the earliest currently detectable human influence. They thus consider what plausibly might have been the case if people had not intervened, and use this to give a measure of what has been lost. Taking the example of lions, the contraction of their range, arguably associated with the spread of humans, is equivalent to reducing their populations from between 1.42 and 14.2 million to the current 20,000 or so – a loss on a par with the recent anthropogenic devastation of the North Sea fish stock where it is estimated that c.97 % of the 'natural' biomass of large fishes weighing 4–16 kg has been lost to fisheries exploitation (Jennings & Blanchard 2004). The difference is that the impact on lions has taken about ten millennia whereas that on fish has taken about ten decades. This 'natural benchmark' has the merit of offering a quantitative assessment of the current status of a species independently of the risk of extinction, and thereby draws attention to the need to think about how to set our aims. Suggesting that the 'natural benchmark' is a useful yardstick of success for conservation does not imply a deluded belief that all, or even any, species or ecosystems can be restored to the state that prevailed before human intervention. But it does provide an often chilling, and non-arbitrary, perspective for taking-stock, considering goals and remaining alert to the corrosive effects of a shifting baseline.

Preservation, conservation, the primacy of process

The idea of natural benchmarks brings us to the need for honesty in our starting points. Much effort, much of it conspicuously laudable, is devoted to retaining, or re-establishing, points that are oddly arbitrary. For example, in Britain inspirational effort has been devoted to conserving (and indeed recreating) the traditional agricultural landscape and organisms such as rare arable weeds and corncrakes whose existence is threatened by the passing of these landscapes (Chapter 16). This effort may be widely misunderstood in so far as whatever value these species may have (and we think it is considerable), neither they nor their environment is 'natural'. They are entirely

man-made, and seeking to preserve them is a consumer choice, loosely analogous to taking pleasure in retaining steam engines (also products of a by-gone technology). This highlights two separate points. First, and most generally, the reality is that throughout much of the world biodiversity conservation is almost entirely about consumer choice of what people want of Nature. Second, for several decades it has been fashionable for conservationists to be dismissive about 'preservation', which was ridiculed as static affection for the status quo, and distinct from something more dynamic that was conservation of process. Emphasis on process is indeed surely a good thing. However, we are less sure that preservation merits ridicule. Some of these rare arable weeds are very interesting (and, for what its worth, stunningly beautiful), so even though their existence is a sort of anachronism, to abandon them would surely be no less philistine than it would be to throw away Da Vinci's Mona Lisa (or indeed the Lascaux cave drawings) because they do not represent contemporary art.

Although conflict in biodiversity conservation is often thought of at the level of species (or individuals of a species)(Chapter 17), it can arise just as fiercely in the conservation of processes. This is vividly illustrated by the conservation of coastal ecosystems which are perpetually in a process of flux. This flux, with altered coastlines either threatening or creating brackish marshlands, has always been dynamic through natural geological processes of erosion and sedimentation (although may now be exacerbated by sea-level rise through 'unnatural' climate change; Chapter 6). Should a treasured saltmarsh, perhaps gazetted as a reserve, be defended, so to speak in aspic, or be allowed to disappear below the sea, even if its disappearance is natural? Hitherto, the answer has often been 'preservationist' – albeit mainly motivated to protect people and industry (of 3760 km of cliffs around England, over 1000 km are protected by coast protection works). However, the process of erosion of sedimentary rocks elevated along the north-east Yorkshire coast 6000 years ago generates the silt that produces the estuaries of the Humber and Wash that are vital to England's wading birds – if the erosion were stopped the estuaries would go. Thus those impressed by the importance of the process of coastal change might favour a (managed) retreat from the advancing tide, whereas those who work in the c.200 jobs offered by cliff-top caravan sites might be more inclined to take a stance alongside King Canute and seek to preserve the cliffs.

Change offers the conservationist a moving target, and makes it hard to be consistent. Take the case of the Eurasian badger, *Meles meles*, in lowland Britain where they are very abundant, first because of the lowland agricultural ecosystem (a direct creation of people) and second because of ameliorating climate (an indirect creation of people, Chapter 6) – society rejoices and enacts laws to keep them numerous. That seems an entirely appropriate stance for society as a whole to take, but let it do so acknowledging that it is because we like numerous badgers (an affection not shared by dairy farmers – Chapter 11) and not because we pretend it is a reflection of pristine nature.

Priority, value and compromise

The most chilling word in conservation is Prioritization. Prioritization is at the heart of every point we have discussed. It is difficult, painful and unavoidable, and all the logic and cleverness brought to doing it (Chapter 2) will do nothing to ameliorate the loss of that which is not prioritized. Of course, the identification of hotspots, and their prioritization, has been appropriate and important, but it can (and should) be complemented with other strategies. The costs of putting your eggs in one basket, and associated considerations about the sizes of these baskets (MacArthur & Wilson 1967; Fahrig & Merriam 1994) have been explored in Chapters 4 & 5, and focus attention on the importance of corridors (Soulé & Terborgh

1999). Although corridors may be important at any scale – enabling small mammals to traverse intensive farmland via hedgerows, or plants to seep between reserves (Macdowel et al. 1991) – surely few examples are more ambitious (or more praiseworthy) than Her Majesty Queen Ashi Dorji Wangmo Wangchuck's declaration in 1999, that 9% of Bhutan would be set aside to create a 'gift to the Earth' under the WWF's Living Planet Programme. Protected Areas in Bhutan comprise 26.5% of the country's area, and these are dispersed as nine protected areas. The capacity to move between these is important to such wide-ranging creatures as tigers, and the Queen's announcement ensured that all nine protected areas will be linked by 10-km-wide corridors (a linked initiative is that the Bhutanese Government has committed to ensuring that 60% of the country remains forested in perpetuity). Bhutan uniquely, along with the Seychelles and Costa Rica, has therefore formulated its plan for development within a framework dictated by its commitment to biodiversity and sustainability. Interestingly, in Bhutan a commitment to twenty-first century evidence-based thinking on environmental issues appears to have been accommodated within the value system of tantric Buddhism (Wang & Macdonald, in press).

The ordering of priorities is the output of values. We have already emphasized how the value of Nature is increasingly expressed in terms of benefits to people, and there are compelling reasons for this. However, a rather fierce political correctness bridles at the view, formerly expressed widely, that Nature should be conserved because it has inherent value. Similarly, sneers of 'elitism' are likely to be provoked by the suggestion that Nature's value is understood or appreciated by only a few. Although the moral and pragmatic cases for emphasizing and delivering nature's benefits to humanity are unarguable, the pendulum may have swung too far towards the expectation, and the demand, that this pay-off to people is short-term or always to a majority. It is worth considering what people mean when they say that Nature has inherent value. This could mean that some people (and we are amongst them) treasure even some parts of Nature that apparently have no market worth. Obviously, this does not mean that something has value independently of human judgment – that would be a vacuous notion because value is something attributed by people. That said, it seems to us that much of Nature that **as yet** has no known market value will be shown to be part of the conditions for market activities, and thus to be essential to sustainability. And for the rest – that which might be lost without infringing sustainable development – even if only a tiny minority currently value those elements of Nature that seem expendable to the utilitarian, it is surely that minority's right to advocate its protection as strongly as they are able. In any case, the value of nature to humans lies in so many unpredictable and slowly-accruing impacts on mental and physical health that it cannot adequately be indexed by short-term economic indicators – in practical terms therefore, it may be helpful to act as if nature has inherent value.

The welfare of populations and individuals

The views of conservationists are often population-based and sometimes rather technical – and other essays in this book reveal them to be both important and compelling. The importance of populations as a unit for consideration does not diminish the importance of views based on the welfare of individual organisms, and the latter may sometimes be more intuitive and appreciated by a wider public. Indeed, an erstwhile tendency of some conservationists to dismiss the importance of individuals may not only be logically shaky but also politically ill-advised. We argue that, at the least, welfare science is one amongst several important tools in conservation biology (Chapter 8), but going further

than this, the need to integrate the cherishing of biodiversity and its processes with that of respecting individuals is surely a priority for holistic twenty-first century conservation. Of course, there will always be hard decisions regarding the fates of individuals in the face of needs to manage populations – but the common ground between those whose emphasis is on populations and those who emphasize individuals is immense. The fact that populations, and ultimately ecosystems, are emergent properties of individuals surely makes irresistible the call for a more unified framework of ideas for tackling all these components of the natural world. At a practical level the synergies are obvious – for example, a 4-year-old network of ape sanctuaries in Africa (the Pan-African Sanctuary Association) is already a superb resource of information about conservation practicalities. The questions hitherto the preoccupation of the animal conservation sector, although mind-bendingly complex, have lent themselves more to quantification than have those previously in the province of the 'care' community, but both sets of questions are open to logic and analysis and neither community should disparage the other because of its 'emotional' attachment to one level of nature. A wide public, having been fascinated and captivated by insights into the lives of individual wild animals, is surely all the more likely to come to value the breadth of nature and its processes. It may therefore be easier to promote conservation to publics (as, famously, the British) that treasure and respect individual animals, than it is to get the same message across to publics that do not value individual animals, and therefore attribute value to populations and processes on purely utilitarian or intellectual grounds. Once again, the future is a quest for alignment.

Advocating, as we do, a continuity between concern for processes, populations and individuals, does not mean we are offended by death or that we are unaware that suffering is commonplace in Nature. Those connected with nature are aware that few animals die peacefully in their beds at the end of a full life-span.

Conclusions

Other essays in this book deliberately took the view from different hilltops in the conservation landscape; it may have been foolhardy even to attempt the climb, but we have tried to glimpse the horizon from an even higher pinnacle. The particular mountaintop from which we have surveyed the state of conservation is too far removed from the hard every-day work of environmental action to allow us to celebrate in this essay the many local achievements that keep habitats and populations in better shape than they would otherwise be (hopefully the people wrestling to achieve these actions will sense in our words a greater respect for them above all others). We can accept that the world may have resources and technology to support even the eight billion people it must soon accommodate – but time is running out to fulfil this capacity against the flow of accelerating economic growth, massive habitat loss, and a global population facing unresolved problems of food distribution. Meanwhile, although it is (or at least should aspire to be) fundamentally evidence-based, all conservation ultimately involves politics, and in practice it is managed globally too much by fragmented rather than united institutions, by advocates rather than politicians, by competition rather than by cooperation, by the weaknesses of single-discipline thinking, and by optimism rather than by acknowledgement of the hard realities and by unspecified value structures. In short, it lacks alignment. To say that conservationists need to become better politicians, business people and leaders is per-

haps too much to ask of these heroic people – better perhaps to say that politicians, business people and leaders need to become conservationists!

Recognition of these problems suggests some solutions. The extraordinary wealth of wisdom held by the 50 other contributors to this book has stimulated our overview, and we hope they, and this final essay, will encourage the next generation to think big. Conservation needs a global platform, disciplinary integrity, a spiritual philosophy, and a newly cooperative spirit. These may seem impossibly distant goals. We must be careful not to use their enormity as excuses for inaction. Rather, we urge that these goals propel us to more creative efforts. Is it too much to hope that this century will bring us billion dollar Institutes of Conservation, a World Environmental Organization, the international adoption of an eco-ethic or politicians with the remit and capacity to deliver a sustainable future for Nature? Such things are possible. It is time for conservation to be more ambitious.

Acknowledgements

We are grateful for comments from Tom Burke, Adam Dutton, Clive Hambler, Brian Harding, Mark Leighton, Andrew Loveridge, Andrew Marshall, James Marsden, Matt Prescott, Greg Rasmussen, Claudio Sillero, Barry Supple, Tom Tew, David Wilkie and Nobby Yamaguchi. It is not to diminish our thanks to others of the foregoing to thank especially Tom Burke for bringing his unequalled grasp of environmental matters to our rescue.

The sportsman who shot the last (Passenger) pigeon thought only of his prowess.
The sailor who clubbed the last auck thought nothing at all.But we, who have lost our pigeons,
mourn the loss.
Had the funeral been ours, the pigeons would hardly have mourned us.
In this fact . . . lies objective evidence of our superiority over the beasts".
(Aldo Leopold, *Death of a Species* – *A Sand County Almanac* 1994)

References

Balmford et al. (2002) *Science* **297**: 950–3

Berry, T. (1999) *The Great Work: Our Way into the Future*. Random House, New York.

Burke, T. (2005) *Speech Celebrating 25th Anniversary of The Green Alliance*, London.

Dobbs, T. & Pretty, J.N. (2004). Agri-environmental stewardship schemes and 'multifunctionality'. *Review of Agricultural Economics* **26**: 220–37

Fahrig, L. & Merriam, G. (1994) Conservation of fragmented populations. *Conservation Biology* **8**: 50–9.

Harcourt, A.H. (2000) Coincidence and mismatch of biodiversity hotspots: a global surveyfor the order, primates. *Biological Conservation* **93**: 163–75.

Hargrove, E.C.(1989) *Foundations of Environmental Ethics*. Prentice-Hall, Englewood Cliffs, NJ.

HMSO (1947) *Conservation of Nature in England and Wales: Report of the Wild Life Conservation Special Committee*. His Majesty's Stationery Office, London.

Hoffman, J. (2004) Social and environmental influences on endangered species: a cross-national study. *Sociological Perspectives* **47**: 79–107.

Jennings, S. & Blachard, JL. (2004) Fish abundance with no fishing: predictions based on macroecological theory. *Journal of Animal Ecology* **73**: 632–42.

MacArthur, R. & Wilson, E.O (1967) *The Theory of Island Biogeography*. Princeton Unvierstiy Press, Princeton, NJ.

Macdonald, D.W. (2000) Bartering biodiversity: what are the options? *Economic Policy: Objectives, Instruments and Implementation* (Ed. D. Helm), pp. 142–71. Oxford University Press, Oxford.

Macdonald, D.W. (2001) Postscript: science, compromise and tough choices. In *Carnivore Conservation*, Vol. 5 (Eds J.L. Gittleman, S.M. Funk, D.W. Macdonald & R.K. Wayne), pp. 524–38. Cambridge University Press, Cambridge.

Macdonald, D. W. & Sillero-Zubiri, C. (2004a). Conservation: from theory to practice, without bluster. In *The Biology and Conservation of Wild Canids* (Eds D.W. Macdonald & C. Sillero-Zubiri), pp. 353–72. Oxford University Press, Oxford.

Macdonald, D.W. & Sillero-Zubiri, C. (2004b). Wild canids - an introduction and dramatis personae. In *The Biology and Conservation of Wild Canids* (Eds D.W. Macdonald & C. Sillero-Zubiri), pp. 3–36. Oxford University Press, Oxford.

Macdonald, D.W. & Tattersall, F.H. (2001). *Britain's Mammals: the Challenge for Conservation*. People's Trust for Endangered Species, London.

Macdowel, C.R., Low, A.B. & Mackenzie, B (1991) Natural remnants and corridors in Greater Cape town: their role in threatened plant conservation. In *Nature Conservation 2: The Role of Corridors* (Eds D.A. Saunders & R.J Hobbs), pp. 27–39. Surrey Beaty and Sons, Chipping Norton.

Mittermeier, R.A., Myers, N., Thomsen, J.B., da Fonseca, G.A.B. & Olivieri, S. (1998) Biodiversity hotspots and major tropical wilderness areas: approaches to setting conservation priorities. *Conservation Biology* **12**: 516–20.

Naess, A. (1986) Intrinsic value: will the defenders of nature please rise? In *Conservation Biology: the Science of Scarcity and Diversity* (Ed. M.E. Soulé), pp. 504–516. Sinauer, Sunderland, MA,

Oates, J.F. (1999) *Myth and Reality in the Rain Forest: how Conservation Strategies are Failing in West Africa*. University of California Press, Berkeley, CA.

Olson, D.M. & Dinerstein, E. (2000) The Global 2000: a representation approach to conserving the earth's most biologically valuable ecoregions. *Conservation Biology* **12**: 502–15.

Pauly, D. (1995). Anecdotes and the shifting baseline syndrome of fisheries. *Trends in Ecology and Evolution* **10**: 430.

Pimm, S. (2000) Biodiversity is us. *Oikos* **90**: 3–6.

Soulé, M.E. & Terborgh, J. (1999) *Continental Conservation: Scientific Foundations of Regional Reserve Networks*. Island Press, Washington, DC.

Terborgh, J. (1999) *Requiem for Nature*. Island Press, Washington, DC.

Waldau, P. (2002) *The Specter of Speciesism: Buddhist and Christian Views of Animals*. Oxford University Press, New York.

Wang, S.W. & Macdonald, D.W. (In press) Livestock predation by carnivores in Jigme Singye Wangchuck National Park, Bhutan. *Biological Conservation*.

Index

Note: Page Numbers in *italics* refer to figures and boxes, those in **bold** refer to tables

genetics 47
  forensic 58–9
genome chip 59–60
Geochelone elephantopus (Galápagos tortoise) *194*
geo-fencing *107*
geographical distance, genetic structure of
      population 53
geographical information systems (GIS) 135
  process-based models 139
geolocators *106*
glaciers, melting 89
Global Environment Facility (GEF) 36, 41, 149–50
global institutions 279
Global Invasive Species Programme (GISP) 199
global positioning system (GPS) *107*, **113**
  limitations 114
global threats 27
global warming *see* climate change
Glossina (tsetse fly) 165
goats *193*, 198
Goedel's proof 287
goose, snow 232
grassland
  agricultural intensification 244–5
  biodiversity 245
grazing 268
Great American Faunal Interchange 189
Great Ape Survival Project (GRASP) 288
Greater Yellowstone Area, wolf reintroduction 262
greenhouse gas emissions 85–6
Greenland ice sheet 89, 91
Grevillea caleyi (understorey shrub) 67–8
grouse, red 156, 162, 163
guardian animals 180, 248–9
Gyps (vulture) 165–6

habitat(s)
  classification 135
  fragmentation and climate change synergy 93
  loss 285
    to agriculture 244
  northward movement 99–100
  protection in sport hunting 231
  static 78–9
habitat conversion 38
  compensation mechanisms 43
  delay 42
habitat corridors 65–6, 100, 168, 291
  dispersal rate 71, 72
  metapopulations 71–3
  water vole population support 250, *251*

habitat islands 100
habitat management
  aquatic systems 75
  sport hunting 231
habitat patches 65
  carrying capacity 70
  extinctions 77–8
  patch-occupancy models 75, 76
  recolonization 77–8
  reserve size limitations 70
  turnover rate 79
  use 77
habitat suitability maps 75
habitat-based approaches 21
Haematopus ostralegus (oystercatcher) 70
hair
  collection 48–50, 60
  genetic identification 50
  scanning 60
hantavirus 194
Hardy Weinberg expected (H_e) 55
harem breeding 232
harvest quotas 67
harvesting 174
  game 211
  wildlife areas 258
hawk-kites 179
headlands, conservation 250
hedgehog 71, 73, 74, 199
hedgerows 245
hens, free-range 128
Heracleum mantegazzianum (giant hogweed) 189
herbivores, crop damage 260–1
herbivory, introductions 191–2
heterozygosity
  excess 55
  levels in subpopulations 51
Himantopus (stilt) 197
Hippotragus niger (sable antelope) 232
HIV/AIDS 165
hogweed, giant 189
honeyeater, helmeted 71
hormone levels
  captured animals 128–9
  see also cortisol
host density 160
  critical threshold 166–7
hosts, spill-over species 165, 166
hotspots, species 9
human error 112
human T cell lymphotropic virus (HTLV) 165